ION

INTRODUCTION TO HEALTH RESEARCH METHODS

A Practical Guide

Kathryn H. Jacobsen, MPH, PhD
Professor of Epidemiology and Global Health
George Mason University
Fairfax, Virginia

World Headquarters
Jones & Bartlett Learning
5 Wall Street
Burlington, MA 01803
978-443-5000
info@jblearning.com
www.jblearning.com

Jones & Bartlett Learning books and products are available through most bookstores and online booksellers. To contact Jones & Bartlett Learning directly, call 800-832-0034, fax 978-443-8000, or visit our website, www.jblearning.com.

Substantial discounts on bulk quantities of Jones & Bartlett Learning publications are available to corporations, professional associations, and other qualified organizations. For details and specific discount information, contact the special sales department at Jones & Bartlett Learning via the above contact information or send an email to specialsales@jblearning.com.

Production Credits
VP, Product Management: Amanda Martin
Director of Product Management: Laura Pagluica
Product Manager: Sophie Fleck Teague
Product Specialist: Sara Bempkins
Project Manager: Kristen Rogers
Project Specialist: Brooke Haley
Digital Project Specialist: Angela Dooley
Senior Marketing Manager: Susanne Walker
VP, Manufacturing and Inventory Control: Therese Connell
Manufacturing and Inventory Control Supervisor: Amy Bacus
Composition: Exela Technologies
Cover Design: Timothy Dziewit
Text Design: Kristin E. Parker
Senior Media Development Editor: Troy Liston
Rights Specialist: Maria Leon Maimone
Cover Image (Title Page, Part Opener, Chapter Opener): © DmitriyRazinkov / Shutterstock
Printing and Binding: LSC Communications
Cover Printing: LSC Communications

Library of Congress Cataloging-in-Publication Data
Names: Jacobsen, Kathryn H., author.
Title: Introduction to health research methods / Kathryn Jacobsen.
Description: Third edition. | Burlington, MA : Jones & Bartlett Learning, [2021] | Includes index.
Identifiers: LCCN 2019047577 | ISBN 9781284197631 (paperback)
Subjects: MESH: Biomedical Research–methods | Research Design
Classification: LCC R852 | NLM W 20.5 | DDC 610.72/4–dc23
LC record available at https://lccn.loc.gov/2019047577

6048

Printed in the United States of America
24 23 22 21 10 9 8 7 6 5 4

Brief Contents

Contents

Preface

Research is the necessary foundation for meaningful improvements in clinical and public health practice. Research helps us learn how to be healthier and how to help our families, friends, patients, communities, and nations improve and maintain their health. We rely on researchers to identify risk factors for infections, noncommunicable diseases, and injuries and to determine which interventions are most effective at preventing adverse health conditions and improving individual and community health status. But it is not just the outcomes of research that make research rewarding. The research process itself—the systematic exploration of the unknown that discovers answers to important questions—can be exciting. The goal of this book is to make the health research process accessible, manageable, and perhaps even enjoyable for new investigators.

This is not a book about the theory of research; it is a book about how to *do* research. The book provides a practical, step-by-step guide to the entire research process. All research projects follow the same steps: identifying a focused research question, choosing a study design, collecting data that will answer the question, analyzing the accumulated evidence, and disseminating the findings. The investigation proceeds through these same basic steps regardless of whether the researchers are surveying community members, running clinical trials, conducting focus groups, analyzing existing data sets, or synthesizing the existing literature through meta-analysis. The same steps are followed whether the researcher is trained in medicine, nursing, public health, dentistry, physical therapy, occupational therapy, dietetics and nutrition, athletic training, health policy, psychology, sociology, counseling, speech-language pathology, respiratory therapy, radiation technology, pharmacy, podiatry, optometry, audiology, or any other clinical or social science discipline. The steps are the same regardless of whether the investigator is an undergraduate student, a master's or doctoral candidate, or a seasoned professional.

Health research is an intentional process that requires fastidiousness and perseverance, but it does not have to be complicated. Anyone who is willing to follow the steps outlined in this guidebook can conceptualize a research project and see it through to completion. Pursuing a research project may lead to the acquisition of new skills, the fulfillment of degree or work requirements, the satisfaction of personal curiosity, and even the opportunity to become a published author. More importantly, every project, no matter how modest, has the potential to contribute to expanding the knowledge base for the health sciences. Researchers who see their projects through to completion may eventually see their findings translated into improved patient care, enriched organizational effectiveness, and enhanced community health. This book is an invitation to make your own contribution to the evidence base that will inform future decisions about promoting healthy behaviors, allocating health resources, and preventing, diagnosing, and treating diseases.

© DmitriyRazinkov/Shutterstock

About the Author

Kathryn H. Jacobsen, PhD, MPH, is a professor of epidemiology and global health at George Mason University in Fairfax, Virginia. Her research portfolio includes a mix of primary studies that use quantitative and qualitative methodologies, secondary studies that analyze existing datasets, and tertiary studies that synthesize the results of prior publications about the population-level burdens from infectious diseases, noncommunicable diseases, mental health disorders, and injuries. She is the author of *Introduction to Global Health* (Jones & Bartlett Learning) and more than 180 peer-reviewed articles.

New to This Edition

In the third edition of *Introduction to Health Research Methods*, every chapter from the second edition has been updated to enhance content and improve clarity. The new edition also features several new chapters and subsections that provide expanded coverage of the clinical and population health research process.

The most important update is the significantly expanded coverage of qualitative research methods. There are now separate chapters on qualitative methodologies (Chapter 13), qualitative data collection (Chapter 23), and qualitative analysis (Chapter 32). Qualitative research methods are also integrated into other chapters, such as the chapters on sampling, ethics, and data management. This update aligns the book with the Council on Education for Public Health (CEPH) accreditation criteria initiated in 2016 that require public health students to demonstrate their ability to collect, analyze, and apply both quantitative and qualitative data. Another major enhancement is the improved integration of glossary terms into the main text. The number of terms in the glossary has been increased from 600 entries to more than 800, and a clear explanation of each key term has been embedded within the main text in the section where the term is first introduced.

All of the information from the second edition remains in the new edition, and the organization of the book remains largely the same. However, new content has been added to nearly every chapter in response to user feedback. The section on Step 1 of the research process (Identifying a Study Question) presents more information about how to generate research ideas, frame research questions, develop testable hypotheses, and define specific aims. Step 2 (Selecting a Study Approach) provides new sections on minimizing bias, conducting person–time analysis, understanding the theoretical paradigms that guide research, using age standardization methods, and other aspects of study design. Step 3 (Designing the Study and Collecting Data) includes new subsections on rigor and reproducibility, ethics, quality assessment, grant writing, grant management, and other themes. Step 4 (Analyzing Data) contains additional coverage of statistical tests, qualitative analysis, and use of analytic software programs. Step 5 (Reporting Findings) includes new sections about using formal sources, critically revising manuscripts, understanding open-access and other publication models, and other topics.

CHAPTER 1

The Health Research Process

© DmitriyRazinkov/Shutterstock

Health research is the process of systematically investigating a single, well-defined aspect of physical, mental, or social well-being.

1.1 Clinical and Population Health Research

Health is defined by the World Health Organization as "a state of complete physical, mental, and social well-being, and not merely the absence of disease or infirmity." **Determinants of health** are the biological, behavioral, social, environmental, political, and other factors that influence the health status of individuals and populations. **Health research** is the investigation of health and disease or any of the factors that contribute to the presence or absence of physical, mental, and social health among individuals, families, communities, nations, or the world population as a whole. Health research encompasses both clinical research and population health research.

Medicine is the practice of preventing, diagnosing, and treating health problems in individuals and families. A clinician is a medical professional who provides direct care to patients or clients. Clinicians include physicians, surgeons, nurses, psychiatrists, physician assistants, midwives, registered dietitians, and other skilled professionals who work in medical settings as well as practitioners in dentistry, clinical psychology, podiatry, physical therapy, occupational therapy, pharmacy, optometry, kinesiology, rehabilitation, and other health-related fields. Clinical practice is contrasted with laboratory work and research that do not involve direct interaction with patients or clients.

Clinical research evaluates the best ways to prevent, diagnose, and treat adverse health issues that adversely affect individuals and families. For example, clinical research projects may examine the progression of a disease over time, compare the effectiveness of various therapeutic regimens, or test the safety and utility of new diagnostic tests, medications, or medical devices.

Clinical research generally uses humans as the unit of investigation, whereas **basic medical research** (also called basic science) studies molecules, genes, cells, and other smaller biological components related to human function and health. **Translational research** bridges basic research and clinical research by applying scientific discoveries to the improvement of clinical outcomes. The aim of translational medicine is to move research from the bench (the laboratory) to the bedside (clinical care settings).

Population health focuses on the health outcomes and the determinants of health in groups of humans. Population health is a function of many factors, including human behaviors, the social and economic environment, the physical environment, access to healthcare services, and many other exposures. **Public health** consists of the actions taken to promote health and prevent illnesses, injuries, and early deaths at the population level. Public health practitioners monitor health status in communities, mitigate environmental hazards, provide health education, support community health partnerships, develop public health policies, enforce safety regulations, and ensure access to essential health services.

Population health research examines health outcomes at the community, regional, national, and worldwide levels. For example, public health researchers assess population needs and capacities; design, implement, and test population-based health interventions; and evaluate population-based health programs, projects, and policies.

Both clinical and population health research studies apply the tools from a diversity of academic disciplines. Clinical research often draws on the tools of the laboratory sciences, such as molecular biology, microbiology, immunology, nutrition, and genetics. Population health research often uses tools from **demography**, the study of the size and composition of populations and of population dynamics, such as birth and death rates. Many clinical and population health research studies use the methods of social science fields such as psychology, sociology, anthropology, and economics.

1.2 The Research Process

Research is the process of systematically and carefully investigating a topic in order to discover new insights about the world. No matter what the goals of a research project are or what methods are used to achieve those goals, the five steps of the research process are the same (**Figure 1-1**). The first two steps are identifying a study question and selecting a general study approach. These two steps are often completed concurrently, because the approach selected may require the refinement of the study question. After the objectives and the approach are set, the last three steps are designing the study and collecting data, analyzing the data, and reporting the findings. These steps apply to nearly every research project. A research project is not finished until all five steps have been completed.

1.3 Text Overview

This text is a handbook for clinical and population health researchers. The chapters are organized according to the five steps of the

FIGURE 1-1 The Research Process

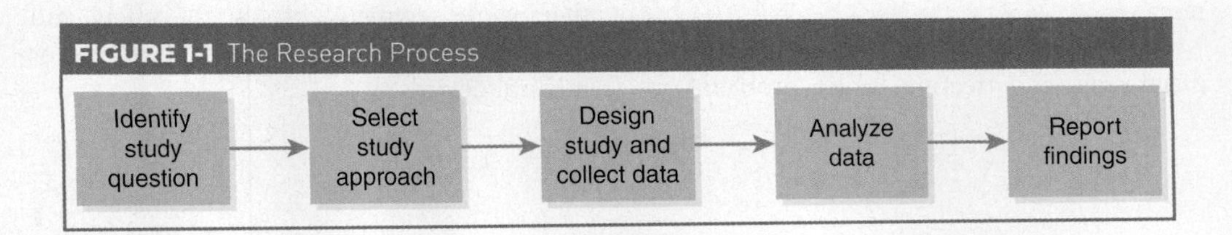

research process. The first section provides suggestions for selecting an appropriately focused research question and establishing good relationships with collaborators and mentors early in a project. The second section presents an overview of each of the main study designs used in the clinical and population health sciences. The third section describes research ethics and the data collection process. The fourth section summarizes common strategies for quantitative and qualitative data analysis. The fifth section presents tips for writing success and a step-by-step guide for preparing a manuscript for review and publication. If the goal is to publish the findings of a study, it may be helpful to write throughout the research process. Thus, some readers may find it helpful to read some of the chapters from the fifth section of the text prior to finalizing their research plans.

The best way to learn about health research is to do actual research and to learn firsthand how the research process works. This guidebook provides a comprehensive overview of the entire research process and details about the most common methods used in clinical and population health research. It is not intended to be a compendium of everything that health researchers know about study design, data collection, and statistical analysis. As a research project unfolds, most researchers will benefit from consulting specialized references. Many excellent books, journal articles, technical reports, and online and library resources contain the advanced information required for complex study designs and analytic techniques. It is also essential for the consulted resources to include human experts—professors, supervisors, colleagues, librarians, statistical consultants, and others—who can provide insights gained from personal research experience and can direct new investigators to the background readings and other information that will be most helpful as they explore their selected research questions.

Anyone who completes a novel, valid research project can contribute to advancing clinical and population health. Health research does not require a license. It does not require a doctorate or a master's degree. It does not even require coursework in research methods, although that is certainly helpful. What research demands is perseverance and patience, honesty and integrity, carefulness and attention to detail, the willingness to learn new knowledge and develop new skills, openness to expert advice and feedback, and the ability to criticize and revise one's own work and writing. These are personal character traits that everyone can cultivate and develop along with the technical skills required for research excellence.

STEP 1

Identifying a Study Question

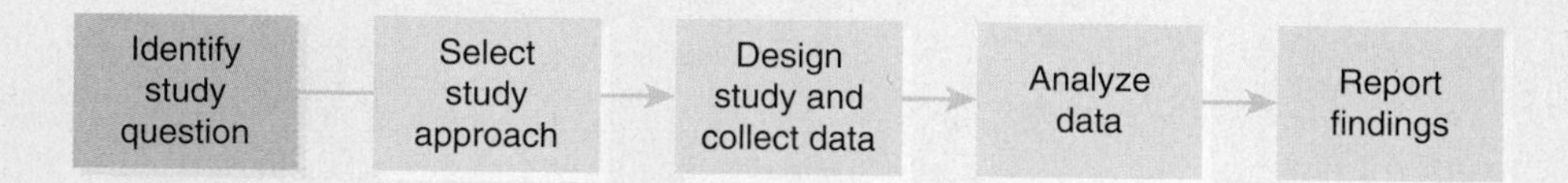

The first step in the research process is selecting the focus of the study. This section describes how to select a research question, review the literature, refine the scope of the project, and work with mentors and collaborators.

- Selecting a research question
- Reviewing the literature
- Defining specific aims
- Professional development
- Coauthoring

CHAPTER 2

© DmitriyRazinkov/Shutterstock

Selecting a Research Question

Selecting one focused study topic is the first step toward completing a successful research project.

2.1 Practical Research

Many research questions in the health sciences arise from observations made during applied practice. Consider the types of questions that different health practitioners might raise about trampoline injuries:

- An emergency room physician: "The trampoline injuries we've been seeing include a mix of limb fractures and head/neck trauma. Are the kids who present with trampoline-related arm fractures being screened for concussions? Are they being adequately treated for multiple injuries?"
- A physical therapist: "It seems like a lot more of the patients coming in for therapy this year are recovering from trampoline-related injuries. Has the rate of injuries really increased, or am I just noticing them more? Should I be telling my patients not to use home trampolines?"
- A health educator: "I'm working at a fitness center that offers trampoline workouts. Is this an effective way to improve cardiovascular fitness? What can we do to ensure the safety of our clients?"

Any of these questions might be worth exploring in a new research project, assuming that a review of the literature shows that there is not yet consensus about the answers to these questions.

A good first step toward selecting a research topic is to think about the various questions about health that have arisen from personal experiences, coursework, clinical or public health practice, and informal reading about subjects of interest (**Figure 2-1**). Practical questions about who, where, when, what, why, and how often point toward unmet demand for descriptive studies, needs assessments, program evaluations, clinical effectiveness studies, and other types of health-related research.

FIGURE 2-1 Brainstorming Questions

Area	Questions
Interests	What are my interests? What health-related conditions have significantly affected me, my family, my friends, my patients or clients, my community, and/or other populations that I care about?
Aptitudes	What knowledge and skills do I already have? What topics am I prepared to study in depth? What methods am I prepared to apply? What methods am I eager to learn?
Applications	What studies would help improve health-related practices or policies? What are the gaps in the literature that I can fill?
Mentors	What are the areas of expertise of my supervisors, professors, and/or other mentors? What source populations and/or data sources might be available to me through professors, supervisors, colleagues, and other personal and professional contacts?

2.2 Brainstorming and Concept Mapping

After thinking about the types of questions that have arisen from personal and professional experiences, the researcher can use a brainstorming session to convert those areas of interest into potential research themes. In this context, **brainstorming** is the process of generating long lists of spontaneous ideas about possible research questions. This is not the stage for eliminating ideas because they do not appear feasible. The ideas do not yet need to be well formed. The goal of brainstorming is to generate a lengthy list of possible topics. In addition to compiling one's own ideas, it can be valuable to check with colleagues, practitioners, and friends about their thoughts. Internet searches, journals, and books might reveal gaps in knowledge that are worth exploring. For example, many research articles end with a call for further research on a particular topic.

A related process is **concept mapping**, a visual method for listing ideas and then grouping them to reveal relationships. The first step of the mapping process is using brainstorming to generate a list of words or phrases that describe topics that might be integrated into a research question, such as the names of health conditions, population groups of particular interest, and the biological, socioeconomic, environmental, or other potential risk factors for various health outcomes. Next, the related ideas that show up several times on the list and appear to be part of a central theme are identified. Circles and arrows are used to group related topics and to visualize the connections between those groups. After some initial decisions about research topics that might be worth exploring have been made, the process of listing words and phrases and then visually grouping them may be repeated. This concept mapping technique can be useful when selecting and refining a study question. (A similar process can be used as part of qualitative data analysis.)

2.3 Keywords

A next step toward refining the areas of interest identified through brainstorming and concept mapping is compiling a list of keywords pertaining to the selected research area. A **keyword** is a word, a MeSH term (described in the following paragraph), or a short phrase that can be used in a database search. For example, a person whose brainstorming and concept mapping processes identify an interest in aging might list keywords like osteoporosis, falls, bedsores, physical therapy, calcium, bone density, home safety, rehabilitation, healthy aging, and prevention. A person who identifies an interest in child health in lower-income countries might list words like children, malaria, bednets, Uganda, measles, vaccination, preschool children, malnutrition, vitamin A deficiency, and community gardens. These keywords can then be explored as potential study foci within the major area of interest.

MeSH (*Medical Subject Headings*) is a vocabulary thesaurus that can be used for searches of MEDLINE and other health science databases. MeSH was developed by the U.S. National Library of Medicine (NLM), which is part of the U.S. National Institutes of Health (NIH). NLM's MeSH database can help a researcher to identify the full extent of a research area and to narrow the scope of a research inquiry. Suppose, for example, that infection is a potential area of interest. The MeSH database suggests a variety of narrower topics related to infection, such as cardiovascular infections, eye infections, sepsis, infectious skin diseases, and wound infections. Within the category of skin diseases, the MeSH database lists a variety of narrower topics, such as cellulitis, dermatomycoses (fungal skin infections), and bacterial skin diseases. Within the category of dermatomycoses, the MeSH database lists yet narrower topics, such as blastomycosis, cutaneous candidiasis, mycetoma, and tinea. Within these categories, MeSH offers even more refined points and still more refined points within successive subcategories.

Searching through the MeSH database can help a researcher in several ways. The researcher can move from a vague interest in infections or skin infections to a more focused interest in fungal skin infections or, even more specifically, ringworm infections. Alternatively, the MeSH database also provides information about broader or related study ideas. A search for preeclampsia, for example, shows that preeclampsia is a type of pregnancy complication. It is related to other forms of pregnancy-induced hypertension, such as HELLP syndrome, which may be an equally interesting study topic for someone with an interest in obstetrics.

After using the MeSH database to expand or focus the list of keywords, the researcher can look for the themes that emerge from those terms. Some potential research topics may be easily eliminated because they do not fit the researcher's interests. Some keywords may stand out as particularly interesting to the investigator.

2.4 Exposures, Diseases, and Populations (EDPs)

Many topics in population health research can be expressed with the following formula: "[exposure] and [disease/outcome] in [population]." Keywords can be filled into this formula to generate possible research questions.

An **exposure** is a personal characteristic, behavior, environmental encounter, or intervention that might change the likelihood of developing a health condition (**Figure 2-2**). Health research often seeks to determine whether an exposure is risky or protective. A **risk factor** is an exposure that increases an individual's likelihood of subsequently experiencing a particular disease or outcome. A **protective factor** is an exposure that reduces an individual's likelihood of subsequently experiencing a particular disease or outcome.

A **nonmodifiable risk factor** is a risk factor for a disease that cannot be changed through health interventions. For example, age is the leading risk factor for many noncommunicable diseases. Although there are many interventions that can promote healthy aging,

FIGURE 2-2 Examples of Types of Exposures

Socioeconomic Status	Health Behaviors	Health Status	Environmental Exposures
■ Income ■ Wealth ■ Educational level ■ Occupation ■ Age ■ Sex/gender ■ Race/ethnicity ■ Nationality ■ Immigration status ■ Marital status	■ Dietary practices ■ Exercise habits ■ Alcohol use ■ Tobacco use ■ Sexual practices ■ Contraceptive use ■ Hygiene practices ■ Religious practices ■ Use of healthcare services	■ Nutritional status ■ Immune status ■ Genetics ■ Stress ■ Anatomy and anatomical defects ■ Reproductive history ■ Comorbidities (existing health problems)	■ Drinking water ■ Air pollution ■ Radiation ■ Noise ■ Altitude ■ Humidity ■ Season ■ Natural disasters ■ Population density ■ Travel

there is no way to reduce age. A **modifiable risk factor** is a risk factor for a disease that can be avoided or mitigated. Identifying modifiable risk factors enables effective preventive interventions to be developed.

Three levels of prevention address modifiable risk factors at different stages of disease progression. **Primary prevention** encompasses health behaviors and other protective actions that help keep an adverse health event from occurring in people who do not already have the condition. Examples of primary prevention actions include nutritious diets, frequent exercise, adequate sleep, vaccinations, and use of seatbelts and other safety equipment. **Secondary prevention** is the detection of health problems in asymptomatic (nonsymptomatic) individuals at an early stage when the conditions have not yet caused significant damage to the body and can be treated more easily. Secondary prevention interventions include cancer screening, blood pressure checks, and routine vision and hearing tests. **Tertiary prevention** consists of interventions that reduce impairment, minimize pain and suffering, and prevent death in people with symptomatic health problems. Tertiary prevention interventions include rehabilitation, palliative care, medications, and surgery.

An **outcome** is an observed event such as the presence of disease in a participant in an observational study or the measured endpoint in an experimental study. For many health research projects, the outcome studied is a **disease**, defined as the presence of signs or symptoms of poor health (**Figure 2-3**). Clinically, a disease is a pathophysiological condition while a **disorder** is a functional impairment that may or may not be characterized by measurable structural or physiological changes. In the EDP framework—exposure, disease, and population—the term "disease" encompasses diseases, disorders, injuries, and other health conditions and outcomes. The particular outcome of interest associated with a disease might be **mortality** (deaths) or might be **morbidity** (nonfatal illnesses). (The term **comorbidity** describes two or more adverse health conditions occurring at the same time. Comorbidities can complicate the management of chronic health disorders, so comorbid conditions are sometimes classified as exposures that affect outcomes associated with the primary disease of interest.) Alternatively, the outcome might be related to quality of life or use of health services.

A **population** is a group of individuals, communities, or organizations (**Figure 2-4**). A population could consist of the patients of a hospital, the clients of a community-based organization, the students attending a school, the employees of a large corporation, the

FIGURE 2-3 Examples of Types of Diseases

Infectious and Parasitic Diseases	Noncommunicable Diseases	Neuropsychiatric Disorders	Injuries
▪ Candidiasis ▪ Cholera ▪ *Escherichia coli* ▪ Hookworm ▪ Influenza ▪ Malaria ▪ Syphilis ▪ Tuberculosis	▪ Asthma ▪ Breast cancer ▪ Cataracts ▪ Diabetes ▪ Hypertension ▪ Osteoporosis ▪ Stroke	▪ Alzheimer disease and other dementias ▪ Autism ▪ Depressive disorders ▪ Posttraumatic stress disorder ▪ Schizophrenia	▪ Bone fracture ▪ Burn ▪ Crush injury ▪ Drowning ▪ Frostbite ▪ Gunshot wound ▪ Poisoning

FIGURE 2-4 Examples of Types of Populations

- Australian children younger than 5 years old
- Women living in rural Ontario
- Adults with diabetes
- Teachers with at least 10 years of classroom experience
- Individuals newly diagnosed with influenza at St. Mary's Hospital in Newcastle
- Nongovernmental organizations working on issues related to HIV/AIDS in South Africa

residents living in a town or county, or any other well-defined set of people.

The keywords compiled after brainstorming and concept mapping often fit into EDP categories. To use a "Mad Libs"–style approach to creating a research question, the final set of keywords are divided into three separate lists:

- One for exposures or interventions
- One for diseases or other outcomes
- One for specific populations

For studies examining the links between two different health conditions, one disorder may be classified as an exposure of interest and a second disorder as the outcome. For experimental studies, the intervention being investigated is the exposure.

These **EDPs**—exposures, diseases (or other health-related outcomes), and populations—can then be combined to form potential study questions using a standard format of "Is [exposure] related to [disease/outcome] in [population]?" For example:

- Are exercise habits [exposure] related to the risk of bone fractures [disease] in adults with diabetes [population]?
- Is reproductive history [exposure] related to the risk of stroke [disease] among women living in rural Ontario [population]?
- Is household wealth [exposure] related to the risk of hospitalization for asthma [disease] in Australian children younger than 5 years old [population]?

A literature review related to the candidate question will assist the researcher in determining what is already known about the topic and what new information a new study could contribute.

2.5 PICOT

Evidence-based medicine (EBM) uses the results of rigorous research studies to optimize clinical decision making. EBM starts with a comprehensive literature search about a particular aspect of risk, prevention, diagnosis, therapy, harm, prognosis, or another aspect of clinical care. The most relevant, high-quality reports are then evaluated and synthesized. Key findings are summarized in practice guidelines and other documents that clinicians can use to enhance patient care and improve outcomes. EBM is not intended to depersonalize the practice of medicine. EBM enables skilled clinicians to integrate the best research into their assessments of the most appropriate ways to care for individual patients. A similar process of **evidence-based practice** is used in a variety of fields to encourage experienced professionals to integrate research into their decision-making processes.

When developing clinical research questions and designing intervention studies that might serve as a foundation for evidence-based practice, one way to operationalize a research question uses the acronym **PICOT**:

- What is the *P*atient (or *P*opulation) group and *P*roblem that will be studied?
- What is the *I*ntervention that will be tested?
- What will the intervention be *C*ompared to?
- What is the *O*utcome of interest?
- What is the *T*ime frame for follow-up?

Sometimes the framework is presented as just PICO, without the *T*. Sometimes a final *S* is added to represent the *Setting* or the *Study design* for the scenario, making this a PICOTS framework.

The PICOT framework operationalizes the EDPs by requiring researchers to define the main exposure (*I*), disease/outcome (*O*), and population (*P*) that will be examined. The framework also prompts the researcher to begin thinking about who the participants will be (*P*), where the study participants will be found (*S*), and when the study will start and end (*T*). For studies that are not experimental, the *I* in PICOT can be replaced with an exposure that is not an intentional intervention. For study designs that do not require a comparison group, the *C* can be ignored.

Publications from previous experimental studies can be helpful for refining the PICOT items for a new study. If a previous study showed that an intervention was successful in one population, a new study might test whether the same intervention is successful in a different population. The *ICOT* can remain the same, with a new *P*. Examining the impact of the intervention in a different type of patient group or with a different time frame might yield a different result. These types of replication studies can provide valuable insights for practice. Similarly, if the previous trial was unsuccessful, testing that same intervention using a different comparison group or a different outcome might reveal ways in which the intervention is valuable in other circumstances.

2.6 From Inquiry to Research

Inquiry is the process of finding answers to questions that arise from personal experiences. Inquiry is grounded in curiosity about a problem or idea. A clinician might ask questions like:

- How well are my colleagues and I preventing, diagnosing, and treating our patients' health concerns?
- How effective are our interventions at improving our clients' health status?
- Does this procedure or process generate the intended outcome?
- What can we do to improve the quality of our services?

A public health practitioner might ask questions like:

- What is the overall health status of this population?
- What are the major health concerns in this population?
- What are the most common risk factors for illness, injury, disability, and death in this population?
- What health-related needs in this population are not being addressed?
- What projects, programs, and policies might improve the health status of this population?
- Is our health education program effective at improving knowledge, transforming attitudes, and changing behaviors?

Inquiry on its own is not synonymous with research, but inquiry processes can grow into research projects.

A personal inquiry project typically involves identifying a practice-related question, thinking about possible answers that align with one's own experiences and observations, searching the Internet for information that might support or challenge these ideas, and then generating hypotheses about solutions. This inquiry process might be sufficient for identifying ways to improve the quality of clinical or public health practice. However, some inquiry projects reveal evidence gaps that are best addressed with more formal research processes. In these situations, there may be opportunities to expand routine practice activities into innovative research endeavors that use structured scientific protocols to test hypotheses and advance generalizable knowledge. Consider three examples of how inquiry projects grow into research projects.

Example 1: The administrators of a physical therapy practice ask their clients to complete a customer satisfaction survey. The survey results reveal several opportunities for quality improvement. The administrative team decides to test a new set of procedures for communicating with clients about scheduled appointments. Rather than just implementing the changes, they decide to conduct a research experiment. They develop a protocol for evaluating whether the new procedures improve client satisfaction and reduce the number of missed and rescheduled appointments. They get approval for the study from a research ethics committee, implement the experimental protocol, and then analyze the data they have collected. Finally, they present their results at a professional conference and publish their findings in the professional society's journal so that other practices can learn about best practices related to scheduling.

Example 2: A clinician is curious about how effective a particular treatment option is for a particular disease. The clinician searches the published literature on the topic and finds a few dozen relevant articles. Some describe the characteristics of the patients who respond best to the treatment and the characteristics of those who do not respond as well. Some compare the treatment of interest with other treatment protocols, describing the circumstances under which one option might be preferred over the other. However, the clinician does not find any review article that summarizes all of this information. The clinician decides to continue exploring this topic by conducting a rigorous synthesis research project. After writing a protocol for a systematic review of the literature, the clinician compiles all of the relevant articles, extracts and organizes the most important information from each article, summarizes the state of current knowledge about the effectiveness of the treatment, writes a paper that explains the clinical relevance of the findings, and submits that manuscript to a peer-reviewed journal.

Example 3: A group of epidemiologists working for a health department track down the source of an outbreak of gastroenteritis. The outbreak investigation team identifies an unusual food item as the likely cause of the outbreak. They realize that this could be an important discovery. If this food product is the culprit, large numbers of people could be at risk of severe illness. They decide to intensify their investigation. They interview the people with gastroenteritis again, using a longer questionnaire. They conduct a case–control study to see whether these people were more likely than healthy individuals to have consumed the suspected food item. They use laboratory testing to confirm their hypothesis that the food item is contaminated with the pathogen. They write a formal report describing their methods and results. They then use a variety of communication strategies to alert consumers to the possibility that the food item could cause illness.

The goal of any single health research project is usually modest: to answer one well-defined question. Some inquiry projects that are very specific to one population at one place during one point in time are not particularly helpful for identifying broader patterns. However, even small research studies—like the preceding examples—can reveal theories, trends, and associations that might be generalizable to other populations, places, and times. When researchers complete the health research process by sharing their findings with others, they are contributing to the evidence base that others will use when making decisions about health policy and practice.

2.7 Testable Questions

Testability is the ability of a research question to be answered using experiments or other types of measurements. A good research question is one that ends in a question mark and is testable.

A research project should not be framed in terms of a value statement like "Mental health is important" or "Tobacco use is bad." Even if these declarations are reframed as questions like "Is mental health important?" or "Is tobacco use bad?," these questions are not answerable because there is no simple scale for measuring concepts such as importance or badness. Values-based assertions may be relevant during the early stages of brainstorming about possible research topics, but they should be set aside when selecting EDPs and framing research questions, hypotheses, and theories. Value-neutral, measurable research questions like "Do older adults perceive mental health to be an important part of their overall health status?" or "Does parental tobacco use increase the risk of bronchitis in children attending elementary school?" should be constructed instead.

A research plan should also not be formulated based on a call to action like "People should exercise more" or "First aid should be a mandatory part of the middle school curriculum." These types of statements often appear in the implications or conclusion sections of research reports when the results of a study directly support the likely effectiveness and impact of the actions. Such claims cannot be supported before a study has been conducted. Reframing these assertions as questions like "Should people exercise more?" and "Should middle school students learn first aid?" does not make the topics measurable, because concepts such as "should" are difficult to quantify. However, calls to action based on the outcomes of previous research studies might point toward interventions that would be valuable to test. A research team could test whether a particular exercise program is effective at improving measurable health outcomes, such as weight, cholesterol levels, or self-reported quality of life, or the researchers could develop a set of lessons for teaching first aid to middle school students and then test the effectiveness of those learning materials in diverse classroom settings. These types of intervention studies reframe calls to action as genuine queries about how best to help people follow through on the desired actions.

2.8 Framing for Generalizability

In order to advance science and practice, research findings must be generalizable beyond the study population. Consider an example of the **Scholarship of Teaching and Learning (SoTL)**, the process of using systematic investigations to improve the quality of education. Suppose that an instructor tries a new active learning method with his or her own students. A manuscript that merely describes the class exercise will be of limited value to readers and is unlikely to be published. However, there are several steps the instructor can take to grow this personal inquiry into a research project that is valuable to other educators.

The instructor can use a validated method for showing that the exercise is effective in enhancing student learning, such as using a pretest and posttest to quantify what students have learned. A study that shows long-term gains in student competencies will be more valuable than one that is tested only at the end of the initial class hour, so a post-posttest could be added that evaluates retention of the core learning objectives from the exercise.

The instructor can add rigor by replicating the experiment. A class exercise conducted with multiple sections of the course would be better than one implemented in just one section. An exercise implemented by multiple instructors would be even better for demonstrating the effectiveness of the exercise and showing that the impact is not dependent on the instructor who is leading the class session. An exercise implemented at several different colleges and universities with diverse student populations would provide even stronger evidence for the effectiveness of the intervention.

The value of the research project could be enhanced by designing the learning exercise based on a recognized educational theory. The manuscript could explain how the theory that informed the particular learning exercise featured could also be used to inform the development of other learning activities. This framing shifts the context of the study from "I did this activity in my class" to "Here is an effective approach that you can use to develop new learning activities for your own classes." This framing moves the study question from a very specific context—one exercise in one class—to a more generalizable case study about a method that can be used in diverse practice settings. A strong research question is one that is likely to yield results that point toward generalizable knowledge or applications.

© DmitriyRazinkov/Shutterstock

CHAPTER 3

Reviewing the Literature

Reading publications related to the selected research question prepares the researcher to begin designing a new study that will fill a gap in the literature.

3.1 Informal Sources

A starting point for learning about potential areas of inquiry is to read nontechnical documents and other files available on the Internet. Many major public health organizations, such as the World Health Organization (WHO) and the U.S. Centers for Disease Control and Prevention (CDC), have online factsheets about various diseases and risk factors for disease. Other government agencies and nonprofit organizations also have factsheets, brochures, and websites that provide basic demographic, political, economic, geographic, and other health-related information about countries and regions. Newspapers and popular magazines may present compelling nontechnical articles about exposures, diseases, and populations that highlight what is interesting and important to know about a topic. The websites of patient advocacy groups, personal websites, and other media may also be helpful in identifying and refining an important and meaningful study question.

These initial background readings can provide a foundation for understanding the more technical scientific literature that will be read later as part of a thorough literature review. However, informal sources that have not been peer reviewed are not part of the formal scientific literature. Researchers must be cautious about any claims in these files that contradict more formal sources of scientific information. Informal sources should never be cited in formal reports and research manuscripts.

3.2 Statistical Reports

When defining specific exposures, diseases, and populations of interest, it may be helpful to identify relevant statistics, such as the estimated prevalence of an exposure in a particular country, the annual global incidence of a disease, or the demographics of a selected population.

For demographic, socioeconomic, and environmental data, reports from national governments are often a good source of up-to-date information. A diversity of statistical estimates are available for every member nation of the United Nations via the World Bank's World Development Indicators reports and the annexes of the annual reports issued by United Nations agencies, such as WHO's *World Health Statistics*.

For health-specific data, national health agencies may publish routine epidemiological updates and reports about **vital statistics**, which are population-level measurements related to births, deaths, and other demographic characteristics. For data about states, provinces, counties, cities, and other smaller governmental units, the relevant public health departments can be contacted. National and subnational estimates of various epidemiological metrics are also available from the Global Burden of Disease (GBD) project. Additionally, the annual reports of nonprofit organizations like the Population Reference Bureau and the American Cancer Society may include up-to-date statistical estimates and projections for relevant population groups. Published scientific articles are often the best place to find very specific information about health-related exposures and diseases.

Although statistics may be readily found on the Internet, few are supported by citations and details about who collected the original data, how and when the data were collected, and how key variables were defined. When possible, trace the statistic back to its original source rather than relying on secondary reports. If the source of data is not clear, the statistic may not be trustworthy.

3.3 Abstract Databases

An **abstract** is a one-paragraph summary of an article, chapter, or book. Abstracts for journal articles in the health sciences usually provide a brief description of the study methods, the study population (such as the sample size and the location of the study), the key findings of the study, and the main conclusion or implication of the investigation. An **abstract database** is a collection of abstracts that allows researchers to search for articles using keywords or other search terms. Online databases allow researchers to quickly search thousands or even millions of abstracts for relevant articles. A careful and comprehensive search of at least one major abstract database is the most important component of a literature search.

Some health abstract databases are available to the public at no cost (**Figure 3-1**). The most popular publicly available health science database is **PubMed**, a service of the U.S. National Library of Medicine that provides access to nearly 30 million abstracts of journal articles. Open-access databases are also available for other world regions. European PubMed Central (Europe PMC) is similar to PubMed but has more extensive coverage of European and Canadian journals. SciELO (the Scientific Electronic Library Online) and LILACS (Literatura Latino Americana e do Caribe em Ciências da Saúde) primarily focus on literature from Central and South America, and they allow searches to be conducted in English, Spanish, and Portuguese. African Journals Online (AJOL) allows searches of partner journals published by African institutions. Other national and regional abstract databases allow for searches to be conducted in other languages.

Many additional tools for finding relevant publications are available through university libraries that pay for subscriptions. Some are discipline-specific databases like CINAHL (nursing and allied health) and PsycINFO (psychology). Some are products provided by companies that produce, manage, and distribute online journal collections, such as:

- EBSCO
- JSTOR
- LexisNexis (which focuses on business and law)
- Ovid (a product of Wolters Kluwer)
- ProQuest

Some large publishing companies also offer databases of the articles published in the journals they own or publish, such as:

- Elsevier, which owns ScienceDirect
- Lippincott Williams & Wilkins (LWW), which is owned by Wolters Kluwer
- SAGE Publishing

FIGURE 3-1 Examples of Abstract Databases

Database	Description	Status
AJOL	African Journals Online	Open
CAB Direct	a product of the Centre for Agriculture and Biosciences International (CABI)	Subscription
CINAHL	Cumulative Index to Nursing and Allied Health Literature, a product of EBSCO	Subscription
Embase	a product of the large publishing company Elsevier	Subscription
ERIC	Educational Resources Information Center, a U.S. Department of Education program	Open
Europe PMC	Europe PubMed Central, similar to PubMed but with more extensive coverage of European and Canadian journals	Open
Google Scholar	a web search engine that indexes scholarly products such as book chapters, journal articles, and abstracts	Open
LILACS	Literatura Latino Americana e do Caribe em Ciências da Saúde, which focuses on literature from Central and South America	Open
MEDLINE	a U.S. National Library of Medicine database that features only journals that have applied for inclusion and passed through a review process	Open
PsycINFO	a product of the American Psychological Association (APA)	Subscription
PubMed	a service of the U.S. National Library of Medicine that includes the MEDLINE database plus additional entries (including articles published about research projects funded by the National Institutes of Health)	Open
SciELO	Scientific Electronic Library Online, which focuses on literature from Central and South America published in English, Portuguese, or Spanish	Open
Scopus	a product of the large publishing company Elsevier	Subscription
SPORTDiscus	a product from EBSCO that focuses on sports and sports medicine research	Subscription
Web of Science	a product from Clarivate Analytics that includes journals from the sciences and social sciences	Subscription

- Springer Nature
- Taylor & Francis
- Wiley-Blackwell, which owns the Wiley Online Library

A librarian can provide advice about the best databases to use for particular research questions.

Even though these health science databases cover many thousands of journals, many peer-reviewed journals are not included in any of the standard databases. Journals that are not published in English are especially likely to be omitted. A supplemental search with a general search engine like Google Scholar may be helpful for identifying additional relevant abstracts. Additional searching is especially important when the search of discipline-specific databases yields only a small or moderate number of hits. Google Scholar has been criticized for prioritizing frequently cited articles over others that might be more relevant and for including low-quality journals that do not use rigorous peer-review processes, but it has a massive database of scholarly products. No matter what databases are searched, researchers need to assess the quality of the articles they find rather than trusting that they are rigorous simply because they have been published in a journal.

Abstract databases can be searched with keywords or MeSH terms using Boolean operators like AND, OR, and NOT. Limits can be set so that results include only abstracts with particular publication years, languages, or other selected parameters. Databases can also be searched by article title, author (often using a last name and first initials format, such as "Baker JD" or "Patel AR"), and journal title. See Chapter 26 for more information about how to successfully search abstract databases.

3.4 Full-Text Articles

Abstracts provide a glimpse into the content of an article. However, the only way to truly understand a study is to read the full text of the article. There are several ways to access the full-text versions of articles identified during a database search.

Some articles are freely available online in their entirety as open-access files on journal websites, in digital archives like PubMed Central, or on the personal websites of the authors themselves. Many open-access articles can be found by searching PubMed or other databases and looking for full-text links. For example, an article that has been added to the PMC repository will be marked as a "Free PMC Article" in PubMed. When an abstract database does not provide a link to the full text, an Internet search tool like Google Scholar can be used to see if the full text is available in another repository.

Most university libraries subscribe to thousands of online journals that allow patrons to access electronic versions of articles. This is true even for universities in low- and middle-income countries, where research institutions can gain free access to hundreds of journals through WHO's Hinari Access to Research in Health Programme. (Most university libraries also have a limited number of journals available in print form on their shelves, but a physical search of the stacks is unlikely to be required unless the article is relatively old.) When a particular article is not part of a university's collection—either because the library does not subscribe to the journal or because the subscription does not cover the years or issues in which the article was published—it may be possible for a librarian to request a copy of the article from a partner institution. These "loans" of journal articles usually take the form of electronic files or photocopies of the article that do not need to be returned. Universities often offer free or low-cost interlibrary loan services to affiliates, but they may restrict the number of requests that one patron can make during any given month or year.

When an article is not freely available online or through a library, another option is to contact one of the authors of the article directly and ask for a copy. The email addresses for coauthors are sometimes included in abstract databases. Sometimes they are available on the publisher's website. For example, some subscription journals show the first page of an article as an image on their websites, and the email address for the corresponding author is often at the bottom of that page. At a minimum, many database entries and publication websites list the institutional affiliations of authors, and it is often easy to find email addresses for the authors by searching the websites of those institutions. There is no risk in writing to an author to politely request an electronic copy of an article. Most authors will be flattered that someone is interested in their work. At worst, the requester will get no response from the author. At best, the author will send an electronic copy of the article and an offer of further assistance. Requests may also be sent through professional networking platforms, but only if the authors have active accounts with those social media sites.

3.5 Critical Reading

Once the researcher acquires a copy of the full-text article, a practical plan of action is to:

- Reread the abstract.
- Look carefully at the tables and figures, which usually display the most important results.
- Read (or at least skim) the entire text of the article.
- Review the reference list for any additional sources that should be read.

All articles should be evaluated carefully. Critical reading involves asking a series of questions about the **internal validity** of a study, the evidence that a study measured what it intended to measure. Checks of internal validity seek to ascertain how well a particular study was designed, conducted, interpreted, and reported so that conclusions can be made about how likely it is that the resulting paper presents the truth about a particular research question in a particular population at a particular place and time (**Figure 3-2**).

Critical reading also requires questioning the **external validity** of the study, which is the likelihood that the results of a study with internal validity can be generalized to other populations, places, and times. **Generalizability** means that the results of one study are considered to be applicable to a broader target audience. Examples of questions about external validity include:

- How well do the findings of this study fit with existing knowledge about the topic? Have replication studies in diverse populations supported the generalizability of the findings?

FIGURE 3-2 Sample Questions for Critical Reading

Section	Key Questions
Introduction/ Background	What was the goal of the study? (For example, was the goal to describe a disease, examine the association between an exposure and an outcome, or test a hypothesis using an experimental design?)
Methods	Were the methods appropriate for the goal? Were the methods used to collect and analyze data scientifically valid? For example, did a study collecting new survey data use an appropriate sample population, recruit an adequate number of participants, define exposure and outcome variables appropriately, use a validated questionnaire, and apply appropriate statistical tests? Was the study conducted ethically?
Results	Was the main study question answered? Do the results seem reasonable?
Discussion	Have the authors acknowledged and discussed the limitations of the study methods? What types of bias in the design, conduct, analysis, and interpretation of the study might have caused some of the results to be inaccurate?
Conclusion	Are all of the implications and conclusions mentioned in the research report supported by the study's results? (For example, if a study claims that an intervention caused an outcome, does the article provide sufficient evidence to support that claim of causality?)

- For experimental studies, how likely is it that the observations from the trial would occur in everyday life outside laboratory conditions?
- To what other populations might the results apply? For example, are results from a study in Canadian men ages 30 to 49 years likely to be applicable to Mexican men ages 30 to 49, Canadian women ages 30 to 49, or Canadian men ages 50 to 69?

3.6 Annotated Bibliographies

One of the common approaches for tracking the articles identified during a literature review is the creation of an **annotated bibliography**, which is a list of related publications that includes, at minimum, a full reference for each document being reviewed, a brief summary of the article or report, and a note about the resource's potential relevance to the new study. Researchers may also find it helpful to make detailed notes about how the studies featured in the annotated bibliography might inform the aims and design of the new research project. The goal is not to replicate a document's abstract. The goal is to summarize the content most pertinent to the new investigation.

Some annotated bibliographies for a new research project are compiled in document files where summary paragraphs can be typed in for each source. Sometimes it is easiest to extract the most relevant details from each source into a spreadsheet, with separate columns for various elements of the research aims, study design, study population, definitions used for exposures and outcomes, statistical results, study limitations, interpretations, and/or the reader's evaluations of internal and external validity. Alternatively, a reference management software program can allow users to add personal notes or other annotations to designated fields in a record for each source.

3.7 What Makes Research Original?

Every researcher is looking for an original topic to study. Originality can be a paralyzing prospect for anyone who thinks that originality requires the discovery of a newly emergent disease in a previously unrecognized group of people who live on a remote island. Such remarkable discoveries are occasionally featured in the news, but even a cursory review of the literature proves that the vast majority of original research is far less dramatic. **Originality** describes the aspects of a new research project that are novel and will allow it to make a unique contribution to the health science literature. For a research project to demonstrate originality, it needs to have one substantive difference from previous work. That could be a new exposure of interest, a new disease of interest, a new source population, a new time period under study, or a new perspective on a field of exploration. It does not require a new exposure and a new disease and a new population.

Figure 3-3 illustrates the concept of originality. An original research project could look at a new potential risk factor (E_2) for a disease (D_1) that is already well studied in a population (P_1). It could look at whether an exposure (E_1) that is known to increase the risk of one disease (D_1) in a population (P_1) also increases the risk of a second disease (D_2). It could see whether the association between an exposure (E_1) and a disease (D_1) observed in one or more parts of the world (P_1 and P_2) is also true in another part of the world (P_3). A research project using a meta-analysis approach could aim to synthesize everything that has already been published on the association between an exposure (E_1) and an outcome (D_1).

For example, a literature review might find several studies showing that older adults (the population) who take 30-minute walks several times a week (the exposure) score higher on memory tests (the outcome) than

FIGURE 3-3 Ideas for New Studies

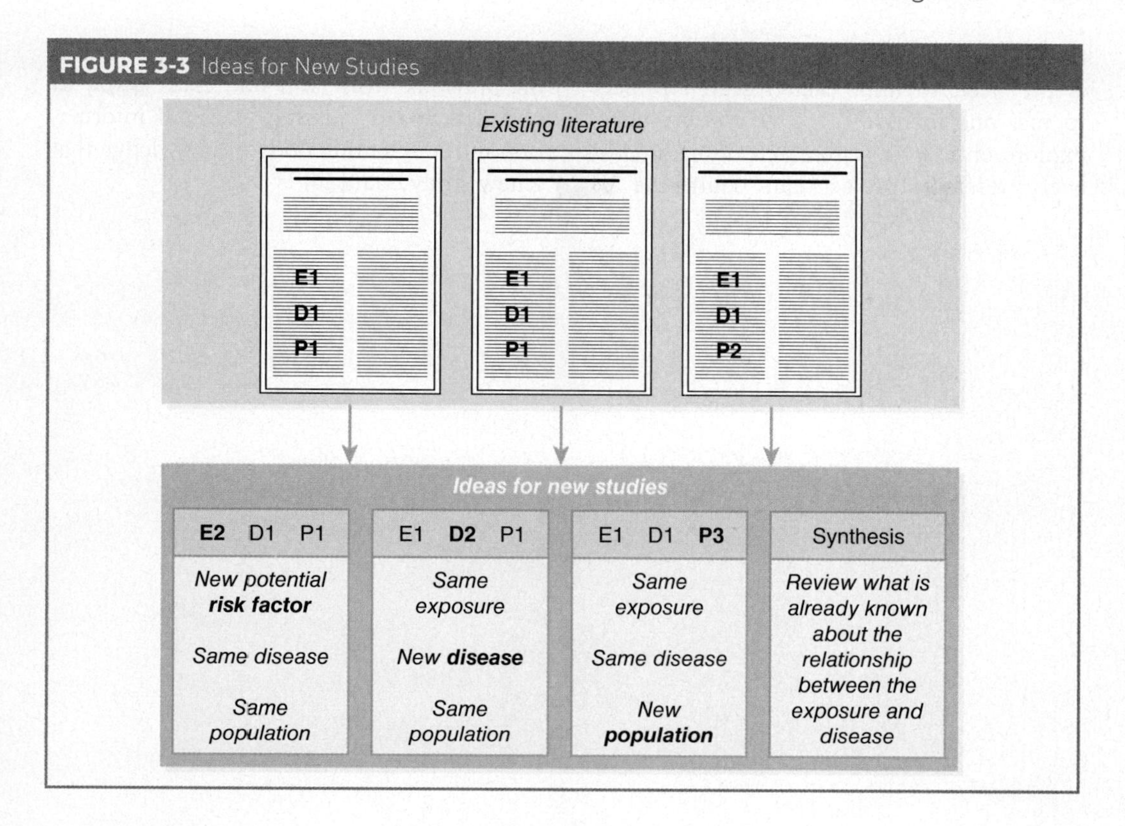

adults who do not routinely walk for exercise. A proposed new study could ask:

- Is playing table tennis (a new exposure) effective at improving memory in older adults (the same outcome and population)?
- Do older adults who walk several times a week (the same exposure and population) improve their balance (a new outcome)?
- Does walking (the same exposure) improve memory (the same outcome) in children (a new population)?

Once a researcher identifies a possibly novel research question, a more complete review of the literature can help confirm that the area has not already been examined.

It is even possible for a replication study to be considered original research. **Replicability** means that a study protocol implemented in a new study population should generate results similar to those of the original study, as long as the exact same protocol is used, including the same recruiting methods and inclusion criteria. **Replication studies** repeat a study protocol in a new population as part of attempting to confirm that the original findings were not due to chance. Replication studies use the same protocol for collecting data about selected exposures and outcomes that a previous study used, but they implement the study in a new population. This is an important way to determine whether observations in one population are generalizable to other populations.

Some new investigators struggle to identify a research topic that has not been previously explored in the literature, but a recognition that most research is about incremental steps forward opens up infinite options for new explorations. The main challenge when selecting a research question is the need to limit each research project to just one focused area. Very few studies explore entirely new research domains, but every research project can contribute to advancing a field of research when it builds on previous work and addresses **gaps in the literature**, missing pieces of information in the scientific body of knowledge that a new study could fill.

CHAPTER 4

Defining Specific Aims

The specific aims for a research project provide a structure for achieving the overall study goal.

4.1 Refining the Study Question

The principle of **purposiveness** states that research projects should be designed to answer one well-defined research question. After identifying a preliminary research area and conducting a review of the relevant literature, researchers are ready to finalize the selection of one very specific study goal that can be achieved via a series of specific aims or objectives. Before moving forward with designing a study and preparing a research proposal, researchers should take the time to think through the answers to four key questions:

- What is the one well-defined research question that the study will answer?
- What specific aims, objectives, or hypotheses will enable the key question to be answered?
- Would a conceptual framework be helpful for guiding the design, analysis, and interpretation of the study and its results?
- Is the proposed study feasible? Is there a high likelihood that the research team will be able to answer the study's main research question?

4.2 One Study Goal

A **study goal** describes the single overarching objective of a research project or the main question that a research project seeks to answer. **Figure 4-1** lists several types of common study goals in the health sciences. A goal or objective statement for a clinical or population health research project usually states the exposures, diseases or outcomes, and/or populations that will be the focus of the study.

As part of the process of focusing and finalizing the study goal, the researcher should be able to answer several critical questions about the context for the study and the justification for the value of the new study, including:

- What is the one key question that this study will answer? (For experimental research, what is the one central hypothesis that this study will test?)
- What is already known about the research topic? What is the gap or limitation in knowledge that needs to be addressed?
- What is the significance of the problem that the study will address? (For example, how severe is the disease of interest and how many people are affected by it?)

FIGURE 4-1 Examples of Types of Study Goals

- To describe the incidence or prevalence of a particular exposure or disease in one well-defined population
- To assess the perceived health-related needs of a community
- To compare the levels of exposure or disease in two or more populations
- To identify possible risk factors for a particular disease in a population
- To measure changes in population health status over time
- To test the effectiveness of a new preventive intervention, diagnostic test, assessment method, therapy, or treatment
- To evaluate whether an intervention shown to be successful in one population is equally successful in a second population
- To understand complex phenomena
- To examine the impact of a program or policy
- To synthesize or integrate existing knowledge

- What will be innovative about the research project? How will the study resolve the current gap or limitation in knowledge?
- What is the likely impact of this research project? If the study is successful, how will it help advance health in relevant populations?

The answers to all of these questions are often included in research proposals, so it is important to answer them early in the process of planning a new study.

4.3 Several Specific Aims

After the overall study goal has been defined, the researcher should identify several specific aims for the study. A **specific aim** (or **specific objective**) is a carefully described action that will help the researcher make progress toward achieving the big-picture goal. (The U.S. National Institutes of Health uses the term "specific aims" to describe the major components of a proposed study, while the National Science Foundation calls those parts of the study "objectives.") Most studies in the health sciences have two to four specific aims, with three the most typical number.

The enumerated items should take the form of a question or a "to" statement that uses an action verb, like "to measure" or "to compare." For observational studies, the specific aims must be testable even if the overall goal is descriptive rather than explanatory. For experimental studies, the specific aims may take the form of a series of hypotheses that will be tested. A **hypothesis** is an informed assumption about the likely outcome of a well-designed investigation that can be tested using scientific methods. A typical hypothesis statement uses an if–then–because format that specifies the independent manipulated variable (the "if") and dependent responding variable (the "then") that will be tested. Hypotheses should be falsifiable and written in neutral, unbiased language that allows the aim to be achieved no matter what the outcome of the experiment is.

There are two common ways to approach writing specific aims for research in the health sciences: sequential and independent. When writing sequential objectives, the specific aims are a chronological list of actions that will achieve the main goal. This is a popular approach in doctoral programs that require candidates for the degree to produce three manuscripts suitable for publication in scientific journals prior to defending the dissertation. For example, suppose a pilot study or a search of the published literature has shown that lead poisoning is a problem in some communities in the Detroit metropolitan area. A researcher could use this information as part of deciding that the overall goal of a new study will be "to assess the impact of lead poisoning on school performance by kindergarten students in southeast Michigan." The specific aims for this study might be:

1. To measure the prevalence of high blood lead levels in a random sample of kindergarten students in southeast Michigan
2. To identify the socioeconomic and environmental exposures that are associated with having high blood lead levels among children in the study population
3. To determine whether children with high blood lead levels have lower scores on academic tests than children with lower blood lead levels after adjusting for other exposures that might affect school performance
4. To estimate the total impact of high blood lead levels on kindergarten performance in southeast Michigan by applying information about risk factors and school performance in the sample population to the total population of the region

Using a sequential approach, aim 1 lays the foundation for aim 2, aim 2 lays the foundation for aim 3, and aim 3 is critical for the success of aim 4. A first journal article could present the results related to aim 1. A second article could cite the first paper and present the results related to aim 2. A third article could cite the first two papers and present results related to aim 3. A fourth article could cite the first three publications and present the results from aim 4.

When writing independent objectives, the specific aims are related but are independent of one another. If one objective cannot be achieved, it is still possible to successfully complete the other objectives. A researcher who wants to examine risk and preventive behaviors related to outdoor activities could select specific aims about insect and tick bites (aim 1), sun exposure (aim 2), use of protective sports gear (aim 3), and recreational water activities (aim 4). Suppose that so few study participants live near a lake where they can fish, swim, and boat outdoors that the study results pertaining to aim 4 lack the statistical power required for significance. The failure to achieve aim 4 will not prevent the success of the other aims. With this independent approach, separate manuscripts could be published about the results related to aim 1, aim 2, and aim 3. The researcher would not need to wait for the first paper to be published before submitting the second paper, because the context for the paper about aim 2 is not dependent on the results generated from aim 1.

The characteristics of a good goal statement—or, in research, a specific aim—are often summarized using the acronym **SMART**:

- *S*pecific
- *M*easurable
- *A*ttainable (or *A*chievable)
- *R*elevant (or *R*ealistic)
- *T*imely (or *T*ime-bound)

After drafting the specific aims for a project, the researcher should confirm that each one is SMART.

4.4 Conceptual and Theoretical Frameworks

Many research projects benefit from the development of a conceptual model or the selection of a theoretical framework that will inform the design, implementation, and interpretation of the study. A **conceptual framework** is a model that a researcher sketches using boxes and arrows to illustrate the various relationships that will be evaluated during a study. For example, a **directed acyclic graph (DAG)** uses nodes (also called vertices, the plural of vertex) and arrows (also called edges) to illustrate hypothesized causal pathways from distal exposures (determinants) to proximal exposures (immediate outcomes) to outcomes. Directed means there is at least one arrow connecting each variable in the model to another variable. Acyclic means the flows are all unidirectional. No loops are allowed, because nothing can cause itself. DAGs can also include variables and arrows representing biases or confounding factors that might distort or hide causal relationships between other variables. Early in the process of designing a study, the creation of DAGs may provide insight about what data need to be collected from study participants. Later on, DAGs provide a roadmap for hypothesis testing and other types of data analysis and a framework for using causal inference to interpret statistical results.

A **theoretical framework** is a set of established models in the published literature that can inform the components and flows of the conceptual framework for a new research study. Theoretical frameworks are especially common in the nursing, social science, and educational research literature. For example, several popular theories describe the factors that influence individual health beliefs and health behaviors, including the health belief model and the social ecological model.

The **health belief model** (HBM) considers health behavior change to be a function of perceived susceptibility to an adverse health outcome, perceived severity of the disease, perceived benefits of behavior change, perceived barriers to change, cues to action, and self-efficacy. An intervention study drawing on the HBM may seek to raise awareness of the risk factors for a disease and the severity of the condition, increase knowledge about the effectiveness of preventive actions related to the disease, or encourage participants to identify and overcome barriers to improved health practices. The research team will collect data about participant demographics, beliefs and behaviors before the intervention, and beliefs and behaviors after the intervention. The analysis will use the components of the HBM to try to understand why the intervention was or was not effective at achieving the desired behavior change in the study population or subgroups within the study population.

The **social ecological model** considers individual health and health behaviors to be a function of the social environment, which includes intrapersonal (individual), interpersonal, institutional (organizational), community, and public policy dimensions. An intervention study applying the social ecological model will seek to influence health behaviors by addressing multiple levels of influence, such as individual risk factors, social relationships, and institutional policies. The research team will collect data about individual, interpersonal, and geographic and sociocultural community characteristics and will use that information to guide the design, implementation, and evaluation of the intervention. When a theoretical framework is selected early in the study planning process, the model can also shape the framing of the study's specific aims.

4.5 Feasibility

An important consideration when narrowing the focus and clarifying the aims of a new research project is the likelihood that the study can be successfully completed. A **feasibility study** is an evaluation of the likelihood that a task can be completed with the time, money, technology, and other resources that are available for the activity. Feasibility studies are valuable for making decisions about whether to pursue a research idea. **Figure 4-2** summarizes some of the critical questions to ask

FIGURE 4-2 Questions Essential to the Success of the Project

Area	Questions
Purpose and significance	■ What will the study contribute? ■ What will be new and noteworthy about the study? ■ Can the importance and necessity of this project be justified? ■ How will the study enhance the body of knowledge in its discipline? ■ Who will benefit from the study besides the researcher? ■ How will the study help individuals and/or communities live healthier lives? ■ How might the study contribute to improving health practices and/or policies?
Scope and feasibility	■ Is the scope of the intended project reasonable and manageable—neither too broad nor too narrow? ■ Can the proposed study question actually be answered? ■ Can the researcher answer the proposed study question?
Capacity and collaborators	■ Does the researcher have the knowledge and skills needed to conduct the study? ■ Does the researcher have access to collaborators who have the expertise needed for the project?
Money and materials	■ Are there adequate financial resources to conduct the study? ■ Does the researcher have access to equipment, space, and other physical requirements? ■ Given the resources available, can the researcher reasonably expect to conduct a scientifically rigorous and valid study?
Time	■ Does the researcher have the time to conduct this study? ■ Does the researcher have the time to make this an excellent study that does not waste health resources?
Population or data	■ If the plan is to collect new data from individuals, does the researcher have access to a reasonable source population and an adequate number of participants? ■ If the plan is to analyze existing data or to write a review paper, does the researcher have access to a reasonable existing data set and/or to an extensive library collection?
Ethics	■ Will the researcher be making good use of the resources available? ■ Has the researcher considered the relevant ethical issues, especially those related to the collection and use of individual-level data? ■ Is the researcher prepared to conduct culturally appropriate and scientifically rigorous research?
Target audience	■ Who is likely to be interested in the findings? ■ Is the resulting paper likely to be publishable?

before committing to a particular project. The characteristics of a viable research project are also captured by the acronym **FINER**:

- Feasible
- Interesting
- Novel
- Ethical
- Relevant

Before moving on to the study design phase, it is helpful to confirm that the scope of the proposed project is manageable for the research team and that the project is likely to be successful.

CHAPTER 5

© DmitryRazinkov/Shutterstock

Professional Development

New researchers should assemble a team of collaborators and mentors early in the research process.

5.1 Research Teams

Scientific research is rarely completed by one person working alone. Although some papers in the health sciences have only one author, the typical paper has about four coauthors, and some have dozens of coauthors. For many projects, the bulk of the work is conducted by one **lead researcher**, defined here as the researcher who will do the majority of the work. (Sometimes the term "lead researcher" refers instead to the **senior researcher**, an experienced researcher who guides the work of a newer investigator.) The lead researcher's work typically is supported by several other contributors. Some of these collaborators may be senior specialists who contribute particular types of expertise to the project. Some may be assistants who are involved in labor-intensive aspects of data collection, data entry, and data cleaning. Those contributions make a project a team effort, even if the lead researcher spends many hours working independently on it.

Once an investigator has committed to doing a research project, it is helpful to assemble a team of collaborators early in the research process who can help ensure that the project will be:

- Scientifically valid
- Ethical and culturally appropriate
- Time- and cost-efficient

For students, the first step is identifying at least one professor or other experienced researcher to serve as a mentor. For early career professionals, one or more senior colleagues may be willing to serve as formal or informal mentors. Mentors can help the new investigator identify and connect with other potential collaborators, such as experts on the exposure or disease being examined, the study design or methods being used for the project, or the study population. Technical experts, such as statisticians, librarians, and laboratory specialists, may also be needed. For international research projects, at least one local researcher at the study site should be a coinvestigator who is involved in every step of the research process, including the selection of the study question, the design of the study, and the collection and interpretation of data.

Some of the individuals the lead researcher communicates with may become core members of the research team and earn coauthorship. Others may play a more limited role as consultants. The lead author should have a conversation with all potential contributors about the amount of time they want to dedicate to the project and their expectations regarding compensation and authorship. For example, a statistical consultant may ask to be paid by the hour to help a researcher think through analysis options as a non-coauthor, the statistician may waive the consulting fee but request coauthorship in return for the development of a data analysis plan, or another arrangement may be requested. The lead author should maintain a record of all the statistical consultants, librarians, laboratory technicians, interviewers, data managers, and others who contribute in a meaningful way to the project. When appropriate, the contributors who do not earn coauthorship can be thanked in the acknowledgments sections of manuscripts that benefited from their contributions. (Always ask for permission to thank people by name, because some people prefer not to have their names published.)

5.2 Finding Research Mentors

Mentorship is a formal or informal relationship in which an experienced mentor offers professional development advice and guidance to a less experienced mentee. It is advantageous for new investigators to seek out a team of several mentors who can provide guidance and advice about research, rather than relying on just one advisor or supervisor. Students writing theses or dissertations typically need to identify a primary mentor and recruit several other established scholars to serve on their supervisory committees, and each of these individuals will bring his or her own expertise and perspectives to the project. New investigators who are not part of a formal research training program may need to seek out their own supervisors and mentors for projects.

New investigators seeking mentorship can identify potential advisors by:

- Asking classmates, colleagues, professors, supervisors, and others about experienced researchers who might be helpful mentors based on shared research interests, the type of mentorship the new investigator is seeking, and whether the communication style of the potential mentor is a good match to that of the mentee
- Searching the profiles of researchers at the new investigator's home institution (or potential collaborating institutions) to see who is actively conducting and publishing research on relevant topics or using relevant methods
- Emailing the individuals identified as potential mentors to share a curriculum vitae (CV) or résumé and request an in-person meeting to learn more about those researchers' current projects and to ask for professional development advice

The new investigator should be prepared for the contacted individuals not to respond or to reply with a message indicating that they are not currently accepting new research assistants, interns, or mentees. Even if a meeting is scheduled, not all conversations will yield a mentor–mentee relationship. An invitation to meet is not an agreement to serve as a mentor. However, all conversations have the possibility of pointing the new investigator to useful resources, including contact information for other individuals who might be well suited to serve as mentors.

5.3 The Mentor–Mentee Relationship

Some formal research mentorship programs require both mentors and mentees to sign an agreement letter that spells out the commitments of both parties, but most mentorships are less formal. A new investigator should not agree to enter into a mentor–mentee relationship before gaining an informed understanding of several key matters, including:

- The potential mentor's time availability.
- The mentor's preferred frequency and style of communication (such as how often emails will be exchanged and how often telephone calls or in-person meetings will be scheduled).
- The roles and responsibilities the mentor agrees to take on.
- The resources the mentor agrees to provide, if the mentee expects the mentor to supply full or partial funding for a project, access to data, access to laboratory facilities or computing equipment, or other types of material support.
- The expectations the mentor has of the mentee.

Once a mentor–mentee relationship is established, there are many things a mentee can do to ensure that the partnership is a productive and pleasant one. Research supervisors appreciate when mentees:

- Communicate often.
- Ask questions.
- Are honest about what they have done and what they plan to do.
- Complete assigned tasks satisfactorily and on time.
- Maintain meticulous research records.
- Are open to receiving constructive criticism.
- Respect the mentor and the mentor's time.

5.4 Professional Development

Professional development is an ongoing and intentional process of establishing short- and long-term professional goals, identifying and completing activities that enable systematic progress toward achieving those goals, and routinely evaluating performance, competencies, and growth. Mentors can help aspiring researchers identify appropriate goals and actions that will help set the foundation for success in a long-term research career. Examples of professional development activities related to research include:

- Completing online or in-person coursework about research methods.
- Participating in journal clubs that read and discuss recently published research articles.
- Working as a research assistant to gain competencies related to methodology, technical skills, and professionalism.
- Becoming active in professional organizations that host research symposia, sponsor workshops, publish academic journals, and/or provide other opportunities for participating in research-related activities.
- Attending and presenting at local, regional, national, and/or international research conferences, and using this time for networking with both early career and established researchers.
- Enrolling in training programs, which may range from half-day workshops to years-long fellowships.
- Seeking out opportunities to practice **interprofessionalism**, the ability to work and communicate well with colleagues in different practice areas in order to achieve a shared goal.
- Using requirements for **continuing education**—the completion of approved learning activities in order to maintain a professional licensure or credential—to understand new discoveries in one's field of interest and to acquire new skills that can be applied to research projects.

5.5 Social Media and Impact Metrics

Several general and discipline-specific social networking platforms are available for researchers to use for networking, having online conversations about methodologies and tools, sharing resources, communicating about recent publications, and building a professional online presence. Users should assume that any content they add to these sites is or could be searchable and readily available to the public. All posts should be respectful and demonstrate professionalism. Negative and poorly written comments may be detrimental to the member's reputation as a health professional or scientist. Authors who want to post PDFs of published articles are responsible for confirming that they are not violating copyright laws. Most articles published in subscription journals are the property of the publisher and cannot be posted in their final form on open social media sites.

Some of these social networking sites rate or rank members based on quantifications of their contributions to the network (such as the number of questions answered on a discussion board) and/or their scholarly productivity. Some of the most common ranking tools are **bibliometrics**, quantitative analyses of publications and citations. Some bibliometrics apply to individual authors. For example, the ***h*-index** is a bibliometric that indicates that an author has at least *h* publications that have each been cited at least *h* times, and the ***i*10 index** is a count of the number of publications by an author that have each been cited at least 10 times. Some bibliometrics apply to the publications themselves, such as ones that quantify the average number of times an article published in a particular journal is cited. Every bibliometric is an incomplete measure of scholarly prowess and impact. For example, if a researcher has only ever published 1 article, the *h*-index and *i*10 index will remain at 1 even if that article is cited 10,000 times. Despite these limitations, bibliometrics are one of the factors used to make decisions about hiring, retention, and promotion in some subdisciplines of the health sciences.

While early career researchers should not be preoccupied with trying to make decisions based on how they might affect performance metrics in the future, an awareness of bibliometrics might influence some aspects of a professional development plan. To increase an *h*-index, *i*10 index, or other type of individual bibliometric, a researcher must publish high-quality articles that will be found, read, and cited by other researchers. Carefully crafted abstracts that use synonyms for key terms can increase the likelihood that other researchers will find an article through database searches. Articles published in widely indexed journals with higher impact factors are more likely to be found than ones published in newer and specialty journals. (Beware of low-quality journals that feature fake metrics on their websites.) Participation in online academic networks, appropriate use of relevant social media platforms, and maintenance of up-to-date profiles on institutional websites can sometimes also increase the visibility of a scholarly product and the likelihood that others will find, read, and cite it. Mentors can offer advice about what content to include on research websites and professional social media profiles.

5.6 Responsible Conduct of Research

Many universities offer training in the **responsible conduct of research (RCR)**, a concept that encompasses research ethics, professionalism, and best practices for collaboration and communication with other researchers. RCR training programs typically spell out expectations and procedures for disclosing conflicts of interest, avoiding research misconduct, reporting research ethics or personnel violations, and otherwise exhibiting professionalism. At their best, RCR programs are not just about avoiding or managing problems. They also explain how to proactively create a healthy, ethical, supportive research environment for all members of a research team.

A typical first experience with research is serving as a research assistant. Being a supporting member of a research team provides a valuable opportunity to become familiar with disciplinary and professional standards, academic writing and publishing, and the habits of good coauthors. Everyone working on a research project is responsible for ensuring the integrity of the project, which includes carefully following the study protocol that has been approved by research ethics committees, protecting study participants from harm, maintaining accurate records of all study data, disclosing potential conflicts of interest, and avoiding plagiarism and other forms of research misconduct.

When a research assistant has concerns about how a study protocol is being implemented or is unable to work productively because of interpersonal communication challenges within the research team, these concerns should be brought to the attention of the lead investigator on the project. If the issues remain unresolved after they are discussed with the lead researcher, it is acceptable to report the concerns up the chain of command. For example, in the academic setting, it would be typical to ask for a meeting with a department chair before seeking counsel from a dean or a senior administrator in the provost's office. If the concerns pertain to potential violations of institutional policies, professional codes of conduct, or government regulations, the appropriate office to consult is likely the institutional review board that approved the research protocol. Many organizations have a designated ombudsman who can provide confidential advice about which individuals or offices to contact about serious concerns when written policies are not clear about institutional roles and responsibilities.

CHAPTER 6

Coauthoring

Decisions about coauthorship should be made early in the research process.

6.1 Coauthorship

Coauthorship is the process of two or more collaborators working together to write a research report. Lead authors should construct the list of coauthors for a report, poster, or paper based on widely accepted disciplinary standards. All decisions about coauthorship should be transparent and should be communicated to all contributors, both those who are expected to earn coauthorship and those who will be acknowledged but not considered coauthors.

Most researchers serve as "middle" coauthors—ones who are not listed first or last in the order of coauthors—before moving into a lead author role for the first time. Students working for professors are often hesitant to ask if they will be listed as coauthors on the manuscripts that result from the projects they are supporting. Similar uncertainty might arise for those who are consulted about a research project because they have special technical or other skills but who are not asked to be involved in drafting the subsequent paper. Anyone who is contributing to a project and wants to be considered for inclusion in the authorship list should have a conversation with the lead author early in the research process so that roles and responsibilities can be clarified.

Good coauthors adhere to the highest ethical and professional standards in how they design studies, interact with collaborators and study participants, analyze data, and report their findings. They ask a lot of questions so that they fully understand the research project's protocols, the roles and responsibilities of all collaborators, and the decisions made about manuscript drafts. They pay attention to details, and they provide valuable feedback to the lead author and other members of the research team. They are committed to developing their technical writing skills. They disclose potential conflicts of interest. They accept responsibility for their own contributions and for the project as a whole. They treat all members of the research team with respect. They respond quickly to research communications, and they never miss a deadline.

6.2 Authorship Criteria

The **International Committee of Medical Journal Editors (ICMJE)** has established criteria for authorship in the health sciences that most journals in the field have adopted. According to the criteria listed in ICMJE's *Recommendations for the Conduct, Reporting, Editing, and Publication of Scholarly Work in Medical Journals* (as updated in December 2018), to earn coauthorship a researcher must meet all four of the following conditions:

1. Making substantial contributions to conception or design of the study and/or to data collection, analysis, or interpretation
2. Drafting the article and/or providing critical revisions of intellectual content
3. Approving the final version of the manuscript that is submitted to a publisher
4. Accepting responsibility for the integrity of the paper

A contributor does not have to engage in all parts of the study—designing the study and collecting the data and analyzing it—to be a coauthor. Participating in a meaningful way in any one of these parts of the study fulfills the first condition. However, participating in design, data collection, and analysis is not sufficient to earn authorship. Authorship requires participation in the writing of the research report. The second ICMJE authorship condition is that all coauthors must make a consequential intellectual contribution to the written product stemming from the research investigation, either by drafting part of the manuscript or by critically revising it. The third condition is intended to ensure that no one is listed as an author against his or her will or without his or her knowledge. A manuscript should not be submitted to a journal until all the coauthors have consented to the submission and agree to accept responsibility for the integrity of their contributions.

Here are some examples about how authorship eligibility is interpreted according to the ICMJE guidelines:

- A person who conducts interviews for the project but does not contribute further would not be eligible for authorship. However, an interviewer who also writes a paragraph for the discussion section might meet authorship criteria.
- A hospital laboratory technician who analyzes blood samples of patients included in a clinical study but makes no further contributions would not be eligible for authorship. A lab tech who analyzes the samples and writes part of the methods section describing laboratory techniques might be a coauthor.
- A data entry assistant who makes no additional contributions to the project would not be considered an author. A data manager who runs statistical tests and creates a results table for the manuscript might meet authorship criteria.
- A technical editor who cleans up the grammar and spelling in a manuscript does not earn authorship. A mentor who volunteers to edit the manuscript draft and then raises important questions about the interpretation of the results and the implications of the work might be eligible for authorship.

These guidelines are not intended to exclude people who have made significant contributions to a study from being included as coauthors. The rules are intended to ensure that everyone who has done the work of a coauthor receives recognition for that work and accepts responsibility for it.

There should be no **ghost authorship**, the failure to include as a coauthor on a manuscript a contributor who has made a substantial intellectual contribution to a research project. If there is doubt about whether an individual has made a significant contribution to a project, it is usually better to err on the side of inviting that individual to be a coauthor on the resulting manuscript. If the invitation is accepted, the individual must fulfill all of the authorship requirements, including participating in the writing process.

There should be no **gift authorship**, which occurs when someone who has not earned authorship according to disciplinary standards, such as those spelled out in the ICMJE authorship criteria, is added to the list of authors of a manuscript. Anyone who contributes to a paper without earning coauthorship can be named in an acknowledgments section but should not be listed as a coauthor. Before a manuscript is submitted to a journal, everyone listed as a coauthor must have explicitly approved the version of the manuscript that is being submitted and have agreed to accept all of the responsibilities of coauthors.

Senior researchers who provide funding and supervision for a project usually qualify for authorship because of their involvement in the research process, but supervision alone—when not accompanied by involvement in study design or interpretation as well as drafting or editing—is not sufficient to justify coauthorship. Just like any other contributor, sponsors and supervisors must make a meaningful intellectual contribution to a project to merit authorship.

6.3 Authorship Order

For most disciplines in the health sciences, the **first author** (or **lead author**) of a manuscript is typically the person who was the most involved in drafting the manuscript. Although this is often the person who took the lead in the whole study process from design through analysis and writing, this is not always the case. Sometimes the person who designed the study and collected the data is unable to conduct the analysis and write up the results, so that person (often a senior researcher) turns the responsibility of drafting the manuscript over to someone else, who is subsequently listed as the first author. Sometimes multiple people are involved in study design and data acquisition, and one person is asked by the group to take the lead on generating a draft of a manuscript. Sometimes organizations make data sets available to researchers for secondary analysis, and the organizations may not request authorship for any of the employees involved in study design or data collection. In all of these situations, the person who does most of the writing is often designated as the first author. When there is any doubt as to who is making the most significant contribution, the decision about who will be first author should be made in consultation with all of the people who took a major role in conducting the study.

The remaining authors—the "middle authors"—are typically listed in order of contribution, which is usually defined in terms of time dedicated to the project as well as intellectual contribution. The person who contributes the second most amount of time and energy to the project is listed as second author, and so on. When many coauthors are involved, it is sometimes difficult to quantify the relative contributions of, say, the seventh and eighth authors. In this situation, the coauthors should be consulted about their preferences, but the best solution may be to list authors with equal contributions in alphabetical order.

The key exception to the rule about listing authors in order of contribution is that the **senior author**, an experienced researcher, often the head of a research group or the primary research supervisor for a student, may choose to be listed last in the order of authors of manuscripts produced under that individual's guidance or supervision. In some health disciplines, the **last author** position designates the senior researcher in whose lab the work was conducted. Senior researchers may prefer to be listed last rather than second even when they were heavily involved in all aspects of the work and might otherwise be listed second. Not every paper has a senior author, but a student's primary thesis advisor or an employee's primary supervisor often serve in this role. The senior author may or may not be heavily involved in the day-to-day details of the study but meets the authorship criteria by providing clarity and direction along the way and by providing critical feedback on manuscript drafts. Additionally, the senior author can serve as a mediator if disputes about authorship or other issues arise. An experienced researcher will be able to provide insight into disciplinary standards and can prevent or resolve many of the issues that might befuddle a newer researcher.

6.4 Decisions About Authorship

In order to avoid last-minute debates over which individuals have made important contributions to a research project, it is helpful to decide ahead of time what the roles and responsibilities of each member of the research team will be and how they will earn coauthorship if that is their intended outcome. There should be no surprises about who is being included or excluded as an author. The lead researcher (or senior researcher) should check with each contributor about expectations. Ideally, this conversation should take place before potential coauthors begin working on project-related tasks. If everyone agrees that a person expected to make only a minor contribution will not earn coauthorship, be sure that individuals in support roles are not asked to write any part of the paper or to provide critical feedback on a draft. If everyone agrees that someone will be a coauthor, be sure that the individual has the opportunity to make an important intellectual contribution to the paper.

Decisions about who will be listed as a coauthor on a report, poster, or paper, as well as the order in which those coauthors will be listed, should be made as early as possible in the research process. Publications are an important marker of success and productivity in the sciences and academia. Authorship is often the only reward for the time collaborators and mentors put into a project. Because publications are so important for professional records, authorship decisions can be very stressful. They can trigger strong emotional responses, and they can sometimes harm relationships among researchers. Lead researchers therefore need to be transparent with everyone involved in the project not only about who will and will not be contributing in ways that merit coauthorship, but also about the roles each person will be playing. A growing number of journals now require a description of how each coauthor contributed to a manuscript and how each met the authorship criteria. It might be helpful to draft that statement before writing any other part of the paper, so that anyone who sees the draft knows what is expected of each coauthor.

If a research supervisor has been assigned by an academic program director or a supervisor, this person will expect to be a coauthor on papers produced with his or her guidance. Research supervisors usually also need to approve the involvement of other contributors who might merit coauthorship. (The supervisor does not need to approve other mentorship roles, such as those related to general professional development.)

Sometimes the list of expected contributors might change during the project. Perhaps a new collaborator is needed to run advanced statistics or to provide an expert's perspective on the policy implications of the work. In such cases, all coauthors need to be informed about the planned addition. When the addition of new collaborators significantly alters another contributor's position in the order of authors, perhaps bumping a person from second to fourth author, the affected person must be consulted and an agreement reached before any promises are made to the new coauthors.

Any disputes over authorship criteria or the order of authors are usually best referred to the senior author on the paper. Coauthors with concerns should speak with the lead author and/or the senior author first, before appealing to anyone outside the research team, such as academic department chairs or research deans. The written guidelines for authorship from ICMJE, relevant professional societies, and the target journal may be helpful resources for amicably resolving disputes.

STEP 2

Selecting a Study Approach

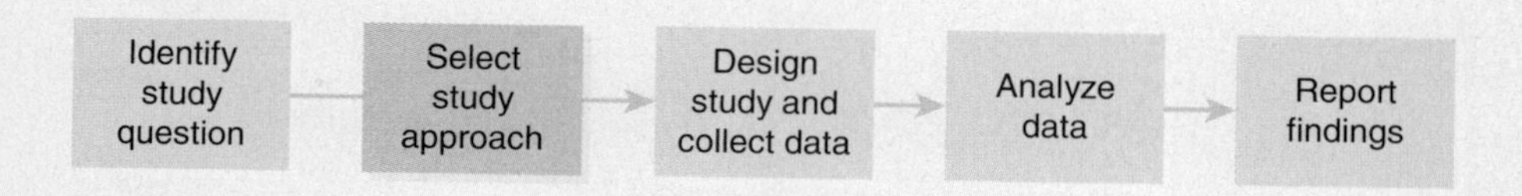

The second step in the research process is selecting a general study approach. This section provides an overview of several of the most common primary study designs.

- Case series
- Cross-sectional studies
- Case–control studies
- Cohort studies
- Experimental studies
- Qualitative studies
- Correlational studies
- Synthesis research

CHAPTER 7

Overview of Study Designs

Many types of observational and experimental study approaches are useful for clinical and population health research.

7.1 Types of Study Approaches

Eight common study designs used for population health research are briefly described in **Figure 7-1**. The figure does not represent a comprehensive list of all types of study approaches. Many research projects use variations of one of these approaches, and in others a hybrid of two approaches might be suitable. A diversity of designs can be valid and helpful for collecting and analyzing new data, analyzing existing data, and reviewing the literature in the health sciences.

The study design selected for a particular project must be appropriate for the goals of the study. A series of questions can help identify the most suitable approach:

- Do new data need to be collected, or are there existing data sources that can be used to answer the study question?
- If new data will be collected, is an experimental study required or can the study question be answered using non-experimental methods?
- Is the research question based primarily on exposure status, disease status, or membership in a particular population?
- Are there time constraints? Some studies allow for rapid data collection, while others require months or even years of follow-up.

7.2 Primary, Secondary, and Tertiary Studies

The first step in selecting an appropriate research approach is deciding whether new data will be gathered. A **primary study** collects new data from individuals. A **secondary study** analyzes an existing data set or existing health records. A **tertiary study** reviews and synthesizes the existing literature on a topic. Each of these three major study approaches has

FIGURE 7-1 Summary of Study Approaches

Study Approach	Goal
Case series	Describe a group of individuals with a disease
Cross-sectional study	Describe exposure and/or disease status in a population
Case–control study	Compare exposure histories in people with a disease (cases) and people without a disease (controls)
Cohort study	Compare rates of a new (incident) disease in people with different exposure histories or follow a population forward in time to look for incident cases of a disease
Experimental study	Compare outcomes in participants assigned to an intervention or control group
Qualitative study	Seek to understand how individuals and communities perceive and make sense of the world and their experiences
Correlational (ecological) study	Compare average levels of exposure and disease in several populations
Review/meta-analysis	Synthesize existing knowledge

FIGURE 7-2 Key Considerations for Primary, Secondary, and Tertiary Studies

Study Approach	Study Plan	Key Questions
Primary	Collect and analyze new data	■ What are possible source populations? ■ Will it be possible to recruit enough participants?
Secondary	Analyze existing data	■ What are possible sources of usable data files? ■ What research questions can be explored with the available data?
Tertiary	Review and synthesize the literature	■ Does the researcher have access to adequate library resources? ■ Can the researcher reasonably expect to access *all* of the needed articles?

its own critical considerations (**Figure 7-2**). Being aware of these likely challenges at the start of the study design process enables the researcher to make informed decisions about how to proceed.

Primary studies allow researchers to design studies that will answer their preferred research questions. A researcher collecting new data gets to design data collection protocols that are optimal for the study goals and resources, including selecting the source population, the sampling and recruiting methods, and the content and wording of the questionnaire. However, there are also some downsides to collecting new data. It may take months for a rigorous protocol to be developed by the research team and then reviewed by one or more research ethics committees.

Some primary studies can collect data quickly after a protocol is approved, but some require months or years of recruitment and data collection. There is a risk of failure if a sufficient number of participants cannot be recruited. Primary studies can become expensive, especially if there is a per-participant cost for laboratory testing or the use of proprietary survey instruments. The types of data that can be collected and the number of people who can participate in a study might be limited by resource constraints.

Data from any type of primary study may be made available for secondary analysis (**Figure 7-3**). The obvious advantage of secondary studies is that a researcher may be able to move very quickly from the definition of the study question to the analysis of related data. The major disadvantage of analyzing existing data is that the available data files might not include the exact variables of greatest interest to the researcher. When a retrospective review of clinical records is conducted, the files might have incomplete information about patient histories or lack confirmation that particular signs and symptoms were not present in cases. These omissions may require the study question to be revised. The data sets generated from national health surveys and made available to researchers might include a very limited number of questions about any particular topic. Some of the questions a researcher might be most interested in exploring may have already been answered by others who had earlier access to

FIGURE 7-3 Primary, Secondary, and Tertiary Study Approaches

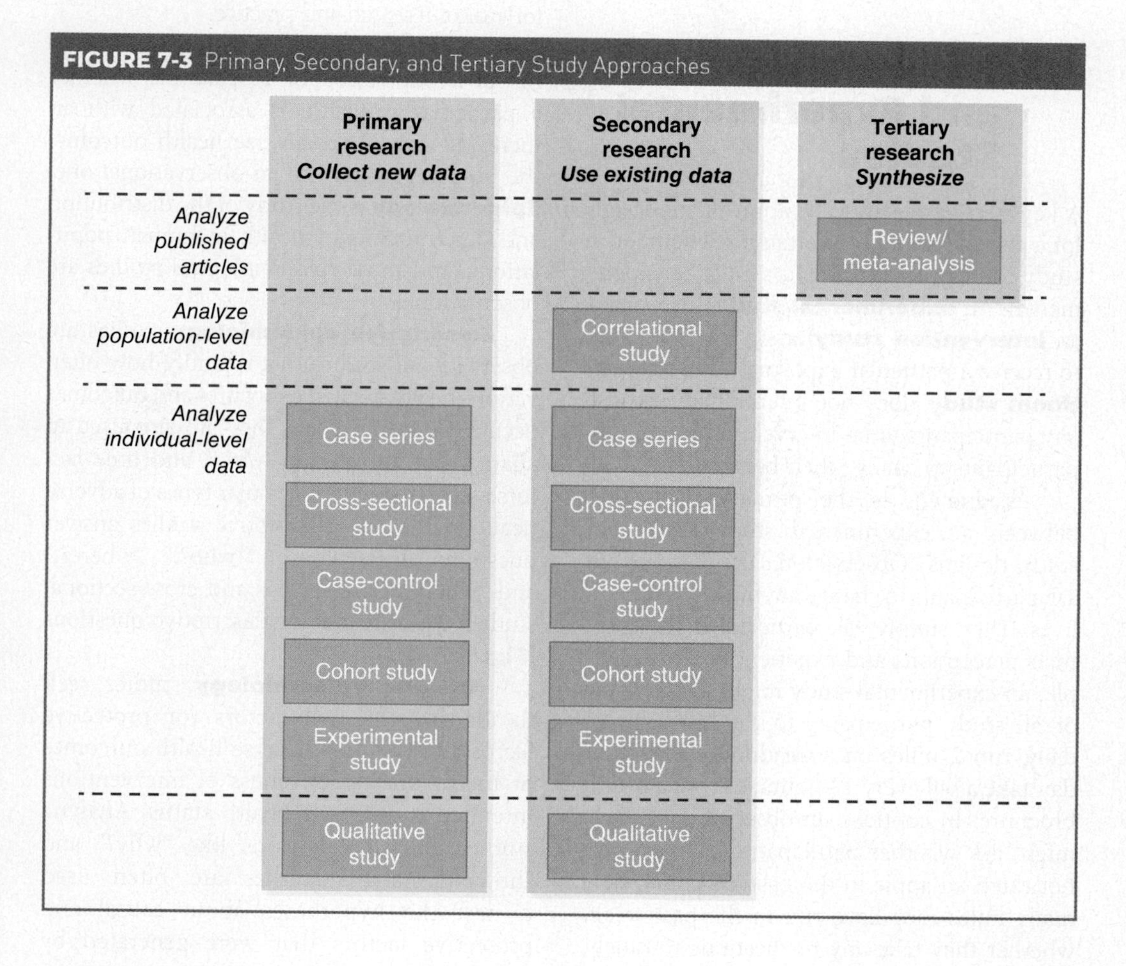

the data. The researcher must identify a valid and accessible source of data, then be prepared to select a study question based on the content of the available data files.

Tertiary studies also allow a researcher to move relatively quickly to the analysis stage, because the data collection process consists of conducting a comprehensive literature review and tracking down the full text of all the potentially relevant articles. Researchers with university affiliations need to check with their institutions' libraries about their policies (and possible fees) for acquiring articles that are not part of their collections. Researchers without university affiliations must consider the costs involved in accessing all of the required articles.

7.3 Observational and Experimental Studies

A key step in selecting the appropriate design for a primary study is ascertaining whether the study can be observational or must be experimental. An **experimental study** (also called an **intervention study**) assigns participants to receive a particular exposure. An **observational study** does not intentionally expose any participants to an intervention or ask any participants to change their behaviors.

Assignment is the primary distinction between an experimental study and other study designs. Observational studies do not ask participants to change anything about their lives. They simply ask participants to report their perceptions and experiences. For example, an experimental study might assign some or all study participants to eat three apples daily, run 2 miles on a treadmill every other day, take a pill every 12 hours, or read a health brochure. In contrast, an observational study might ask whether participants have or have not eaten an apple in the past 24 hours, how many miles they have run in the past week, whether they take any medications routinely, and whether they have seen an ad for a health promotion campaign.

If the goal is to examine whether an intervention is effective, an experimental design is likely to be the only suitable one. Experimental designs are used to test the efficacy of new medications, vaccines, and medical devices. They are also used to test the effectiveness of preventive interventions, diagnostic methods, treatment protocols, and rehabilitative therapies. Most experimental studies randomly assign some participants to an active intervention group and others to a comparison group, so that the results in the two populations can be compared. Experimental studies require careful design and oversight to ensure that they minimize the risks to participants and do not violate other standards for ethical research and practice.

If the goal is to describe the health profile of a population or to examine whether a particular exposure is associated with an increased risk of an adverse health outcome, the best design may be an observational one. **Epidemiology** is the study of the distribution and determinants of health in human populations, and most epidemiological studies are observational.

Descriptive epidemiology studies are observational studies that quantify how often various health-related exposures and outcomes occur in a population. They are also used to characterize the person, place, and time factors associated with particular types of adverse health outcomes. Descriptive studies answer questions like "what?," "who?," "where?," and "when?" Case studies and cross-sectional studies typically answer descriptive questions (**Figure 7-4**).

Analytic epidemiology studies seek to identify the risk factors (or protective factors) for various adverse health outcomes or to test the effectiveness of interventions intended to improve health status. Analytic studies answer questions like "why?" and "how?" Analytic studies are often used to test the hypotheses about causal and protective factors that were generated by

FIGURE 7-4 Epidemiological Study Approaches

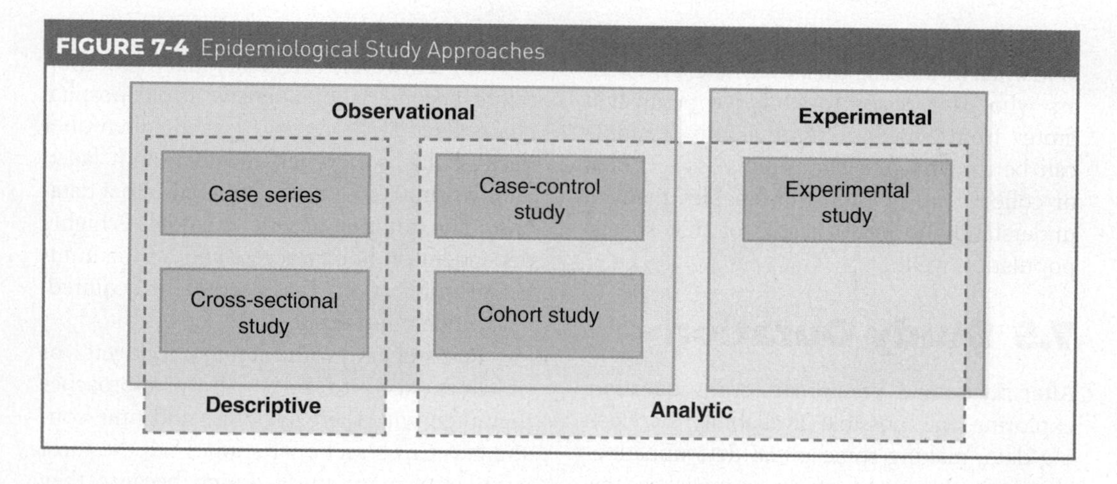

descriptive studies. Case–control, cohort, and experimental studies typically answer analytic questions.

7.4 Exposure, Disease, or Population?

Most clinical and population health studies have a research question that emphasizes one particular exposure, disease, or population (**Figure 7-5**). All interventional studies and many cohort studies focus on a selected exposure, one that is assigned in experimental trials and observed in cohort studies. Case series and case–control studies both focus on "cases," individuals with a particular disease. Cross-sectional studies and some types of cohort studies seek to recruit a study population that is representative of a well-defined group of people.

Researchers who have access to occupational records may preferentially choose a cohort study design that compares health outcomes among exposed and unexposed workers. This type of cohort study approach is especially valuable when an exposure is rare in the general population but high in the occupational population. Researchers who have access to clinical records may preferentially choose a research question that can be answered by a case series or a case–control study.

FIGURE 7-5 Population Selection for Each Study Approach

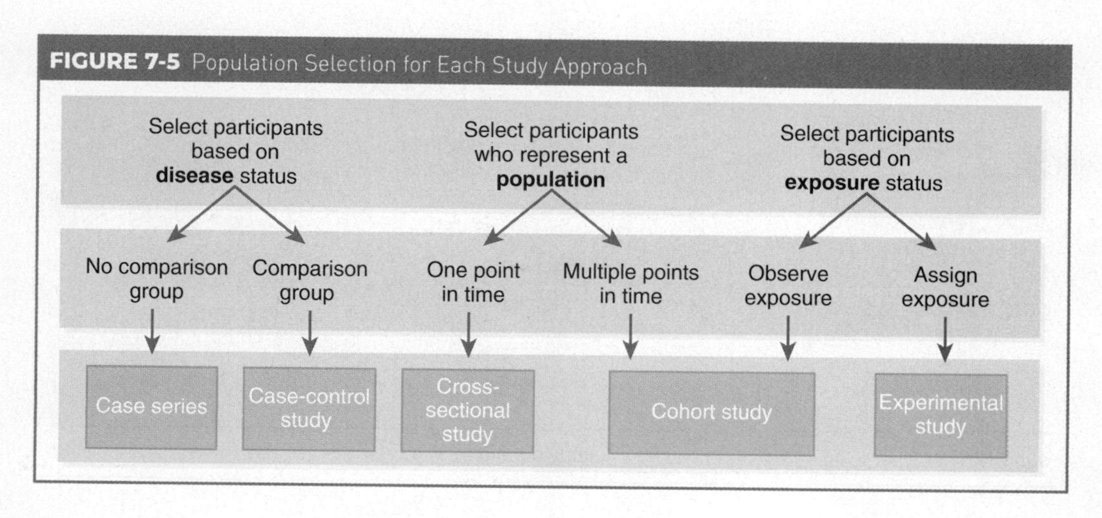

Those study approaches are particularly valuable when the disease is relatively rare. Researchers who have access to a unique population group from which a representative sample can be drawn often choose a cross-sectional or cohort study design that enables them to understand the health status of that special population well.

7.5 Study Duration

After selecting a preliminary study question, exploring the possible availability of existing data, making some initial determinations about the types of study approaches that might be appropriate, and identifying the special populations or data sources that might be accessible to the researcher, the final decision about which study design to use requires an evaluation of the expected duration and cost of the possible approaches.

The time required for collecting and analyzing data varies from study to study. Some primary studies allow for the collection of all needed data from participants at one point in time. Others require participants to be followed for weeks, months, or even years (**Figure 7-6**). The timeline for a secondary study might be very short if an entire data file and the relevant supporting documentation (such as copies of the questionnaire and codebook) can be downloaded from a website. However, secondary data collection might become labor intensive if old hospital charts have to be retrieved, read (often after deciphering somewhat illegible and faded handwriting), coded, and entered into a database. The duration of tertiary studies is highly dependent on library access and on the number of publications that need to be acquired, read, and summarized.

Researchers with limited budgets or timelines need to select study approaches that align with their resources and time constraints. Cross-sectional studies are the most popular primary study design because they collect data from each participant at one point in time, typically using a simple questionnaire. Case series, case–control studies, and qualitative studies may also allow data to be collected quickly. Cohort and experimental studies require data to be collected from each participant at least twice, first during a baseline examination and then at one or more follow-up times, so they typically take longer than other primary study designs. The fastest option for secondary analyses is accessing a data set that is already clean and ready to use, because compiling, cleaning, and coding new data sets from patient records, client records, and other types of information can take a long time. Population-level data (used during

FIGURE 7-6 Time Frame for Primary Data Collection

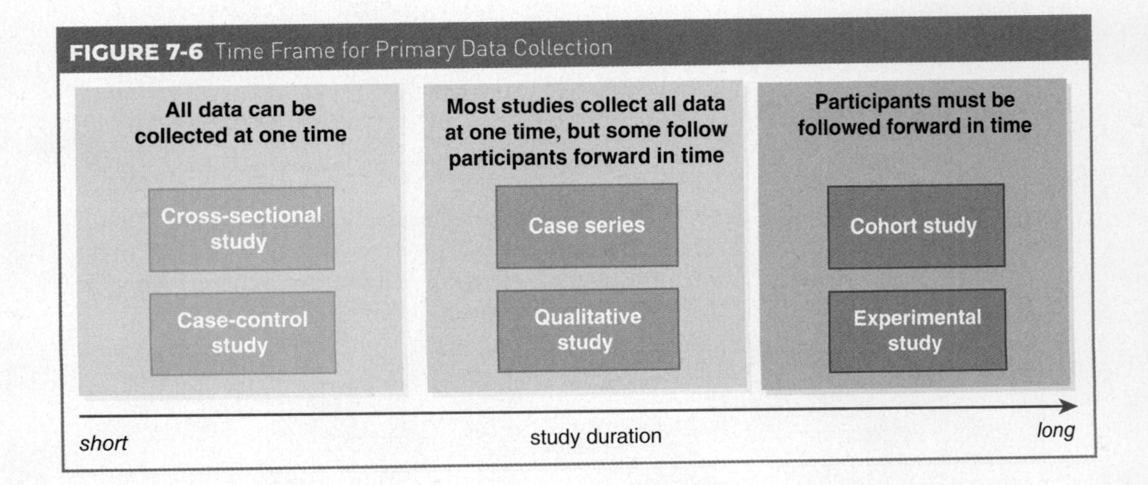
All data can be collected at one time
Cross-sectional study
Case-control study
Most studies collect all data at one time, but some follow participants forward in time
Case series
Qualitative study
Participants must be followed forward in time
Cohort study
Experimental study
short
study duration
long

correlational studies) and publicly available, deidentified individual-level data can be analyzed immediately, whereas all primary studies and any secondary studies using private or identifiable data must undergo review by a research ethics committee prior to analysis.

Flexibility is necessary at this stage of the research planning process, because research questions, specific aims, and the study approach may need to be adjusted to align with the resources available to the research team (**Figure 7-7**).

FIGURE 7-7 Flow Diagram for Selecting a Study Approach

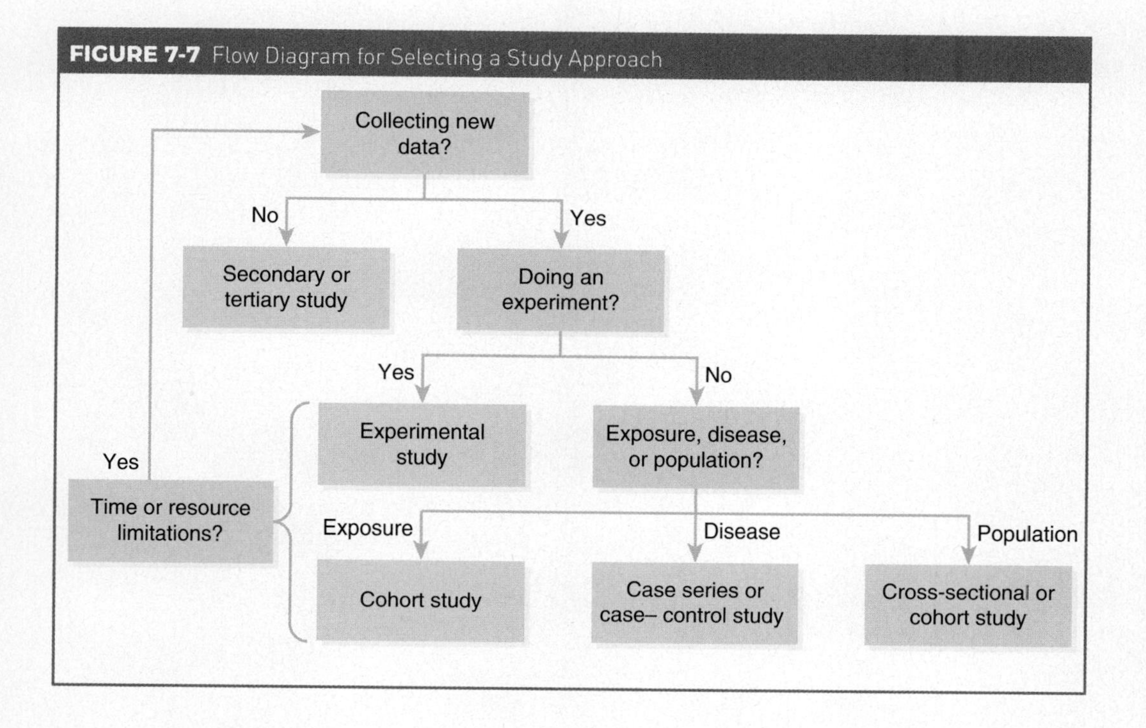

CHAPTER 8

Case Series

A case series describes a group of patients who have the same health condition or who have undergone the same procedure.

8.1 Overview

A **case report** is a report that describes one patient. A **case series** is a report that describes a group of individuals who have the same disease or disorder or who have undergone the same procedure (**Figure 8-1**). A case series can be written and disseminated only when a researcher has access to an appropriate source of cases and there is a compelling reason to write about those cases. This study approach can be useful for a variety of purposes, including:

- Describing the characteristics of and similarities among a group of individuals with the same signs and/or symptoms of disease

FIGURE 8-1 Key Characteristics of a Case Series

Objective	Describe a group of individuals with a disease.
Primary study question	What are the key characteristics of the cases included in the study?
Population	All individuals included in the study must have the same disease or disorder or have undergone the same procedure.
When to use this approach	A source of cases is available, and no comparison group is required or available.
Requirement	An appropriate source of cases is available.
First steps	1. Specify what new and important information the analysis will provide. 2. Identify a source of cases. 3. Assign a case definition. 4. Select the characteristics of the study population that will be described.
What to watch out for	A lack of generalizability
Key statistical measure	Only descriptive statistics are required.

- Identifying new syndromes and refining case definitions
- Clarifying typical progression of a disease or disorder
- Describing atypical presentations of a disease or disorder or unusual complications from a treatment
- Developing hypotheses for future research

Some case series for rare conditions may require only a handful of participants. Others may include several hundred or even several thousand individuals.

8.2 Case Definitions

A researcher conducting a case series must select one disease, disorder, or procedure of interest, determine what will be new and interesting about the study, and identify an appropriate and available source of cases. The next step is to establish a clear case definition. A **case definition** is a list of the inclusion and exclusion criteria that must be met in order for an individual to be classified as a person with the disease of interest in a case series, a case–control study, or another type of study. Case definitions are also essential for any outbreak investigation, no matter which study approach is used to investigate the event.

The first step in writing a case definition is clarifying what constitutes the disease or disorder of interest. A **sign** is an objective indication of disease that can be clinically observed, such as a rash, cough, fever, or elevated blood pressure. Sometimes a case definition can be based solely on these types of clinical observations. However, for many diseases and disorders, clinical observations and laboratory test results are not sufficient on their own to yield a valid diagnosis. A **symptom** is a subjective indication of illness that is experienced by an individual but cannot be directly observed by others. For example, when a patient rates his or her pain on a scale from 0 to 10, no one else can verify that the patient is accurately reporting his or her pain level. A clinician might observe behaviors consistent with severe pain but cannot measure pain. Illness and sickness describe other subjective components of a disease. **Illness** describes how a person perceives his or her own experience of having an adverse health condition. **Sickness** describes how a person with an adverse health condition relates to and is regarded by his or her community. A **syndrome** is a collection of signs and symptoms that occur together.

After a clinician makes a diagnosis based on signs and symptoms, details about a patient's condition are added to the patient's medical record. Diagnostic and procedural codes used as part of this charting process can be a valuable part of a case definition, especially since the codes allow electronic databases of clinical records to be searched for potentially eligible patients. **ICD codes** are diagnostic categorizations based on the International Classification of Diseases (ICD), more formally called the *International Statistical Classification of Diseases and Related Health Problems*. ICD codes can be used to search clinical records for patients with the same diagnosis. **CPT codes** are the Current Procedural Terminology codes published by the American Medical Association. CPT codes can be used to search clinical or insurance records for patients who have undergone the same procedure or received the same service.

The other important step in establishing a case definition is selecting the relevant **PPTs**, which stands for the *p*erson, *p*lace, and *t*ime characteristics that set the context for a case series or for other types of descriptive epidemiology studies (**Figure 8-2**). An ICD code or CPT code alone is not sufficient as a case definition, because it does not provide information about the sources of cases and other eligibility criteria. Will demographic characteristics, such as age or sex, be used to eliminate some patients with relevant diagnoses from the study population? Will only patients of particular clinics or hospitals, or residents of particular cities or counties, be eligible for inclusion? Will the study include 1 year

FIGURE 8-2 Sample Case Definitions

Category	Example 1	Example 2
Disease/procedure	Whooping cough (ICD-10 code A37)	Liver transplantation
Person	Any person with a confirmed case of whooping cough, defined as (1) an acute cough of any duration with isolation of *Bordetella pertussis* from a clinical specimen *or* (2) a cough lasting 2 or more weeks with paroxysms of coughing, inspiratory "whoop," and/or posttussive vomiting in an individual known to have had contact with a laboratory-confirmed case of pertussis	Adult patients (ages 18 years and older at the time of transplant), excluding those who were not receiving their first liver transplant and those who received multiple organ transplants
Place	Residents of River City whose diagnoses were reported to the River City Health Department (which requires notification of all diagnoses of pertussis)	Patients who had transplant surgery at the Oakville Regional University Medical Center
Time	Patients who first sought clinical care for a cough between January 1 and March 31, 2020	Recipients of liver transplants between January 1, 2010, and December 31, 2019, who were followed for a minimum of 2 years post-transplant

of files, or will it include 5 or more years? A comprehensive case definition must include a disease description plus an appropriate set of PPTs that specify additional inclusion and exclusion criteria.

8.3 Data Collection

A case series might be constructed from primary data acquired by interviewing cases about their experiences using a questionnaire and/or qualitative techniques. These data might be supplemented or confirmed with a review of the participants' medical records. Alternatively, a case series can be—and often is—based solely on secondary data, usually acquired from a review of patient charts.

When medical records will be consulted as part of the data collection process, it is often helpful to create a questionnaire that guides the extraction of data from these files. One of the limitations of relying on patient charts is that they usually contain only information deemed at the time of examination to be clinically relevant. The medical information in patient files is not recorded for research purposes, so records are unlikely to contain all the details that researchers would like to know. Many signs and symptoms, patient comments, and clinician observations are not routinely recorded. As a result, the absence of a specific note about a sign, symptom, or history does not necessarily mean that the exposure was not present, just that it was not recorded. A data extraction tool used to compile information about patient characteristics should include at least three response options for each item: "yes" (the symptom was present), "no" (the symptom was absent), and "no information" (the record did not indicate whether the symptom was present or absent). During the analysis and interpretation stage of the research project, the researcher should carefully consider the amount and type of missing information.

8.4 Ethical Considerations

Researchers conducting case series studies must be extremely careful about protecting patient privacy. All researchers using clinical data must strictly adhere to all laws and regulations pertaining to the use of medical records. All case series projects also require approval by at least one research ethics committee.

Institutional review boards usually require evidence that the patients whose records will be accessed have provided informed consent for their health data to be included in the analysis. Some clinical facilities ask all patients to sign forms indicating their general consent to allow their deidentified records to be available to researchers. Those hospitals and clinics may allow affiliated researchers to access patient records that have been stripped of information that could make individual patients identifiable, such as names and addresses. The data administrator will remove the records of patients who have opted out of research participation before providing files to a researcher. When researchers will have direct contact with patients as part of the data gathering process, or when they will be accessing identifiable information, additional consent forms specific to the study may need to be signed by participants.

Patients must provide additional consent before any photographs are taken. The photographer must follow the policies of the medical center where the patient is receiving care while also being aware of the relevant patient privacy laws and regulations. When a photograph may be shown in a public presentation or a published article, the researcher usually must document that a patient approved the use of the image. Many journals require proof of consent from patients before images with potentially identifiable features are published. Documentation of consent may be required even when there are no identifying marks in the image that could reveal a patient's identity.

Researchers must pay careful attention to protecting the identities of participants, especially when the disease or procedure is relatively rare and/or when the place and time characteristics are so narrow that individuals familiar with the source community might be able to identify the participants. In most situations, all potentially identifiable information must be removed prior to publication.

8.5 Analysis

Most case study reports do not require any numbers beyond simple counts and percentages. When the sample size is sufficiently large, statistical tests may be used to compare subpopulations of cases or to compare before-and-after measures of included patients.

Some case series may benefit from the use of various measures of morbidity and mortality. For example, the **case fatality rate (CFR)** is the proportion of people with a particular disease who die as a result of that condition (**Figure 8-3**). The CFR is different from the **mortality rate**, which is the percentage of members of a population who die from any condition (an all-cause mortality rate) or from a particular condition (a cause-specific mortality rate) during a specified time period. Mortality rates are typically expressed in units such as "per 100,000." The CFR is also different from the **proportionate mortality rate (PMR)**, which is the proportion of deaths in a population during a particular time period that were attributable to a particular cause. Both the CFR and the PMR are often expressed as percentages.

Although many case series studies do not have any time dimension, some follow patients for days, months, or even years. In this type of study approach, the case series becomes, functionally, a longitudinal cohort study in which all participants are defined by their disease status. Chapter 11 discusses cohort study approaches.

FIGURE 8-3 Mortality Rate, Case Fatality Rate, and Proportionate Mortality Rate

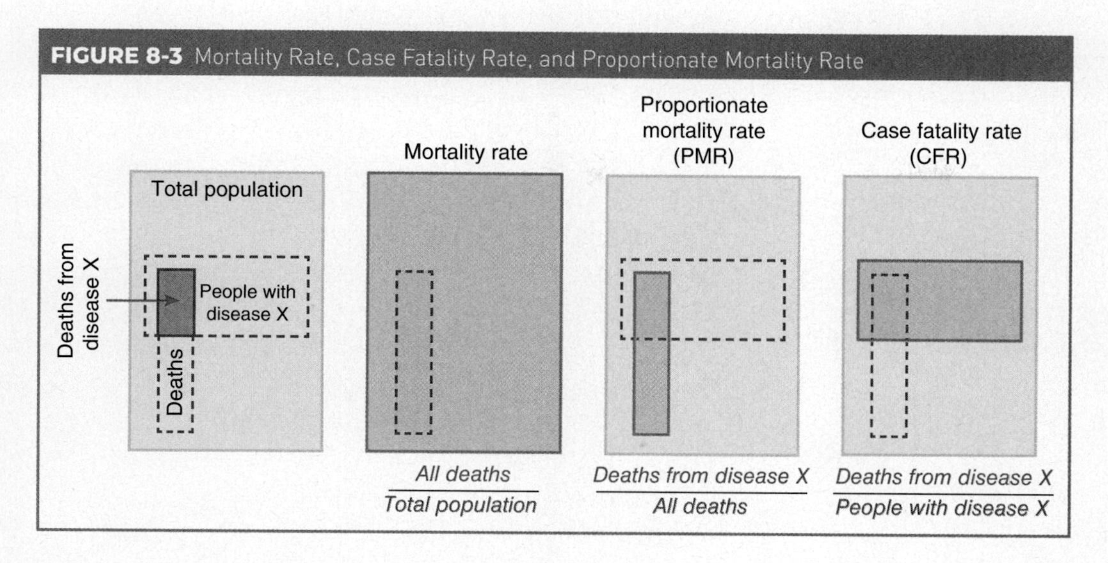

CHAPTER 9

Cross-Sectional Studies

A cross-sectional study provides a snapshot of the health status of a population at one point in time.

9.1 Overview

A **cross-sectional study**, also called a **prevalence study**, measures the proportion of members of a population who have a particular exposure or disease at a particular point in time. This determination is typically made over a short duration of time and must be based on a representative sample of the source population (**Figure 9-1**). Cross-sectional studies are used to:

- Describe communities.
- Assess population needs.
- Support program planning.
- Monitor and evaluate programs.
- Establish baseline data prior to the initiation of longitudinal studies.

Because cross-sectional studies are time- and cost-effective, they are the most popular approach used for descriptive epidemiology.

9.2 Representative Populations

In some ways, cross-sectional studies use the simplest epidemiological study design. The researcher just asks a sufficient number of people—usually a few hundred—to complete a short questionnaire, and then those data are analyzed. However, there is one very important requirement: The participants must be reasonably representative of one well-defined population. A cross-sectional study may be conducted with a representative sample of the patients of a hospital system or clinical practice, the clients of a community organization or business, the students or employees of a school district, the residents of a neighborhood or city, or the members of some other carefully defined population.

FIGURE 9-1 Key Characteristics of Cross-Sectional Studies

Objective	Describe the exposure and/or disease status of a population.
Primary study question	What is the prevalence of the exposure and/or disease in the population?
Population	The study participants must be representative of the source population from which they were drawn.
When to use this approach	Time is limited and/or the budget is small.
Requirement	The exposures and outcomes are relatively common, and the researchers expect to be able to recruit several hundred participants.
First steps	1. Define a source population. 2. Develop a strategy for recruiting a representative sample. 3. Decide on the methods to be used for data collection.
What to watch out for	Non-representativeness of the study population
Key statistical measure	Prevalence

Representativeness is the degree to which the participants in a study are similar to the source population from which they were drawn. Researchers implementing a cross-sectional study protocol cannot simply ask friends, the fans attending a youth football game, or individuals attending one chiropractic clinic to complete a survey and then assume that the results of the survey will be generalizable to all town residents. If the results are intended to reflect the profile of a particular population group, then the study's sampling strategy must recruit a population that is as diverse as the source population. Chapter 19 has more detailed information about populations for a cross-sectional study, and Chapter 20 explains how to estimate sample size requirements.

9.3 KAP Surveys

For research conducted with human populations, a **survey** is the gathering of data from individuals using a list of questions. Survey methodologies are widely used by health and social scientists, including researchers in psychology, political science, and economics. One commonly used tool for collecting data during a cross-sectional study is a **KAP survey**, a survey instrument that asks participants about their *k*nowledge, *a*ttitudes (or beliefs or perceptions), and *p*ractices (or behaviors). KAP questionnaire items can be particularly helpful for identifying gaps between what people know and how they act on that knowledge. For example, the adults who complete a KAP survey form might demonstrate high knowledge about the benefits of exercise on cardiovascular health but at the same time indicate that they exercise rarely because a variety of perceived barriers prevent them from being as physically active as they know they ought to be for maximum fitness.

9.4 Repeated Cross-Sectional Surveys

A **repeated cross-sectional study** is a series of cross-sectional studies that resample and resurvey representatives from the same source population at two or more different time points. A repeated cross-sectional study design does not track the same individuals forward in time. Instead, a new set of participants is sampled from the source population for each round of data collection. Some people may happen by chance to be selected for more than one round of surveying, but their answers to the different surveys are not linked. This is different from a longitudinal study that follows the same people forward in time. Repeated cross-sectional surveys can reveal trends in population-level status over time, but they do not allow for the examination of individual-level changes.

Repeated cross-sectional surveys are often conducted annually or every few years as part of national health surveillance programs. **Surveillance** is the process of continually monitoring health events in a population so that emerging public health threats can be detected and appropriate control measures can be implemented quickly. Health officials can use surveys of randomly sampled population members as part of monitoring long-term trends in public health. Population-based surveys are used for many of the largest studies conducted by the U.S. Centers for Disease Control and Prevention (CDC), including the Behavioral Risk Factor Surveillance System (BRFSS), the National Health and Nutrition Examination Survey (NHANES), and the U.S. National Health Interview Survey (NHIS).

Besides cross-sectional studies, many additional types of surveillance are used by governments to track population health status. **Active surveillance** is the process of public health officials contacting healthcare providers in their jurisdictions to ask how often the clinicians are diagnosing particular types of disease. **Passive surveillance** is the compilation of reports of notifiable disease diagnoses submitted by medical laboratories. The **case detection rate (CDR)** is the proportion of people with a disease who are diagnosed as having that disease. The CDR will be low if few patients are tested for suspected infectious diseases or other conditions that can be verified through various types of medical tests. **Syndromic surveillance** is the process of tracking potential outbreaks or other disease events based on reports of symptoms rather than relying solely on counts of laboratory-confirmed diagnoses. Crowd-sourced data culled from social media sites and other platforms can assist with the recognition of growing population health concerns. **Sentinel surveillance** is the continuous collection and analysis of high-quality data from a limited number of clinics or hospitals so that public health officials will be able to detect changes in health status occurring in the larger population from which the sentinel sites were sampled.

9.5 Prevalence

Cross-sectional studies measure the prevalence of various exposures or exposure histories, diseases and disorders, and demographic characteristics in one well-defined population at one point in time or over a short duration of time, with all data collected within a few days, weeks, or months. The most common result reported for a cross-sectional survey is the **prevalence**, the percentage of members of a population who have a given trait at the time of a study.

A **point prevalence** measures the proportion of a population with a particular characteristic at one point in time. A point prevalence is a "snapshot" of population status, such as the percentage of 18- to 64-year-olds in a city who were current smokers as of July 1, 2020. A **period prevalence** measures the proportion of a population with a particular characteristic during a defined time period, such as several weeks or several months. A period prevalence might ask what percentage of eighth-grade students in a school district have ever been told by a doctor that they have asthma or what percentage of those students have had a dental checkup during the past 6 months.

Comparative measures can also be used as part of the analysis of data from cross-sectional studies. For example, a **prevalence ratio (PR)** compares the prevalence of a characteristic in two independent populations (or independent subpopulations of study participants) by taking a ratio of their prevalence rates. Populations are independent when no individual study participants are members of more than one of the populations being compared. A PR could compare the prevalence rates for males and females or compare the prevalence rates for people who identify as current smokers and those who identify as never having used tobacco products.

Because a cross-sectional study has no time dimension, it cannot be used to assess causality. An exposure can be said to be "associated with" or "related to" a disease, but a cross-sectional study cannot show that an exposure caused a disease.

CHAPTER 10

Case–Control Studies

© DmitriyRazinkov/Shutterstock

A case–control study compares the exposure histories of people with and without a particular disease in order to identify likely risk factors for the disease.

10.1 Overview

A **case–control study** is a study that compares the exposure histories of people with disease (cases) and people without disease (controls). A **case** is a study participant with the infectious or parasitic disease, noncommunicable disease, neuropsychiatric condition, injury, or other disease, disorder, disability, or health condition of interest to the researcher. A **control** in a case–control study is a participant who does not have the disease being examined. (The term "control" has a different meaning in experimental studies. In an experiment, a control is a participant who is assigned not to receive the active intervention.)

Individual participants in a case–control study are selected for inclusion in the study based on their disease status (**Figure 10-1**), then both cases and controls are asked the same set of questions about past exposures (**Figure 10-2**). A case–control study is often the best study approach for identifying possible risk factors for a disease. This is especially true when the disease is uncommon, so a study of the general population would be unlikely to yield more than a few cases. A special type of statistic—an odds ratio—is used to identify likely risk factors.

10.2 Finding Cases and Controls

Because case–control studies require an adequate number of cases in order to be valid, the first step in designing a case–control study is to identify an appropriate and accessible source of individuals who have the disease or disorder of interest. Hospitals, specialty clinics, physicians' offices, public health agencies, disease registries, and disease support groups may be able to assist researchers in identifying individuals who are likely to meet the study's case definition.

Most organizations will not release any information about patients or members until after a research protocol has been approved by the appropriate ethics oversight committee. Researchers accessing health data from cases and controls must exercise extreme care to protect the privacy of potential participants

FIGURE 10-1 Key Characteristics of Case–Control Studies

Objective	Compare exposure histories of people with a disease (cases) and people without that disease (controls).
Primary study question	Do cases and controls have different exposure histories?
Population	Cases and controls must be similar except for their disease status.
When to use this approach	The disease is relatively uncommon, but a source of cases is available.
Requirement	A source of cases is available.
First steps	1. Identify a source of cases. 2. Assign a case definition. 3. Decide what type of control population will be appropriate for the study. 4. Decide whether cases and controls will be matched.
What to watch out for	Recall bias
Key statistical measure	Odds ratio (OR)

FIGURE 10-2 Framework for a Case–Control Study

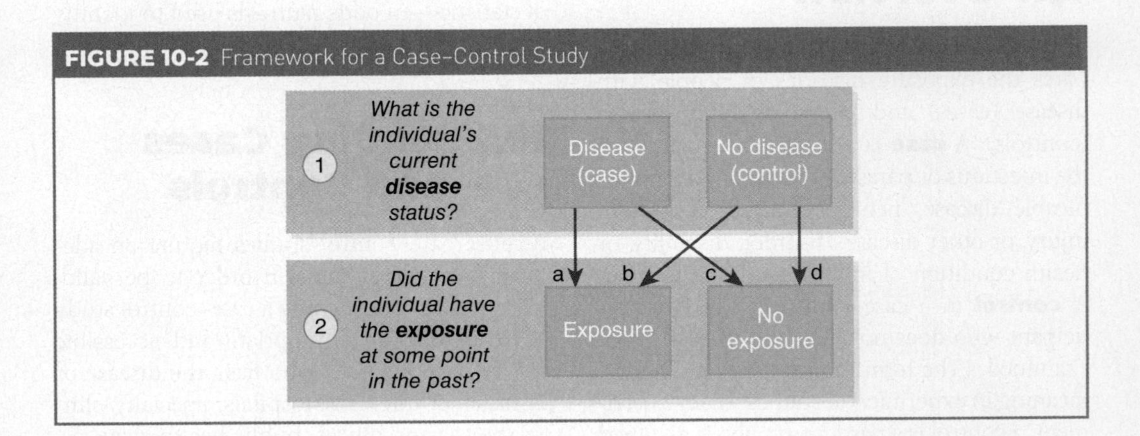

and the confidentiality of their personal information.

All cases must have the same disease, disorder, disability, or other health-related condition, and the study's case definition must specify exactly what characteristics must be present or absent for a person to be deemed a case. Clinical manuals and reports about previous studies of the disease can be helpful references for drafting and refining the inclusion and exclusion criteria. The case definition should include person, place, and time (PPT) characteristics.

Next, an appropriate source of controls must be selected. Depending on the goals of the study, controls may be recruited from, among other sources:

- Friends and relatives of cases
- Hospital or clinic patients without the disease of interest
- The general population

A **control definition** is a list of all of the eligibility criteria for inclusion in a comparison population. Controls must be similar to cases except for their disease status, so the inclusion and exclusion criteria for cases that do not specifically relate to the disease must also apply to controls. For example, if cases must be males between 25 and 39 years of age, controls must also be men in this age group.

Individuals who do not meet the case definition or the control definition must be excluded from the study. Excluded individuals may not meet one of the PPT criteria for inclusion, or they may have an intermediate or indeterminate disease status that prevents them from meeting either the case or the control definition. Chapter 19 provides additional details about the selection of cases and controls for case–control studies.

10.3 Matching

Matching in a case–control study describes the process of recruiting one or more controls who are demographically similar to each case. (For a cohort study, matching describes the process of recruiting one or more unexposed individuals who are demographically similar to each exposed person participating in the study.) Early in the study design process, a decision must be made about whether and how to match cases and controls. There are three often-used options for matching: no matching, frequency (group) matching, and matched-pairs (individual) matching.

Many case–control studies use no matching. They simply assume that similar inclusion and exclusion criteria for cases and controls will result in case and control populations that have similar distributions according to sex, age group, socioeconomic status, and other characteristics that may be confounders of the association between the key exposure and the disease.

Frequency matching (also called **group matching**) is a sampling design that ensures that cases and controls in a case–control study have similar group-level demographic characteristics. (In cohort studies, frequency matching ensures that exposed and unexposed participants have similar group-level profiles.) For example, suppose a study is using hospitalized cases and controls. For each case, the researcher may select one control from the hospital registration files who was admitted the same week as the case, is the same sex as the case, and is within ±3 years of the age of the case, but does not have the disease of interest. Frequency matching can be used to identify one, two, or several controls for each case. For group matching, the goal is to recruit a control population that is similar to the case population. Individual cases are not linked to individual controls during analysis, so the analysis uses the same approaches that are used for unmatched case–control studies.

Matched-pairs matching (also called **individual matching**) is a sampling design that links each case in a case–control study to one or more controls with similar characteristics, such as genetic siblings or community members with the same date of birth. (In cohort studies, matched-pairs matching links each exposed individual to one or more unexposed individuals.) This approach is common in genetic studies that link each case to a close genetic relative. When this type of matching is used for a study, the pairs are linked for analysis. The unit of analysis is the pair, not the individuals. This approach requires the use of special statistical tests.

For both frequency matching and matched-pairs matching, it is important not to overmatch. **Overmatching** describes the recruiting challenges and possible statistical bias that can result from matching too many characteristics of the cases and controls (or exposed and unexposed participants). The demographic and exposure variables used as matching criteria cannot be evaluated as possible risk factors for the disease. For example, suppose that cases and controls are frequency matched based on the date of hospital admission, sex, and age. The case and control populations will, by design, have the same proportion of admissions in April, the exact same percentages of males and females, and nearly identical mean ages. As a result of this forced similarity, the study will not be able to examine whether cases are more or less likely than controls to require hospitalization in a certain month, to be males, or to be octogenarians. Additionally, when there are more matching characteristics, it can be difficult to find controls who meet all of the matching criteria. The study population may end up being quite different from the general population because of the strict eligibility requirements, and this may limit the usefulness of the study. Overmatching may also result in a statistical bias that obscures the relationship between an exposure and the disease.

10.4 Minimizing Bias

Once the preliminary decisions about the study design are finalized, planning for data collection may begin, as described in the third section of this text. As the protocol is developed, researchers should seek to minimize the likelihood of various types of bias occurring. **Bias** is a systematic flaw in the design, conduct, or analysis of a study that can cause the results of the study not to accurately reflect the truth about the source population. Each type of study design has particular kinds of bias that are especially likely to be problematic, but careful study design and implementation can prevent or minimize the risk of bias occurring.

A key word in the definition of bias is "systematic," which means methodical, orderly, and routine. Any study might experience random error, like a random sample of people recruited for a population-based study happening, by chance, to have a mean age that is much younger or much older than the population from which the sample was drawn. If a second sample of people is drawn from the same population, it is very unlikely that the new sample will have a mean age that is far from the population mean age. By contrast, bias is not something that happens randomly or by chance. A flawed research protocol will introduce bias into any sample drawn from the same source population, because bias is a systematic error.

A strong survey instrument for a case–control study will ask each participant questions that confirm whether the respondent is a case, a control, or neither. The disease status for each participant may need to be confirmed by clinical or laboratory testing or other types of secondary verification. The researcher must ensure that only confirmed cases and confirmed controls are included in the analysis. Adhering to strict definitions for what constitutes a case and what constitutes a control minimizes the risk of **misclassification bias**, which occurs when participants are not correctly categorized, such as when some controls in a case–control study are incorrectly classified as cases or some cases are incorrectly classified as controls due to a systematic problem with the case definition or the control definition.

The protocol for a case–control study must also seek to minimize **recall bias**, which occurs when cases and controls systematically have different memories of the past. Participants are often asked to remember events from the distant past that cannot be confirmed by documents created around the time when the exposure would have occurred. Cases may be searching for answers to questions about why they have become ill. As a result, they may have more vivid memories of participation or lack of participation in activities perceived to be risky or beneficial. Although there is no way to prove that recall bias is occurring because of systematically different memories among cases and controls, the results of case–control studies must be interpreted cautiously in light of the possibility that differential recall may have influenced the findings.

For example, adult cases in a study of night blindness may systematically report that they rarely ate carrots as children. They may say this not because they are sure that they never ate carrots, but because their memories are fading and they unconsciously assume that they would have good vision in adulthood if they had eaten a lot of vegetables high in vitamin A when they were younger. Alternatively, cases may systematically overestimate childhood carrot intake. They may wonder why they developed night blindness when they have such fond memories of happily munching on carrot sticks every day during lunch in grade school. The reality may be that they ate carrots only once per month. Controls, on the other hand, are unlikely to have spent much time thinking about risk factors for poor eyesight. They may recall eating carrots sometimes rather than rarely or often. A study of night blindness affected by recall bias might find a significant difference in the reported childhood consumption of carrots by cases and controls even if in reality there was no difference in the average diet of the two groups. Alternatively, the survey may fail to capture a true difference in dietary history. There is no way for the researcher to determine whether one scenario is more likely than the other. The limitations section for a report about this study would need to acknowledge and discuss the possible errors that might be attributed to recall bias.

10.5 Odds Ratios

Probability is the likelihood that an event will happen. The probability of an event can be as low as 0 (0%) or as high as 1 (100%). Probability can be written mathematically as p. **Odds** are a single number calculated as the ratio of the likelihood of an event happening (p) to the likelihood of that event not occurring ($1 - p$). The odds can be written as $p/(1 - p)$. Odds are most familiar from their connection with betting. A horse with an equal chance of winning a race (50% likely to win) or losing a race (50% likely to lose) is said to have "even odds," or odds of 1 (50%/50%). A case–control study examines the likelihood of participants having had a particular exposure or not having had that exposure (**Figure 10-3**). If 50% of the participants in a study report a history of exposure and 50% report no exposure, then the odds of exposure are 50%/50%, or 1. If 25% report having the exposure and 75% do not, then the odds are 25%/75%, or 0.33. If 2% report being exposed in the past and 98% report not being exposed, then the odds are 2%/98%, or 0.02.

For most health research studies, the term **measure of association** refers to a number that summarizes the relationship between an exposure and a disease outcome. The measure of association for case–control studies is the **odds ratio (OR)**, a ratio of odds in which the denominator represents the reference group. For a case–control study, the OR is the ratio of the odds of exposure among cases (in the numerator) to the odds of exposure among controls (in the denominator). Researchers considering using a case–control study approach must become familiar and comfortable with odds and odds ratios, because the OR is the measure of association that readers will expect to be reported for a case–control study.

A **contingency table** (also called a **crosstab**) is a row-by-column table that displays the counts of how often various combinations of events happen. A **two-by-two (2×2) table** is a contingency table displaying two variables that have been divided into dichotomous (yes/no) categories. In epidemiological analysis, the columns typically display disease status (yes/no) and the rows typically display exposure status (yes/no).

Figure 10-4 shows a sample 2×2 table for a case–control study. In the 2×2 table for an unmatched case–control study, the columns are for disease status (case = yes, and control = no) and the rows are for exposure status (exposed = yes, and unexposed = no). All of the participants in the study are assigned to one of the four resulting boxes: *(a)* cases with an exposure history, *(b)* controls with an exposure history, *(c)* cases with no exposure history, and *(d)* controls with no exposure history. The total number of cases in the study is $a + c$, the total number of controls in the study is $b + d$, and the total number of participants is $a + b + c + d$.

FIGURE 10-3 Odds

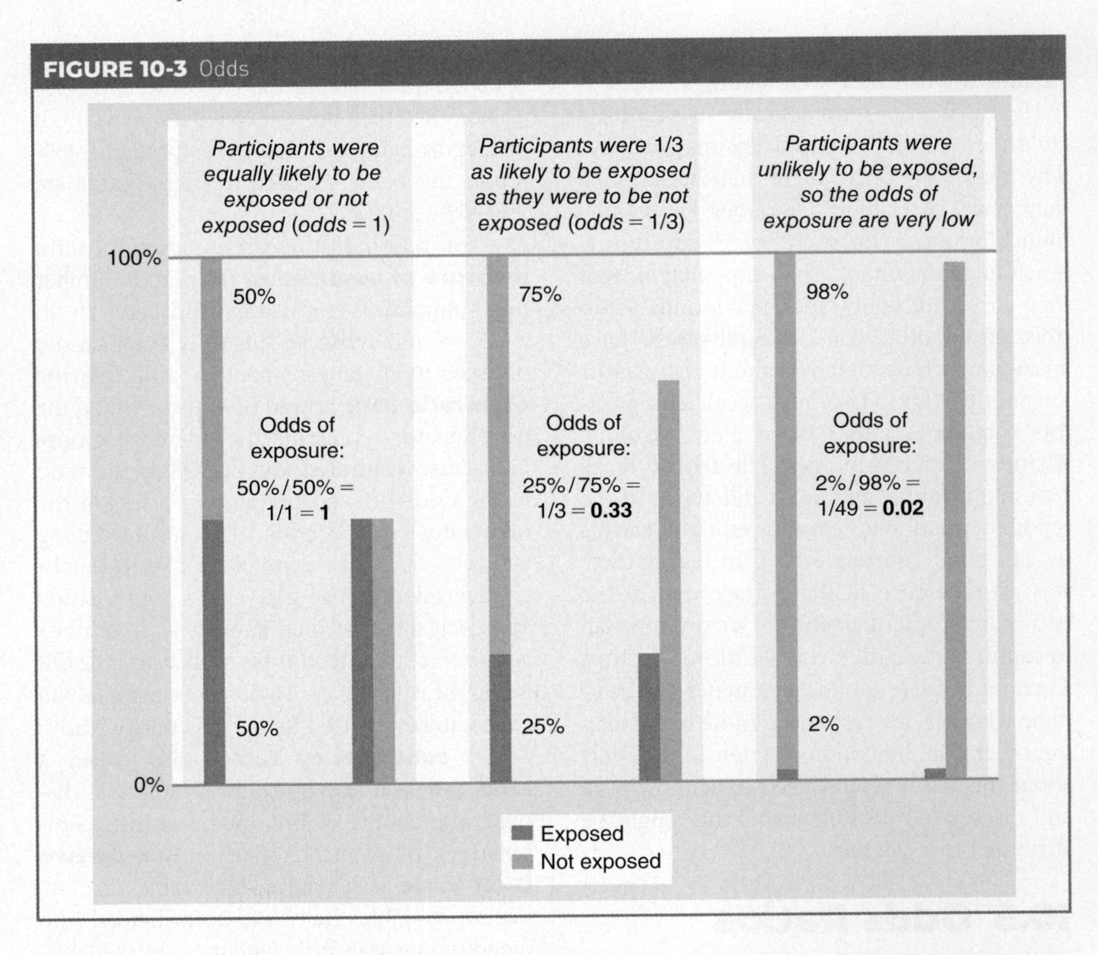

FIGURE 10-4 Odds Ratio (Point Estimate)

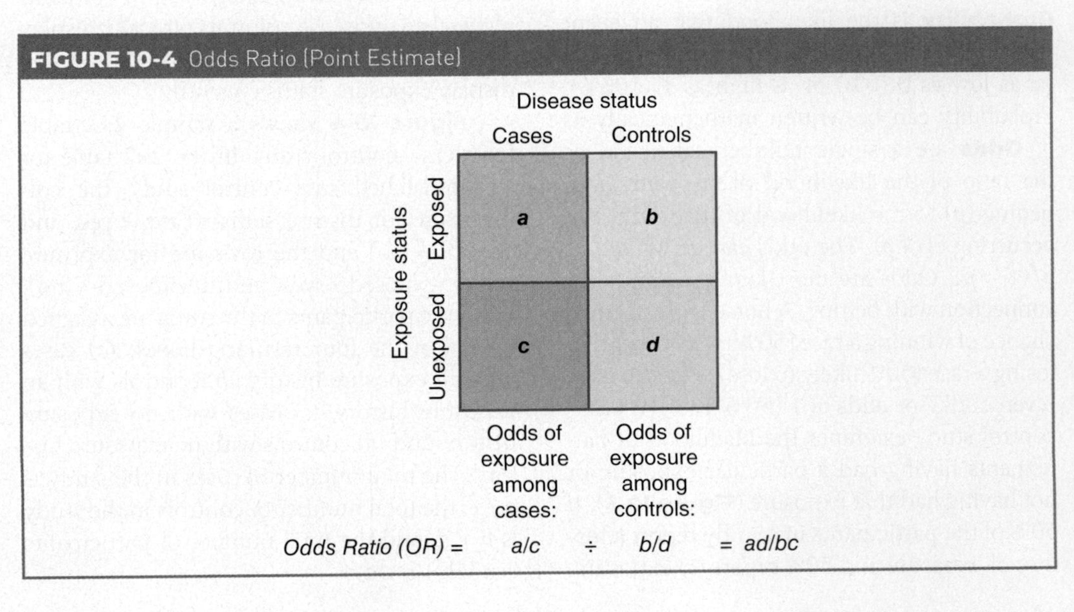

The odds of exposure among cases are the number of cases with the exposure (*a*) divided by the number of cases without the exposure (*c*). The odds of exposure among controls are the number of controls with the exposure (*b*) divided by the number of controls without the exposure (*d*). Basic algebra shows that the equation for the OR of $(a \div c)/(b \div d)$ can be simplified to $a/c/b/d = ad/bc$. The **point estimate** for any statistic is the value of the statistic in a study population. For an OR, the equation *ad/bc* calculates the point estimate. Point estimates of ORs provide a starting point for understanding the relationship between the disease and exposure status in the study population.

- OR = 1 (or close to 1): The odds of exposure were the same (or about the same) for cases and controls.
- OR > 1: Cases had greater odds of exposure than controls, implying that the exposure was risky.
- OR < 1: Cases had lesser odds of exposure than controls, implying that the exposure was protective.

Point estimates are typically presented along with a corresponding 95% confidence interval (CI) that provides additional information about the likely value of the statistic in the source population from which study participants were drawn. The 95% CI shows whether an OR is statistically significant (**Figure 10-5**):

- When the entire 95% CI is less than 1, the OR is statistically significant and the exposure is deemed to be protective in the study population.
- When the entire 95% CI is greater than 1, the OR is statistically significant and the exposure is deemed to be risky in the study population.
- When the 95% CI overlaps OR = 1, the OR is not statistically significant in the study population. The lower end of the CI is less than 1, suggesting protection, while the higher end of the CI is greater than 1, suggesting risk. In this situation, the exposure and disease are deemed to have no association in the study population.

A non-significant CI may reflect a true absence of a relationship between the exposure and the disease, but it may also indicate that the sample size was too small. Power calculations can be

FIGURE 10-5 Interpretation of the Odds Ratio Based on Its 95% Confidence Interval

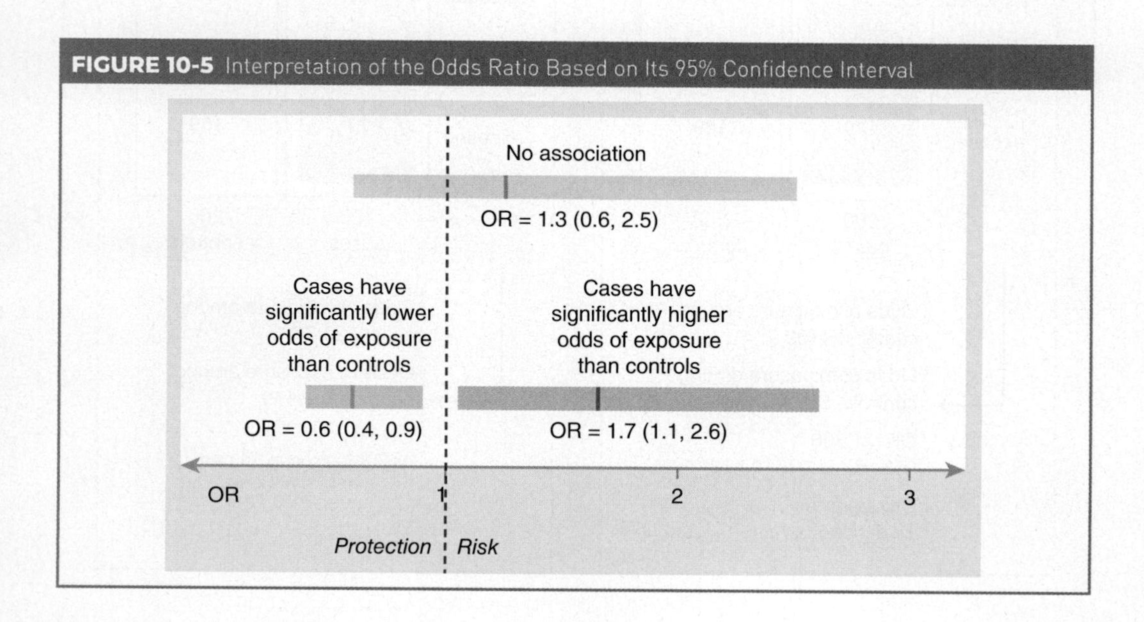

used to verify whether the sample size was sufficient to detect differences in the odds among cases and controls if a difference really did exist.

Chi-square tests can be computed from the same 2×2 tables used to calculate ORs. When the 95% CI does not overlap 1, the p value for the chi-square test will be $p < .05$, which is statistically significant and indicates a difference between cases and controls. When the 95% CI for an OR overlaps the number 1, the p value for the chi-square test will be $p > .05$, which indicates no association or no difference between cases and controls.

Computer- and Internet-based statistical programs (such as statistical software packages or the OpenEpi.com website) can use the counts for a, b, c, and d to calculate the point estimate for the OR (the value of ad/bc) along with its corresponding 95% CI. Sample output is shown in **Figure 10-6**. One example has an OR of 1.606 and a 95% CI of (1.05, 2.48). Because the entire 95% CI is greater than 1, this implies that the exposure was risky. The chi-square p value of $p < .05$ confirms this conclusion. The other example has an OR of 1.16 (0.65, 2.07). Since the 95% CI overlaps 1, the association is not statistically significant. The correct conclusion in this example is that there is no association between the exposure and the disease. The chi-square p value of $p > .05$ confirms this conclusion. Logistic regression models can be used to calculate ORs that adjust for possible confounding variables.

For a case–control study, it is incorrect to say that "the exposed had a higher (or lower) rate of disease than the unexposed" because the rates of disease in the exposed and unexposed groups are not known. Case–control studies recruit participants because they have or do not have a disease. Usually about 50% of participants in a case–control study are cases even if cases make up less than 1% of

FIGURE 10-6 Examples of Odds Ratio Calculations

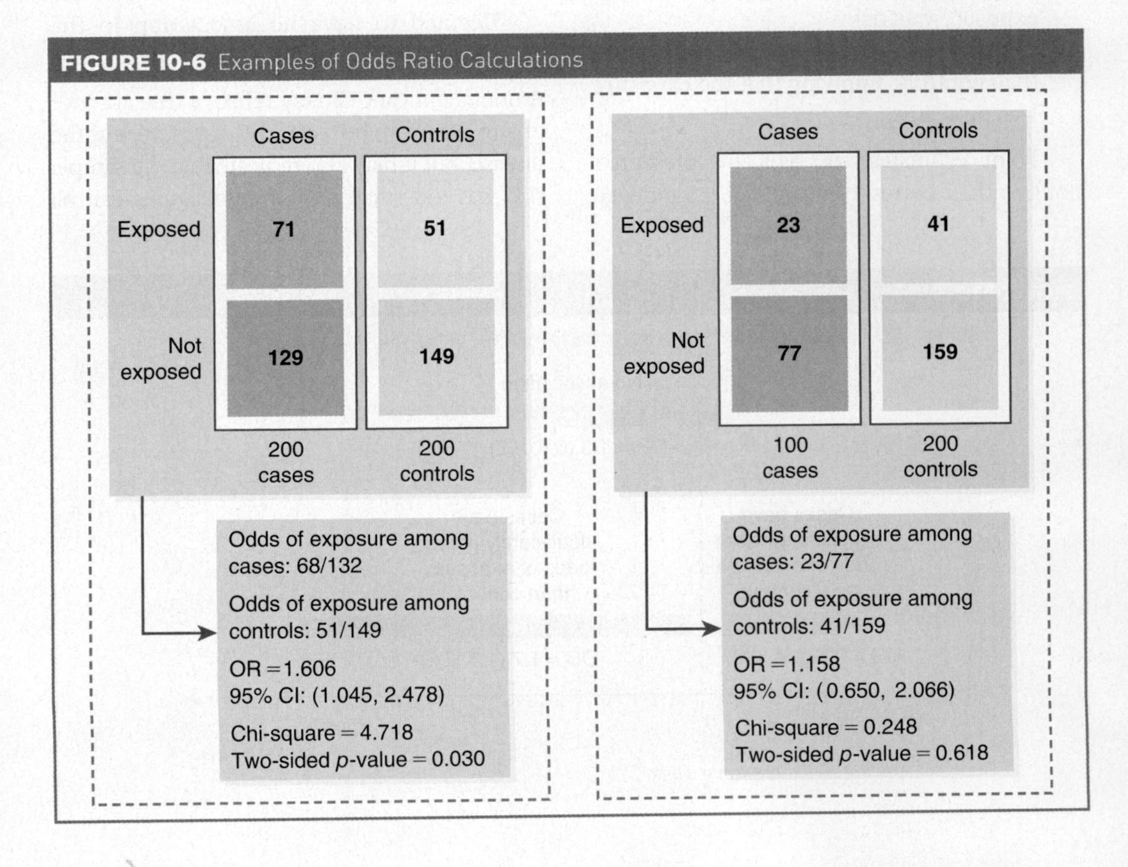

the community from which the study population was drawn. Cases are oversampled. As a result, the proportion of exposed and unexposed people in the study who are cases will be much higher than 1%. Because the study population is usually not representative of the community as a whole, case–control studies are unable to estimate rates of disease among the exposed and unexposed.

Case–control studies are, however, able to examine odds of exposure among the diseased and the not diseased. For case–control studies, the orientation should always be from disease status to exposure history, and from odds rather than risks or rates. Accordingly, the results should always be phrased to indicate that cases had greater (or lesser) odds of exposure than controls.

10.6 Matched Case–Control Studies

Individually matched case–control studies require the calculation of a matched-pairs OR that uses a special kind of 2×2 table that shows how often pairs of cases and controls had the same or different exposure histories (**Figure 10-7**). **Concordance** occurs when there is agreement; **discordance** occurs when there is disagreement. When both the case and the control in a matched pair have the same history of exposure or no exposure, their experiences are **concordant** (cells *a* and *d*). Concordant pairs do not provide much useful information about the potential relationship between the exposure and the disease. However, when the exposure histories for a pair are discordant (cells *b* and *c*), they provide an indication about whether the exposure is likely to be risky or protective. A **matched-pairs odds ratio (OR_{mp})** is a special kind of OR for a matched-pairs case–control study that compares the number of pairs in which the case had the exposure and the control did not (*b*, in the numerator) to the number of pairs in which the control had the exposure and the case did not (*c*, in the denominator). The matched-pairs OR is typically calculated as $OR_{mp} = b/c$.

FIGURE 10-7 Matched-Pairs Odds Ratio

		Controls: Exposed	Controls: Unexposed
Cases	Exposed	a	b
	Unexposed	c	d

$OR_{mp} = b/c$

- When $b/c > 1$ and the 95% CI (which is calculated using all four categories in the figure, including the concordant pairs) does not overlap 1, cases were more likely than controls to have been exposed. This implies that the exposure is risky.
- When $b/c < 1$ and the 95% CI does not overlap 1, cases were less likely than controls to have had the exposure. This implies that the exposure is protective.
- When the 95% CI for b/c includes 1, there is no statistically significant association between the disease and the exposure.

Most 2×2 tables present individuals, so the total count across the four cells is the total number of *individuals* in the study. For a matched-pairs 2×2 table, the sum of the four cells is the total number of *pairs* in the analysis. A matched-pairs 2×2 table can be converted to an individual-level 2×2 table (**Figure 10-8**). The total number of individuals in the individual-level exposure-by-disease 2×2 table should be twice the number of pairs in the matched-pairs 2×2 table. The point estimate for the OR_{mp} should be similar to the standard OR calculated for the same data.

FIGURE 10-8 Example of a Matched-Pairs Odds Ratio

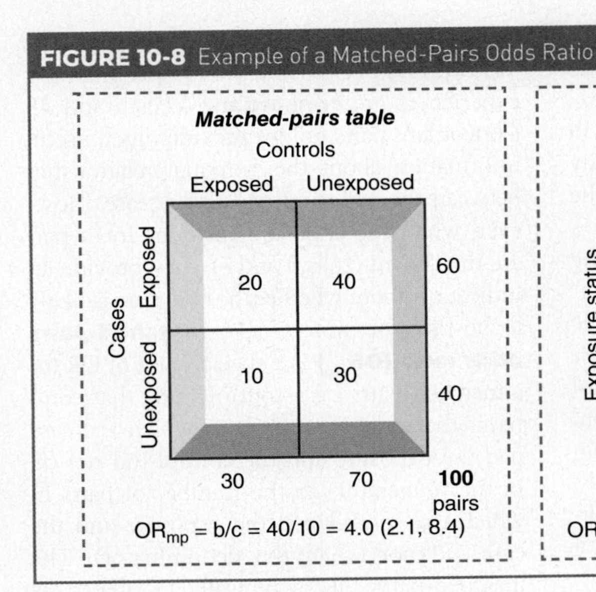

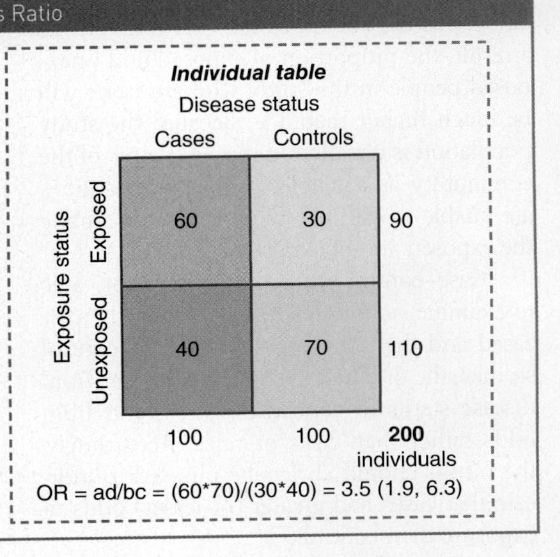

CHAPTER 11

Cohort Studies

A cohort study follows participants through time to calculate the rate at which new disease occurs and to identify risk factors for the disease.

11.1 Overview

A **cohort** is a group of similar people followed through time together. A **cohort study** is an observational study that follows people forward in time so that the rate of incident (new) cases of disease can be measured. Cohort studies can take many forms. For simplicity, this chapter will group cohort study designs into three categories: prospective, retrospective, and longitudinal (**Figure 11-1**). The published literature is somewhat inconsistent in how

FIGURE 11-1 Key Characteristics of Cohort Studies

Approach	Prospective or Retrospective Cohort	Longitudinal Cohort
Objective	Compare rates of new (incident) disease over time in people with and without a particular well-defined exposure.	Follow a representative sample of a well-defined population forward in time to look for new (incident) diseases associated with a diversity of exposures.
Primary study question	Is exposure associated with an increased incidence of disease?	Is exposure associated with an increased incidence of disease?
Population	Participants must be similar except for exposure status.	Participants must be available for follow-up months or years after enrollment.
	Because the goal is to look for incident disease, no one can have the disease of interest at the start of the study.	The study participants must be reasonably representative of the population from which they were drawn.

(continues)

FIGURE 11-1 Key Characteristics of Cohort Studies *(continued)*

Approach	Prospective or Retrospective Cohort	Longitudinal Cohort
When to use this approach	An exposure is relatively uncommon, but a source of exposed individuals is available.	The goal is to examine multiple exposures and multiple outcomes, and time is not a concern.
Requirement	A source of individuals with the exposure is available.	The research team has adequate time and money for the study.
First steps	1. Identify a source of individuals with the exposure. 2. Decide what type of unexposed individuals will be an appropriate comparison group.	1. Select a source population. 2. Select the exposures and outcomes that will be assessed. 3. Decide how often data will be collected. 4. Develop a strategy for minimizing the burden of participation and maximizing benefits and incentives.
What to watch out for	Loss to follow-up (prospective studies) or missing records (retrospective studies)	Loss to follow-up
	Information bias in which the exposed participants are more thoroughly examined for disease than unexposed participants	Potential data management challenges if a lot of information is collected at many points in time
Key statistical measure	Incidence rate ratio (RR, also called the relative risk)	Incidence rate ratio (RR, also called the relative risk)

these terms are used, but the distinct epidemiological approaches used for these three types of cohort studies are widely recognized.

All cohort studies have at least two measurement times. First, an initial survey determines the baseline exposure and disease status of all participants. A **baseline** is an initial measurement used as a benchmark for examining changes over time. Later on, one or more follow-up assessments determine how many participants who did not have the disease of interest at baseline have developed incident disease during the study (**Figure 11-2**). Because data for a cohort study are collected from individuals at multiple points in time, researchers can know with certainty which exposures were present in individual participants before the onset of new disease. This information allows for the identification of potentially causal exposures.

FIGURE 11-2 Framework for a Cohort Study

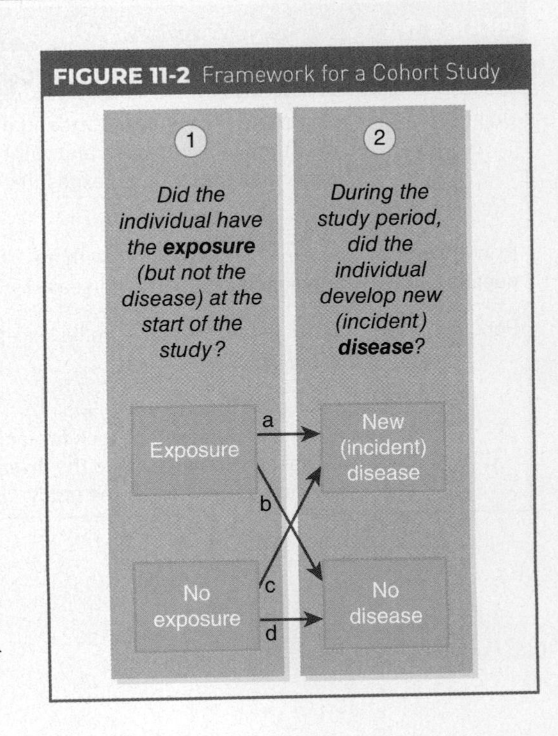

11.2 Prospective Cohort Studies

The word "prospective" means future-oriented. Any study that follows participants forward in time—including all types of cohort studies—can be considered to be a **prospective study**. However, this term is usually used to refer to cohort studies that recruit participants based on their exposure status and follow them into the future. A **prospective cohort study** recruits participants because they have or do not have an exposure of interest at the time of the baseline survey and then follows both exposed and unexposed people forward in time to look for incident cases of disease. Recruiting based on exposure status makes prospective cohort studies an excellent option for studying uncommon exposures.

The first step in designing a prospective cohort study is identifying two accessible source populations, one for individuals who are known to have had the exposure of interest and one for people who are known not to have been exposed. The members of the two comparison groups should be similar except for their exposure status. A study examining the health effects of a particular industrial chemical should compare workers at plants that use the chemical with workers at plants that do not use the chemical. It would usually not be valid to compare factory workers to office managers. A prospective cohort study comparing health outcomes in children with high blood lead levels and low blood lead levels might recruit all of its participants from the same elementary school. It would not be as helpful to examine the impact of blood lead levels if the exposed students were from one primary school and the unexposed were from another school, because observed differences in health status might be due to differences in geography or socioeconomic status rather than lead exposure.

11.3 Retrospective Cohort Studies

A **retrospective cohort study** (sometimes called a **historic cohort study**) recruits participants based on data about their exposure status at some point in the past and typically also measures outcomes that have already occurred (but happened after the baseline exposures were established). The key difference between retrospective and prospective studies is when the baseline measurements are established (**Figure 11-3**). Retrospective cohort studies use documented baseline data collected at some point in the past and follow the cohort to another point in the past or to the present. Prospective and longitudinal cohort studies collect baseline data in the present and follow the cohort to some point in the future.

For a retrospective study, the first step in study design is identifying a source of existing records that can provide baseline data of adequate quality. Retrospective studies compile

FIGURE 11-3 Times of Baseline and Follow-Up Data Collection for Cohort Studies

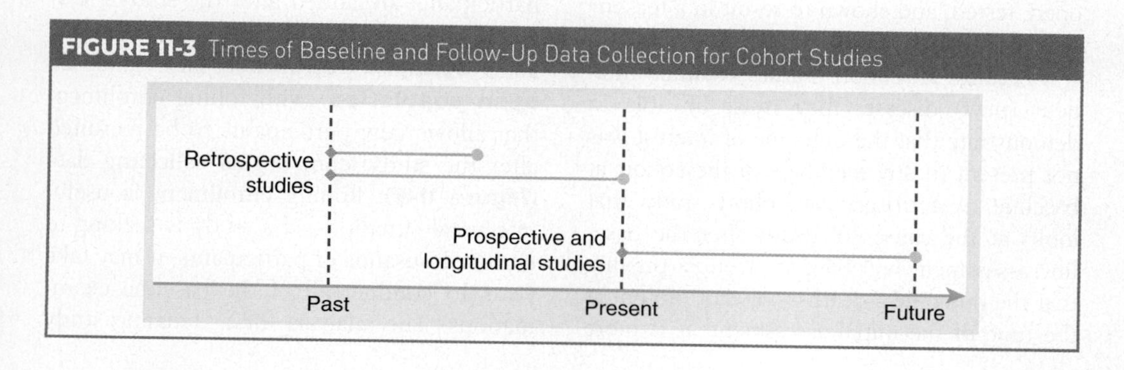

baseline data from birth records, school records, medical files, occupational records, or other sources that may be decades old. These baseline records are then matched to files compiled in subsequent years.

In some situations, existing records may provide all required follow-up data, and no contact with the individuals whose files are being examined is required. For some historic cohort studies, this is the only option available because all of the individuals included in the analysis are deceased. For example, a retrospective study might track down the causes of death after discharge from the armed services of soldiers whose military records indicate whether they served or did not serve in a particular deployment zone during World War I. None of those soldiers would still be living, but death records might be available to researchers. (In many jurisdictions, access to death records is restricted for a few decades, but the files then become part of the public record.)

Sometimes the optimal way to gather information about health outcomes occurring after a baseline in the past is to contact the sampled individuals in the present. For example, a research team exploring how birthweight influenced adult health status might seek to track down two groups of middle-aged adults: those born at a particular hospital in a particular year who had low birthweights and those born in the same hospital in the same year who had normal birthweights. If direct communication with participants will be required, a method for contacting the people identified in historic records must be developed, tested, and shown to result in a reasonable participation rate.

Because all cohort studies examine incident (new) disease, they must be able to demonstrate that the outcome of interest was not present in any members of the cohort at baseline. A retrospective cohort study that looks at the causes of death after the baseline assessment will have no trouble proving that the outcome—death—was not present at the time of the initial assessment. It is more challenging to conduct a retrospective study when the outcome of interest is a condition that may have been present at baseline but not documented.

11.4 Longitudinal Cohort Studies

A **longitudinal cohort study** follows a group of individuals who are representative members of a selected population forward in time. Longitudinal studies do not recruit participants based on exposure status. Instead, participants are recruited based on membership in a well-defined source population. Longitudinal cohorts may follow a representative sample of the residents of one town, a cohort of randomly sampled students recruited from the same university, or all members of one professional organization.

Individual participants in longitudinal studies are usually assessed at baseline for several exposures and diseases. They are then followed forward in time to determine the incidence rates for several outcomes of interest. A participant with a history of breast cancer at the baseline exam would need to be excluded from any analyses of breast cancer incidence. However, that person could be included in studies of heart disease incidence if she did not have heart disease at the baseline exam.

Longitudinal studies may use a **fixed population** in which all participants start the study at the same time and no additional participants are added after the study's start date. Alternatively, longitudinal studies may use a **dynamic population** (also called an **open population**) with rolling enrollment that allows new participants to be recruited after the study team begins collecting data (**Figure 11-4**). Rolling enrollment is useful in several situations. If a study is seeking to recruit thousands of participants, it may take years to conduct all of the baseline examinations. The analysis for a dynamic study

FIGURE 11-4 Fixed and Dynamic Populations

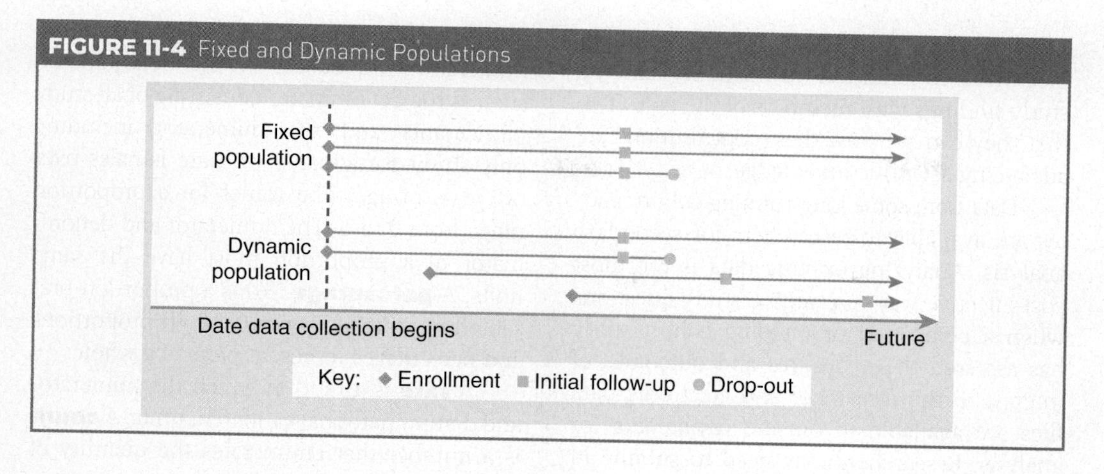

adjusts for early enrollees being followed for several years more than later enrollees. During analysis, the follow-up time can be based on individual participants' dates of enrollment rather than on a fixed calendar date. If researchers expect that many of the original participants will die or drop out of the study—as is likely to happen when studies are following older adults—a dynamic study allows lost participants to be replaced by new recruits. This design allows the total number of participants being followed at any given time to remain relatively stable even if the particular individuals enrolled in the study change over time.

Several variants of longitudinal studies measure the same individuals repeatedly over time. In some disciplines, the terms **time series study** or **panel study** describe research studies that measure participants or samples of participants at multiple points in time. Surveillance systems that are designed to monitor whole populations over an extended period of time, often using continuous data collection rather than discrete time points, may also use a cohort approach. However, repeated cross-sectional studies that measure individuals randomly sampled from the same population at different points in time are not using a cohort study approach, because they do not necessarily capture the same individuals in each round of questioning.

11.5 Data Collection

Once source populations have been identified, plans for data collection can be made. Survey instruments and other assessments for cohort studies must establish exposure and disease status for all participants at baseline and at follow-up. All participants must complete the same assessments in order to prevent the information bias that might result when exposed participants are more thoroughly examined for disease than unexposed participants. A strong data management system must be established to link baseline and follow-up data while maintaining the confidentiality of the information provided by participants.

For prospective and longitudinal cohort studies, decisions must be made about how often follow-up data collection will take place and how long the study (or at least the first wave of the study) will continue. **Loss to follow-up** is the inability to continue tracking a participant in a prospective or longitudinal study because the person drops out, relocates, dies, or stops responding to study communication for another reason. Because loss to follow-up is a major concern of studies that follow participants forward in time, researchers must develop strategies that minimize the burden of participation while maximizing interest in continuing to participate. Some studies may increase retention rates by offering participants free medical tests or other incentives in addition to regularly sharing study findings with members of the cohort so that they can see how their contributions are advancing scientific knowledge.

Data from some long-running cohort studies are available to researchers for secondary analysis. Analyzing existing data is the most cost-effective way to examine study questions when a completed or ongoing cohort study has assessed the exposures and outcomes of interest to the researcher and electronic data files are available to outside researchers for analysis. Researchers may need to submit an application to the data collection team that includes a detailed protocol describing the specific research question that will be explored and the exact variables required. Approval from relevant research ethics committees is often required before data are released to the secondary analyst. A fee may be charged to cover the expenses associated with compiling the requested variables into a useable data file.

11.6 Ratios, Rates, and Risk

A **ratio** is a comparison of two numbers. A ratio can be displayed as *A*:*B*, calculated as a fraction in the form of *A*/*B*, or reduced to a single number using division. The **numerator** is the top number in a ratio—the *A* in the ratio *A*/*B*. The **denominator** is the bottom number in a ratio—the *B* in the ratio *A*/*B*. A **reference population** is a group that is used as a comparison for another population. In rate ratios, the rate for the reference population is placed in the denominator of the ratio. A rate ratio comparing males to females that makes females the reference population will place the rate for males in the numerator and the rate for females in the denominator of the ratio. If males are selected to serve as the reference population, then the rate for females should be the numerator for the ratio and the rate for males should be the denominator.

A **proportion** is a ratio in which the numerator is a subset of the denominator, such as the denominator consisting of all study participants and the numerator including only study participants who are females over 60 years of age. The values for a proportion range from 0 to 1. The numerator and denominator of a proportion must have the same units. A **percentage** (%) is a proportion presented in units of "per 100." Both proportions and percentages represent parts of a whole.

A **rate** is a ratio in which the numerator and denominator have different units. A **count** is a number that enumerates the quantity of similar items, such as the number of people who have been classified as cases according to a case definition. Counts are often used as the numerator for the calculation of rates. The denominator of a rate typically expresses a measure of time, so rates are often presented in units like "cases per year" or "miles per hour." **Risk** is the probability of an individual in a population becoming a case during a defined period of time. The terms "rate" and "risk" have distinct meanings even though they are often incorrectly used interchangeably in epidemiology reports. *Risk* implies causality in a way that the more mathematical *rate* does not, so the term "risk" should be used judiciously.

11.7 Incidence Rates

The goal of cohort studies is to observe the incidence of new disease or other new health-related outcomes. **Incidence** is the number of new cases of disease in a population during a specified period of time. Incidence is typically reported in terms of an incidence rate or an incidence proportion.

The **incidence rate** is the number of new cases of disease in a population during a specified period of time divided by the total number of people in the population who were at risk during that period. The incidence rate is sometimes called the **absolute risk** to emphasize that the number is a measured value in one population. Absolute risk is contrasted with relative risks that compare several observed values. The **population at risk** consists of people who do not already have the disease being tracked in a cohort study. Individuals who already meet the case definition at the start of the study period are not at risk of getting new disease. Because they cannot be included among the incident cases tallied for the numerator of the incidence rate, they are removed from the denominator when the incidence rate is calculated (**Figure 11-5**). For example, suppose a cohort study examined the incidence of disease over 1 year in a population with 50 members. If 7 of those 50 people already had the disease at the start of the year, the denominator should be 43 rather than 50. If 4 of those 43 are diagnosed with the disease during the year, then the incidence rate is 4/43.

Incidence rates are usually converted to units of "per 1000," "per 10,000," or the like so that they can be more easily compared. An incidence rate of 4/43 might be reported as a rate of 93 cases per 1000 people per year. This numerator is calculated by dividing 4 by 43 and then multiplying that value (0.093) by 1000. Alternatively, the numerator for a rate with units of "per 1000" can be calculated by dividing 1000 by 43 and then multiplying that value (23.3) by 4 to quantify the numerator (93).

The incidence rate is typically the preferred measurement of incidence. The **cumulative incidence** (also called the **incidence proportion** or the incidence risk) is the percentage of people at risk in a population who develop new disease during a specified period of time. The cumulative incidence is a measurement of risk during a stated time frame, such as the probability of a neonate dying before the first birthday or the probability

FIGURE 11-5 Incidence

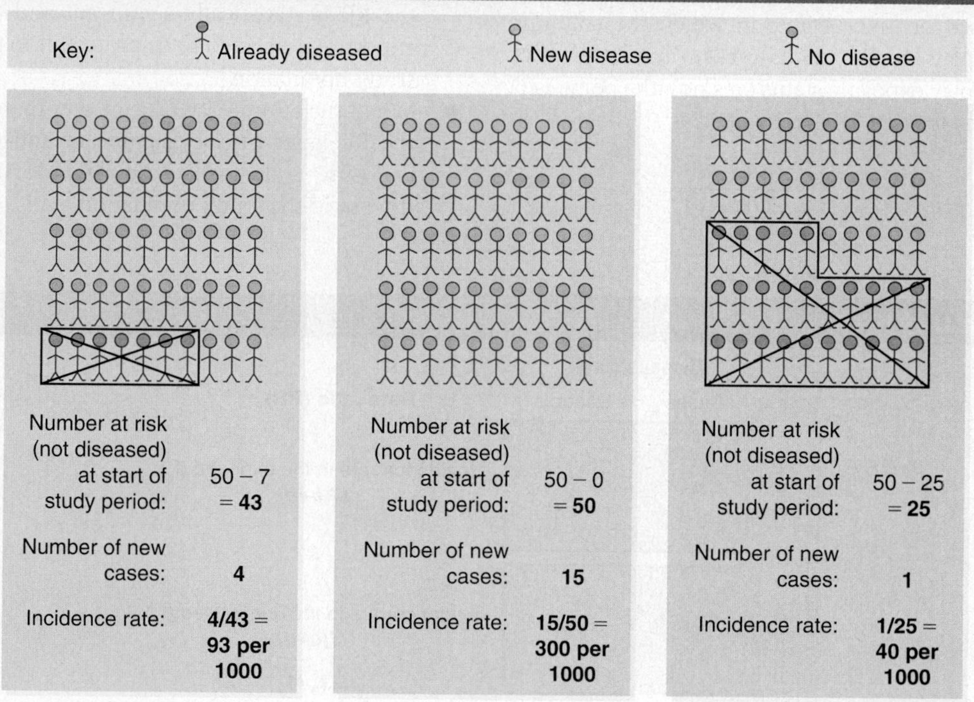

Key: Already diseased | New disease | No disease

Number at risk (not diseased) at start of study period:	50 − 7 = **43**	Number at risk (not diseased) at start of study period:	50 − 0 = **50**	Number at risk (not diseased) at start of study period:	50 − 25 = **25**
Number of new cases:	**4**	Number of new cases:	**15**	Number of new cases:	**1**
Incidence rate:	**4/43** = **93 per 1000**	Incidence rate:	**15/50** = **300 per 1000**	Incidence rate:	**1/25** = **40 per 1000**

of an individual contracting a novel strain of influenza during the first 3 months of an epidemic. Cumulative incidence is typically only calculated when the time period for the calculation is relatively short and the risk is not negligible. The **attack rate** is the cumulative incidence of infection during the course of an epidemic.

11.8 Incidence Rate Ratios

A **rate ratio (RR)**, also reported as a **relative rate** or sometimes as a risk ratio or relative risk, describes any ratio of two rates in which the reference (comparison) group is placed in the denominator. For cohort studies, the most common measure of association is the **incidence rate ratio**, which compares the incidence rate among the exposed to the incidence rate in the unexposed (**Figure 11-6**). Although this term would technically be called an "IRR," it is usually just reported as an "RR." Based on a 2×2 table in which the columns display disease status (yes/no) and the rows display exposure status (yes/no), the point estimate for this RR is calculated as:

$$\text{RR} = \frac{a/(a+b)}{c/(c+d)}$$

The point estimate provides an initial interpretation for the RR:

- RR = 1 (or close to 1): The incidence rate was the same (or about the same) in the exposed and in the unexposed.
- RR > 1: The incidence rate was greater in the exposed than in the unexposed, suggesting that the exposure was risky.
- RR < 1: The incidence rate was lesser in the exposed than in the unexposed, suggesting that the exposure was protective.

The 95% confidence interval (CI) for the point estimate indicates whether the calculated RR is statistically significant (**Figure 11-7**). For the risky example in the figure, the report could state that "participants with the exposure were twice as likely to develop the disease as participants without the exposure." For the protective example in the figure, it would be accurate to report that "participants with the exposure were half as likely to develop the disease as those without the exposure."

- When the entire 95% CI is less than 1, the RR is statistically significant and the exposure is deemed to be protective in the study population.
- When the entire 95% CI is greater than 1, the RR is statistically significant and the exposure is deemed to be a risk for the disease in the study population.

FIGURE 11-6 Rate Ratio (Point Estimate)

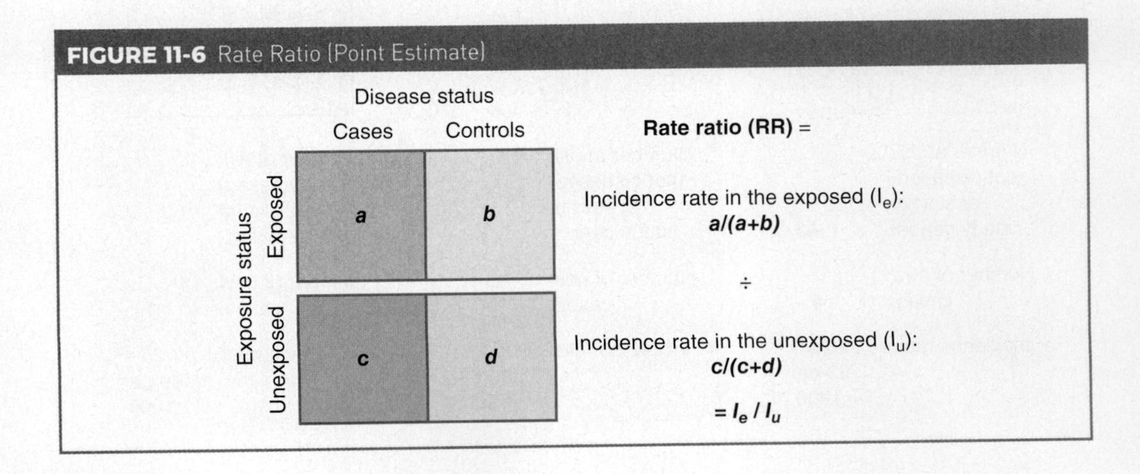

FIGURE 11-7 Interpretation of the Rate Ratio Based on Its 95% Confidence Interval

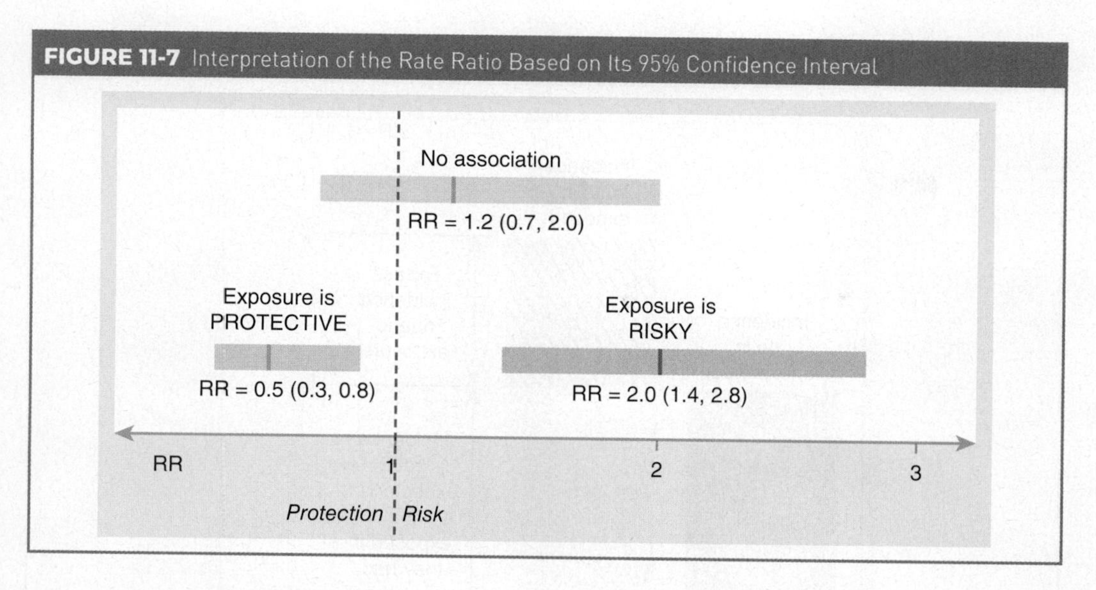

- When the 95% CI overlaps RR = 1, the association between the exposure and the outcome is not statistically significant. The lower end of the CI is in the protective range (RR < 1) and the higher end is in the risky range (RR > 1), so there is no clear association between the exposure and the disease. The appropriate conclusion is that there is no evidence for an association between the exposure and the disease in the study population.

A 95% CI showing no association may indicate that there truly is no relationship between the exposure and the disease, but it may also result from a small sample size. A study with a small number of participants might not have the statistical power to yield a statistically significant result even if the exposure truly is risky or protective in the source population. Power calculations can be used to examine whether the sample size for a study was sufficient to allow conclusions about non-significance.

It is also possible to calculate an odds ratio (OR) for a cohort study. For a cohort study, the OR is the ratio of the odds of disease among exposed people (in the numerator) to the odds of disease among unexposed people (in the denominator). Basic algebra shows that the equation for the OR of $(a \div b)/(c \div d)$ can be simplified to ad/bc, which is the same equation for the point estimate that is used for case–control studies. The value of an OR for a cohort study might be quite different from the value of the RR. Suppose that 50% of exposed people and 25% of unexposed people developed the disease of interest. The RR for this study is 50% ÷ 25% = 2.0. The odds of disease among the exposed are 50:50 = 1. The odds of disease among the unexposed are 25:75 = 0.33. The odds ratio is OR = 1/0.33 = 3.0. Exposed individuals have double the rate of disease compared to unexposed individuals, but triple the odds of disease. Because odds are more difficult to interpret than rates, the RR, rather than the OR, is generally the preferred measure of association for cohort studies.

11.9 Attributable Risk

In addition to the RR, there are several other ways to compare the rates of new disease in the exposed and unexposed members of a cohort. The **rate difference** is the absolute difference in the incidence rate between the exposed group and the unexposed group (**Figure 11-8**). The rate difference is also called the **excess risk**, the **attributable risk (AR)**,

FIGURE 11-8 Attributable (Excess) Risk

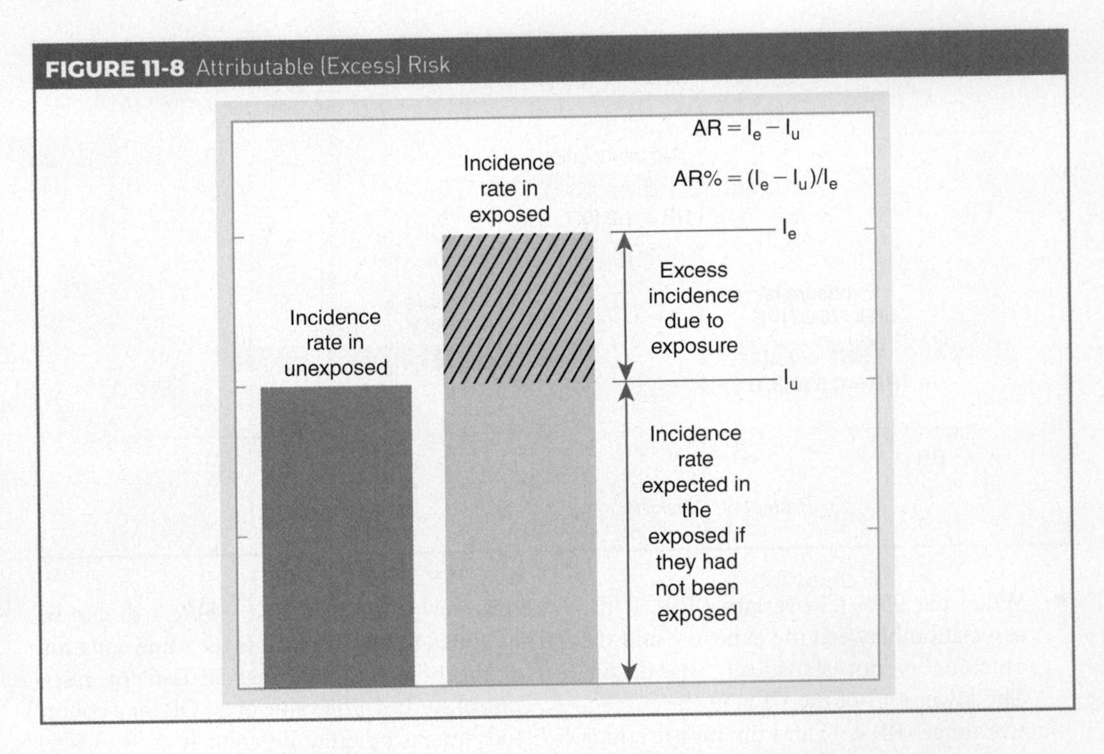

and the risk difference. For example, if 10% of the unexposed and 15% of the exposed became ill during the study period, then the excess risk in the exposed was 15% – 10% = 5%. This number represents the additional rate of disease in the exposed that can be attributed to the exposure. The calculation assumes that the exposed would have had the same rate as the unexposed if they had not had the exposure. This assumption is one of the reasons why the exposed and unexposed populations in a cohort study must be similar except for their exposure status.

The **attributable risk percentage (AR%)** is the proportion of incident cases among the exposed population in a cohort study that is due to the exposure. The AR% is sometimes called the etiologic fraction in the exposed. The percentage is calculated by comparing the excess risk to the incidence rate in the exposed. For the preceding example, the AR% is 5% ÷ 15% = 33%. The interpretation of this result is that one-third of the cases of disease in the exposed could have been prevented if the exposure had been removed.

The **population attributable risk (PAR)** is the rate of new disease in a population that can be attributed to some people in the population having an exposure. The **population attributable risk percentage (PAR%)** is the proportion of incident cases in the total population that can be attributed to some people having the exposure. The PAR% is sometimes called the etiologic fraction in the population. If 100% of cases occur among exposed people, the PAR% is 100%. If some cases occur among both exposed and unexposed people and the prevalence of exposure is low in the population, then the PAR% will be small compared to the AR%. The PAR and PAR% can be calculated only when the prevalence of the exposure in the total population is known. Longitudinal cohort studies that recruit a cross-section of a population often allow for the calculation of the PAR and PAR%. Prospective cohort studies that preferentially recruit exposed individuals do not allow the population prevalence rate to be estimated, so the PAR and PAR% cannot be calculated.

FIGURE 11-9 Examples of Rate Ratio Calculations

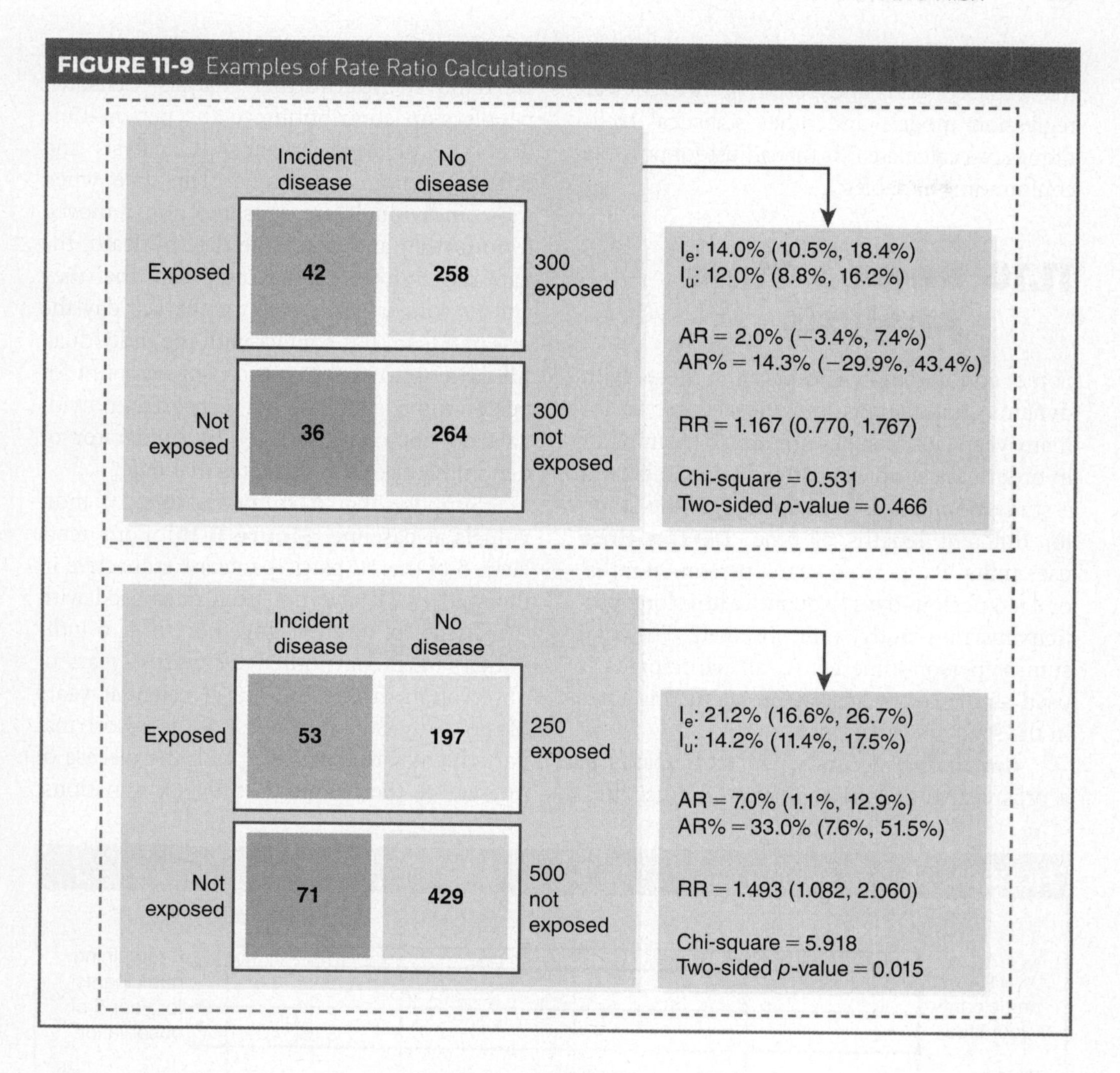

Computer- and Internet-based statistical programs are available for the calculation of a variety of statistics that can be derived from a 2×2 table (**Figure 11-9**), such as the:

- Incidence in the exposed
- Incidence in the unexposed
- Excess risk or attributable risk (AR)
- Attributable risk percentage (AR%)
- Incidence rate ratio (RR)
- Confidence intervals for the incidence rate, AR, AR%, and RR
- Chi-square statistic and its associated *p* value

Sample results from a computer-based statistical program are shown in Figure 11-9. The top example in the figure shows an RR of 1.17, a 95% CI of (0.77, 1.77), and a *p* value of $p = .47$. This RR is not statistically significant because the 95% CI overlaps 1. The chi-square *p* value of $p > .05$ confirms the conclusion of no association or no difference between the exposed and the unexposed groups. The bottom example in the figure has an RR of 1.49 (1.08, 2.06) and a *p* value of $p = .02$. This is a statistically significant result that implies that the exposure is risky, because the entire 95% CI is greater than 1. The chi-square *p* value of $p < .05$ confirms that there is a statistically significant difference between the exposed and unexposed groups. Linear regression models and other statistical techniques can calculate RRs that adjust for possible confounding variables.

11.10 Person–Time Analysis

Some cohort studies, especially those with dynamic populations and those that run for many years, use person–time as a denominator in order to account for individual members of a study population participating in the study for different lengths of time. **Person–time** uses units like person–years, person–months, or even person–days to quantify how long participants in a study were observed. The total sum of person–time across all participants is used as the denominator for calculating rates in the study population.

Censoring occurs when participants in a prospective or longitudinal study die, drop out, are lost to follow-up, or for another reason are removed from further analysis. Censored people stop contributing to the person–time denominator used in survival analysis and other outcome evaluations. The date when those individuals are censored and removed from further analysis is the date of death, the date they inform the research team that they will no longer participate, or the last day the research team has contact with the individual. All days up to the date of censoring count for person–time analysis, but censored individuals do not contribute to the numerator or denominator of analysis after that date.

Suppose that a study recruits 10 individuals at baseline (**Figure 11-10**). Four years later, 6 of the 10 participants are still active in the study and have not been diagnosed with the disease of interest. Together, these 6 individuals have contributed 24 person–years of observation during the first 4 calendar years of the study. Suppose that 2 of the 10 original participants are diagnosed with the disease of interest at their annual study examinations.

FIGURE 11-10 Person–Year Analysis

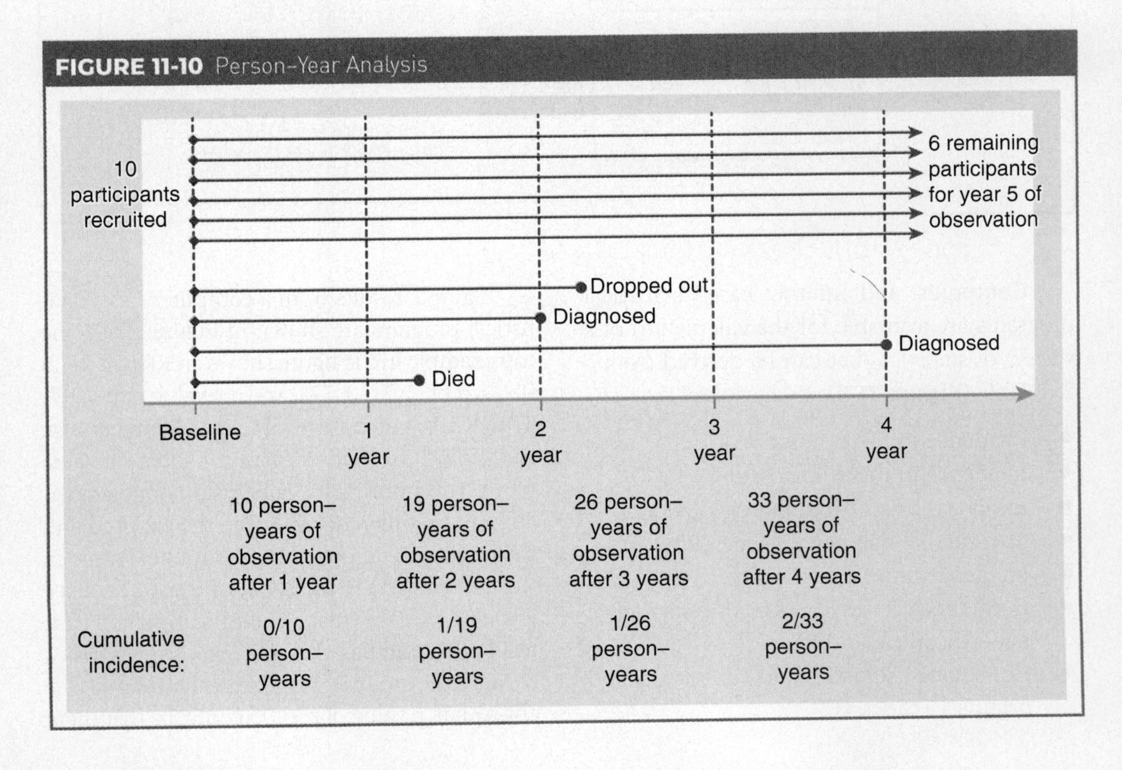

One person is diagnosed 2 years into the study, and the other 4 years into the study. Together, these 2 individuals contributed 6 person–years of observation to the study. However, once they are diagnosed and no longer able to develop incident disease, they are no longer able to contribute further person–years to the denominator for the calculation of incidence. Two other participants also leave the study and are censored and removed from analysis. One drops out of the study after the second year but before the third year. That participant is considered to have contributed 2 person–years of observation. Another dies after the first year and contributes only that 1 person–year of observation. In total, over 4 calendar years, the 10 original participants experience 2 incident cases of disease over 33 person–years of observation. For the calculation of RRs and other measures that rely on the comparison of incidence rates, it does not matter whether the incidence rates are measured per 1000 participants (Figure 11-5) or per 1000 person–years (Figure 11-10), as long as all incidence rates in the equation use the same units.

CHAPTER 12

Experimental Studies

An experimental study assigns participants to intervention and control groups in order to test whether an intervention causes an intended outcome.

12.1 Overview

Experimental studies are the gold standard for assessing causality (**Figure 12-1**). They are used for clinical trials of new therapies for individuals with various illnesses, field trials of preventive interventions like vaccinations, and community trials of public health and environmental interventions. Because the researcher assigns participants to receive a

FIGURE 12-1 Key Characteristics of Experimental Studies

Objective	Compare outcomes in participants assigned to an intervention or control group.
Primary study question	Does the exposure cause the outcome?
Population	Similar participants are randomly assigned to an intervention or control group.
When to use this approach	Assessing causality
Requirement	The experiment is ethically justifiable.
First steps	1. Decide on the intervention and eligibility criteria. 2. Define what will constitute a favorable outcome. 3. Decide what control is an appropriate comparison for the intervention. 4. Decide whether blinding will be used to prevent participants and/or the researchers who will assess outcomes from knowing whether a participant has been assigned to the intervention or the control group. 5. Select the method for randomizing participants to an intervention or control group.
What to watch out for	Noncompliance
Key statistical measure	Efficacy

particular exposure, the exact dose, duration, and frequency of the exposure are known. The researcher knows when the exposure occurred, so the health status of each participant before and after the exposure can be compared. The researcher can therefore assess whether the exposure may have caused a particular outcome.

A **controlled trial** is an experiment in which some of the participants are assigned to an intervention group and some are assigned to a nonactive comparison group. A very common experimental study design used in the health sciences is a **randomized controlled trial (RCT)** in which some participants are randomly assigned to an active intervention group, the remaining participants are assigned to a control group, and all participants from both groups are followed forward in time to see who has a favorable outcome and who does not (**Figure 12-2**). RCTs and all other types of experimental study designs require careful descriptions of:

- The intervention
- The type of control that will be used and why it is appropriate
- How participants will be assigned to exposure groups
- The end point that will constitute a favorable outcome for the trial

FIGURE 12-2 Framework for an Experimental Study

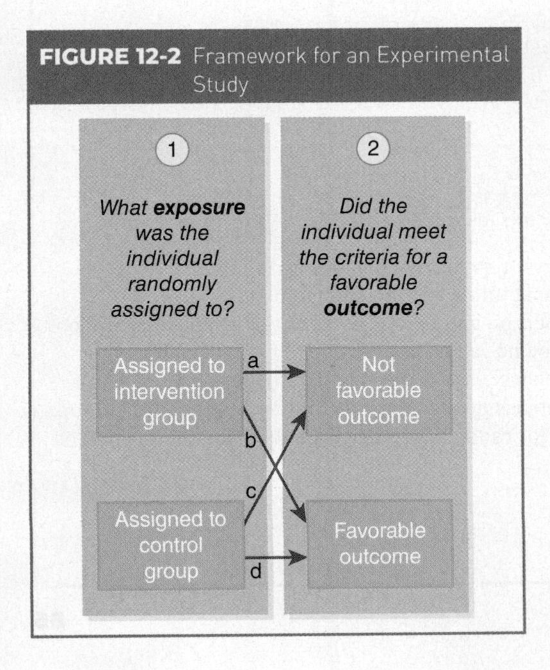

Experimental studies also require careful consideration of the ethical challenges associated with assigning participants to an exposure, even if that exposure is expected to improve health status.

12.2 Describing the Intervention

An **intervention** is a strategic action intended to improve individual and/or population health status. Experimental medical studies typically focus on testing the effectiveness of interventions intended to diagnose, treat, cure, or prevent adverse health conditions. Experimental public health studies typically focus on testing the effectiveness of interventions designed to prevent health problems at the population level. **Prevention science** is the scientific study of which preventive health interventions are effective in various populations, how successful the interventions are, and how well they can be scaled up for widespread implementation.

The first step in an experimental study is to carefully define the intervention that participants assigned to the active intervention group will receive and to decide on the person, place, and time (PPT) criteria for the study. The description should state exactly:

- What the intervention will be
- The eligibility criteria for participants
- Where and how participants will receive the intervention
- When, how often, and for what duration participants will receive the intervention

For example, a trial of a new drug will impose very strict requirements for the composition of the pill to be ingested, the eligibility criteria for volunteers, and the schedule for how often the pills will be taken and for how many weeks the trial will last. A new strength-building intervention will provide detailed descriptions of the exercise procedures and how they will change in intensity over the study period, the inclusion and exclusion criteria that will apply to potential volunteers, how participants will be coached or supervised, where participants will engage in the exercises, and how many months the study will run.

12.3 Defining Outcomes

Most experimental studies are superiority trials. A **superiority trial** aims to demonstrate that a new intervention is better than some type of comparison, not merely as good as the comparison (**Figure 12-3**). "Better" could mean that an intervention is more effective than a current therapy at curing existing disease, or it could mean that a new intervention is more effective than a placebo at preventing new disease from occurring.

For some experimental studies, the goal is to show that the intervention yields outcomes that are as good or no worse than the comparison. For example, an experimental study may test whether a cheaper intervention is as good as, or no worse than, a more expensive intervention. If both of the interventions yield similar outcomes, that is helpful evidence to support scaling up the use of the more cost-effective approach. An **equivalence trial** aims to demonstrate that a new intervention is as good as some type of comparison. A **noninferiority trial** aims to demonstrate that a new intervention is no worse than some type of comparison.

FIGURE 12-3 Types of Success

Goal	Success
Superiority trial	The intervention is better than the comparison.
Noninferiority trial	The intervention is not worse than the comparison.
Equivalence trial	The intervention is equal to the comparison.

Because the term "better" can be defined in so many ways, the researcher must carefully define what constitutes a favorable outcome for an individual participant and for the experimental study as a whole. These measures of success must be stipulated prior to the initiation of the study. For example, an individual participant's success in a weight-loss program could be defined as the loss of at least 10% body weight and the maintenance of the lower weight for at least 6 months (**Figure 12-4**). Alternatively, success could be defined as the loss of at least 15 pounds over a 2-month intervention period or as achieving a body mass index of less than 30 by the end of the study period. These measures of individual success can then be translated into measures of study success. For example, the weight-loss trial could be considered successful if the proportion of participants with favorable individual-level outcomes is significantly greater in the intervention group than in the control group.

For many studies, the outcome can be objectively measured. If the goal is to evaluate whether a new vaccine prevents infection, laboratory tests can be used to confirm that the participants did not have immunity prior to the infection but did have immunological markers for the relevant pathogen after vaccination. Some studies allow only subjective measures of outcomes, like those that ask participants to describe the effects of a drug intended to alleviate pain or anxiety. If the goal of an experiment is to test whether a new counseling intervention is better than another type of psychosocial intervention at improving the quality of life of people with a particular health condition, the survey instrument that will be used to measure quality of life before, during, and after the trial must be carefully constructed and validated.

FIGURE 12-4 Examples of Favorable Outcomes

Intervention	Intended Outcome	Favorable Outcome for an Individual	Unfavorable Outcome for an Individual	Favorable Outcome for the Study Population
New diet- and exercise-based weight-loss program	Significant weight loss	The loss of ≥ 10% body weight and maintenance of lower weight for ≥ 6 months	The loss of < 10% body weight or failure to maintain weight loss of ≥ 10% or more for ≥ 6 months	The proportion of those who lose at least 10% of their body weight and maintain that loss for at least 6 months is higher in the intervention group than in the control group.
New drug therapy	Improvement of the quality of life for those with a particular disease condition	Improvement in quality of life	Failure to demonstrate improvement in quality of life	The rate of improvement in the drug therapy (intervention) group is higher than the improvement rate in the placebo (control) group, according to a carefully defined and validated set of criteria for what constitutes improvement.
New preventive vaccine	The prevention of infection	Incident infection does not occur	Incident infection occurs	The incidence of infection in the vaccinated (intervention) group is lower than the incidence of infection in the unvaccinated (control) group, as confirmed by laboratory testing.

12.4 Selecting Controls

Experimental studies usually assign some participants to the active intervention and the remainder to a control group (**Figure 12-5**). One commonly used type of control is a **placebo**, an inactive comparison that is similar to the therapy being tested. Examples of placebos are a sugar pill used as a control for a pill with an active medication, a saline injection used as a control for an injection of an active substance, and a sham procedure that is designed to look and feel like a real clinical procedure used as a control for that active procedure. The mere act of taking a pill or receiving some other form of therapy, even if it is inert or inactive, is often enough to make recipients feel better. Placebo-controlled studies allow the effect of the active therapy to be examined separately from the boost in health status that people may experience simply by participating in a clinical trial or receiving some other intervention.

Not all experimental studies use placebos. When the goal of the experiment is to test whether a new therapy is better than (or at least equivalent to) a current one, it is appropriate to compare the new therapy to some **standard of care**, an existing therapy

FIGURE 12-5 Examples of Types of Controls

Type of Control	Active Intervention	Comparison
Placebo/inactive comparison	Active pill	Inactive pill that is indistinguishable from the active pill in terms of appearance, odor, taste, texture, and delivery mechanism
	Injection of an active substance	Injection of saline solution
	Acupuncture needles inserted at acupuncture points	Acupuncture needles inserted at locations in the body that are not acupuncture points (sham acupuncture)
Active comparison/ standard of care	New therapy	Current best therapy for the condition being studied
	New therapy	Current standard therapy
	New therapy	Some other existing therapy
	Current therapy plus new therapy	Current therapy alone
Dose-response	Some dose of a medication	Alternate doses of the medication
	Some duration of a therapy	Alternate durations of the therapy
No intervention	New intervention	Participants assigned to the control group are asked to maintain their usual routines.
Self	New intervention	Each participant's status before the intervention is compared to his or her own status after the intervention.
	New intervention	Each participant receives the new intervention for some duration and the comparison for some duration, preferably in a random order.

that is used as a comparison for a new therapy being experimentally tested. Some experimental studies use the best therapy currently available as the comparison group. Others use the therapy that is used most often in the location where the study is being conducted. Sometimes the new therapy may be given in addition to the existing therapy.

Sometimes the goal is to determine how much of an intervention is required. For example, should the dose of a substance be changed? Is 100 mg of a medication as effective as 200 mg? Or should the duration of therapy be reconsidered? Is 4 weeks of physical therapy as effective as 8 weeks? In such cases, varying doses and durations may be tested and compared to one another. Some RCTs use a **factorial design** that tests several different interventions in various combinations within one trial (**Figure 12-6**).

Experimental studies sometimes include a control group of participants who are randomly assigned to maintain their usual routines, but this method is usually not preferred. The approach raises ethical concerns about discouraging the adoption of healthier lifestyles during the course of the study. It also raises concerns about the **Hawthorne effect**,

FIGURE 12-6 Example Approaches to Randomized Control Trials

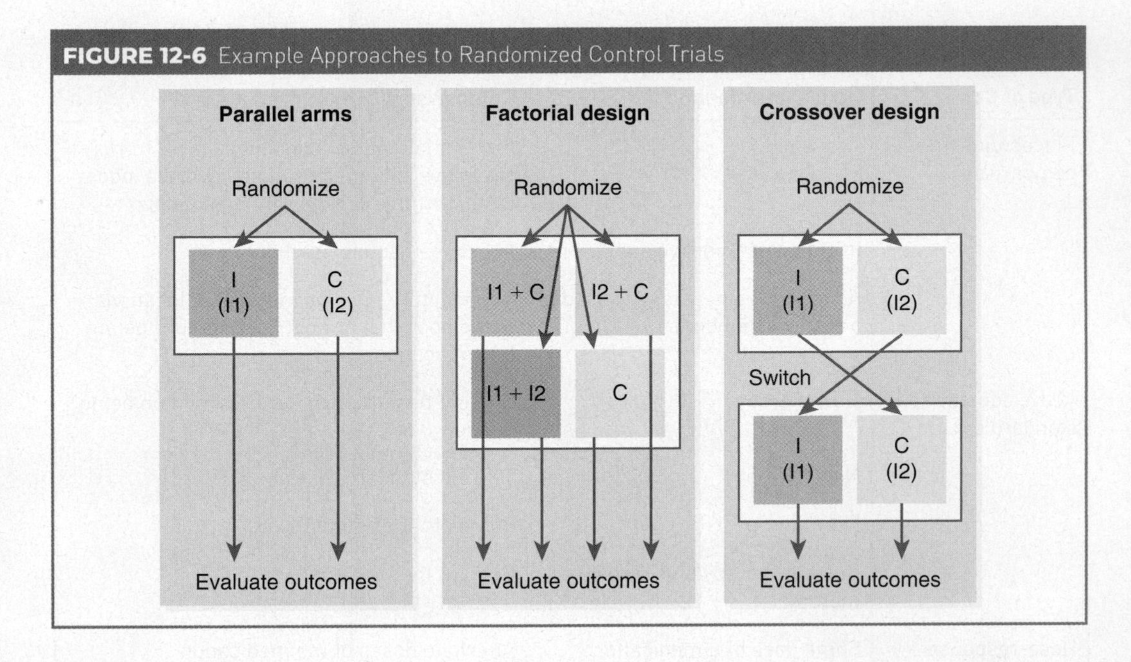

a type of bias that occurs when participants in a study change their behavior for the better because they know they are being observed. For example, suppose a researcher is initiating a study of a new weight-loss program and plans to randomly assign participants either to the new therapy or to a usual-routine group. In this situation, simply informing the controls that they will be weighed at the start and end of the study period will be enough to spur a sizable proportion of the control group to initiate an exercise program, start eating a healthier diet, or take other steps to lose weight. These changes may interfere with the accurate measurement of the impact of the new intervention.

When there are ethical concerns about withholding a potentially lifesaving intervention from some participants, it may be possible for participants to serve as their own controls. A **before-and-after study** is a nonrandomized experimental study that measures the same individuals before and after an intervention so that each participant's "before" status can serve as that individual's control. Some experimental studies in which each participant serves as his or her own control use a **crossover design** that randomly assigns some participants to receive the active intervention first and then the control, and assigns the others to receive the control first and then the active intervention. The term **arms** describes the treatment and nontreatment groups of an experiment. A **washout period** for an experimental study is a time between arms of the study when patients receive no treatment. In a crossover study, both groups may take a break between the two arms of the experiment. This break is intended to reduce **carryover effects**, the residual effects from the first part of an experimental study that may bias the results of the second part of a crossover study if a sufficient washout period between the two arms of the study is not implemented. Each participant's status before the intervention is compared to his or her own status after the intervention. However, the results of crossover experimental designs may not be as clear-cut as placebo studies because time alone can lead to significant improvements or declines in health status, especially among those who are severely ill.

12.5 Blinding

Blinding, sometimes called **masking**, is an experimental design element that keeps participants (and sometimes some members of the research team) from knowing whether a participant is in the active intervention group or the control group. In a **single-blind** experimental study, participants do not know whether they are in an active group or a control group. In a **double-blind** experimental study, neither the participants nor the researchers assessing the participants' health status know which participants are in an active or control group.

Blinding is intended to minimize **information bias**, bias in an epidemiological study that arises due to systematic measurement error. There are many types of information bias, such as recall bias, reporting bias, detection bias, and observer bias. **Reporting bias** occurs when members of one study group systematically underreport or overreport an exposure or outcome. For example, cases in a case–control study of the outcomes of drug abuse might systematically underreport their histories of drug use, while controls might provide a more complete accounting of their drug exposure histories. **Detection bias**, also called **surveillance bias**, occurs when a population group that is routinely screened for adverse health conditions incorrectly appears to have a higher-than-typical rate of disease because more frequent testing enables a higher case detection rate in that population than in the general population. **Observer bias** occurs when an observer (a researcher) intentionally or unintentionally evaluates participants differently based on their group membership, such as systematically evaluating cases and controls in a case–control study differently.

Blinding prevents participants and assessors from being able to evaluate outcomes differently based on the results they expect for an exposure. Blinding ensures that participants in the active intervention group will not report more favorable results simply because they expect a positive outcome. It also keeps assessors from intentionally or unconsciously recording more favorable results for participants in the active intervention group.

Blinding is usually possible only when all participants are assigned to similar exposures. If participants in both the active intervention group and the control group are taking pills (of the same color, shape, size, and taste) or if both are getting injections, a blinded study may be possible. In contrast, if the active intervention is a special diet and the controls eat their usual diets, if the active group will participate in exercise classes and the controls will be on their own, or if the active intervention will include both diet and exercise components and the control only a diet plan, then a blinded study may not be possible. To minimize the risk of bias in studies that are not blinded, it is helpful to identify objective outcome measures such as laboratory tests rather than relying on subjective outcome measures such as participants' self-reported feelings.

12.6 Randomization

Randomization is the assignment of participants to an exposure group in an experimental study using a chance-based method that minimizes several types of possible bias. For example, randomization minimizes the problems that could occur if participants were able to choose the intervention or control group they preferred. Some people would prefer to know they were getting the active intervention, while others with less risk tolerance might prefer to be in a control population. Self-selection might significantly alter the results of the trial. Randomization also mitigates the **allocation bias** that might occur as a result of nonrandom assignment of participants to experimental study groups, such as when people with different exposure histories are not equally distributed across treatment arms.

A variety of approaches can be used to randomly allocate participants to an active intervention group or a control group. These approaches

FIGURE 12-7 Examples of Types of Randomization

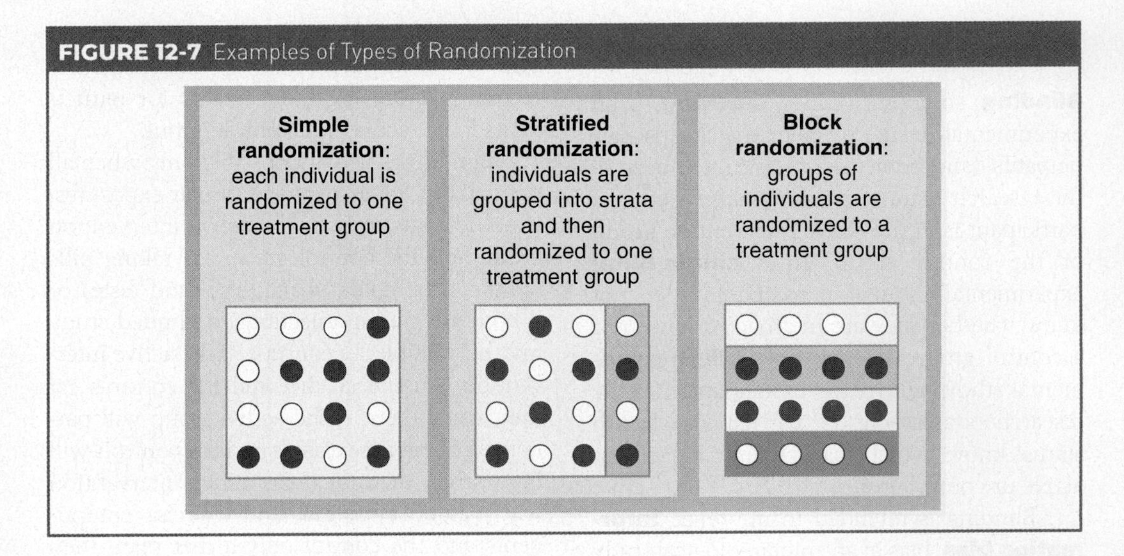

include simple randomization, stratified randomization, and block randomization (**Figure 12-7**). **Simple randomization** is the use of a coin toss, a random number generator, or some other simple mechanism to randomly assign each individual in an experimental study to one of the exposure groups.

Stratified randomization is the division of a population into subgroups prior to randomly but systematically assigning each individual within each subgroup to one of the exposure groups in an experimental study. Stratified randomization is used when it is important for members of certain subpopulations to be distributed evenly across the treatment arms of a trial. For example, suppose that 75% of the volunteers for a study are female and only 25% are male. Stratified randomization can ensure that enough males are assigned to the intervention group. The list of female volunteers can be sorted into alphabetical order by last name, and then every other individual in the ordered list can be assigned to the active group. This same process can be repeated with the male volunteers. This stratified process will ensure that half of the females and half of the males are assigned to the intervention group.

Block randomization is an allocation method that randomly assigns groups of people to an intervention group and other groups of people to a control group. In this method, randomization occurs at the group rather than individual level. For example, if there were 10 elementary schools in a county, schools could be randomly assigned to be intervention or control schools. All of the students in the 5 schools randomly assigned to the intervention group would receive the intervention. All of the students in the other 5 schools would be assigned to the comparison.

Some experimental studies use nonrandomized approaches because randomization is unethical or is not feasible. A **quasi-experimental design** is an experimental study that assigns participants to an intervention or control group using a nonrandom method. Other than using a nonrandom method to assign participants to exposure groups, most quasi-experimental studies use methods similar to those of randomized studies. Most quasi-experimental studies use both pre- and post-intervention tests to compare the two arms of a controlled study. However, some quasi-experimental studies have no control group, and some use only a post-intervention assessment (with or without a control group).

Some research studies are considered to be natural experiments because the researchers do not have any control over the interventions. A **natural experiment** is a research study in which the independent variable is not manipulated by the researcher but instead changes due to external forces. For example, a researcher may seek to understand the impact of a devastating tornado on the health of residents of the affected community by comparing residents in the damaged areas to residents of neighboring areas who were not directly harmed by the twister. The researcher does not cause the tornado or select which community it affects, but the outcomes can still be examined and compared. Or suppose that a hospital announces that it will implement a new infection control policy. A **policy** is a set of principles and procedures defined by governments or other groups to guide decision making and resource allocation. A researcher would not have the authority to assign some patients to one infection control strategy and other patients to a different policy. However, inpatients hospitalized during the year after the policy update could be considered to be the active intervention group for a study examining the effectiveness of the new policy in reducing healthcare-associated infections, and inpatients who stayed at the hospital during the year before the policy change could be considered to be the control group. These are not true experimental studies because the "interventions"—a natural disaster and a policy change—are not ones that can be manipulated by a researcher, but they can be evaluated using analytic methods similar to those used for true experiments.

12.7 Ethical Considerations

All research with human participants or their identifiable personal data raises ethical concerns that researchers must address, but experimental studies involve a particularly high level of ethical risk. In experimental studies, the researcher assigns participants to exposures that the participants do not choose and may have been unlikely to encounter had they not volunteered to participate in a research project. Thus, a number of issues, such as the following, must be considered before initiating an experimental study (**Figure 12-8**):

- The principle of **equipoise** states that experimental research should be conducted only when there is genuine uncertainty about which treatment will work better.
- The principle of distributive justice necessitates that the source population be an appropriate one and that the research study not exploit individuals from populations that are unlikely to have continued access to the therapy if it is found to be successful.
- The principle of respect for persons requires that all participants volunteer for a study without being unduly influenced by the prospect of being compensated for their participation. Respect also requires that all participants understand what it means to be a research subject, including the possibility of being assigned to a control group instead of the new intervention.
- The principles of beneficence and nonmaleficence require that researchers balance the likely benefits and risks of the study.

Researchers must make careful decisions about when to use a placebo or another type of control, must put in place a system for monitoring adverse reactions, and must identify the conditions under which an experiment would be discontinued early either because the exposure proves to be risky or because the new intervention appears to be so beneficial that keeping it from the control group would be unethical. An **adverse reaction** is a negative side effect of a medication, vaccination, or other exposure, or another bad outcome related to a study. An **adverse event** is a negative outcome that may be the direct result of a study-related exposure or may be a coincidental occurrence that is not directly related

FIGURE 12-8 Examples of Ethical Issues in Experimental Studies

Study Stage	Examples of Questions to Ask
Study topic selection	■ Is the study really necessary (equipoise)? ■ Is an experimental design truly necessary?
Recruitment	■ Is the source population an appropriate and justifiable one? ■ Is the inducement to participate appropriate and not coercive?
Randomization	■ Do participants truly understand that they might not receive the active intervention? ■ Is it appropriate to use a placebo? Is it appropriate to use some other control?
Data collection	■ How will adverse outcomes be monitored and addressed? ■ When might an experiment need to be discontinued early?
Follow-up	■ What happens if a participant experiences study-related harm after the conclusion of the study? ■ Will participants have continuing access to the therapy if it is shown to be successful?

to the study but happens after an individual receives a study-related exposure. Research ethics committee review what is required for all experimental studies. Adverse events that occur during a research study must be immediately reported to the appropriate institutional review board. See Chapter 17 for discussions of additional ethical principles that must be considered when planning and conducting research with human subjects. Chapter 18 explains the ethical review process.

12.8 Efficacy

Experimental studies use many of the same measures of association that cohort studies do, including rate ratios, attributable risks (excess risk or risk reduction), attributable risk percentages, measures of survival, and various types of regression models. Cohort studies use these measures to examine the impact of an unassigned exposure on the incidence of disease. Experimental studies use the statistics to quantify the impact of an assigned exposure on the likelihood of having a favorable or an unfavorable outcome.

There are also several measures that are specific to experimental studies. **Efficacy** is a measure of the success of an intervention that is calculated as the proportion of individuals in the control group who experienced an unfavorable outcome but could have expected to have a favorable outcome if they had been assigned to the active group instead of the control group (**Figure 12-9**). A high efficacy is an indication that an intervention is successful.

The **number needed to treat (NNT)** is the expected number of people who would have to receive a treatment to prevent an unfavorable outcome in one of those people (or, alternately stated, to achieve a favorable outcome in one person). A small NNT indicates a more effective intervention. If a drug intended to prevent stroke has an NNT of 5, then 5 people have to take the drug for 1 year (or some other specified time period) to prevent 1 of the 5 from having a stroke. If the drug has an NNT of 102, then 102 people have to take the drug to prevent 1 of the 102 from having a stroke. The **number needed to harm (NNH)** is the number of people who would need to receive a particular treatment in order to expect that one of

FIGURE 12-9 Efficacy and Number Needed to Treat

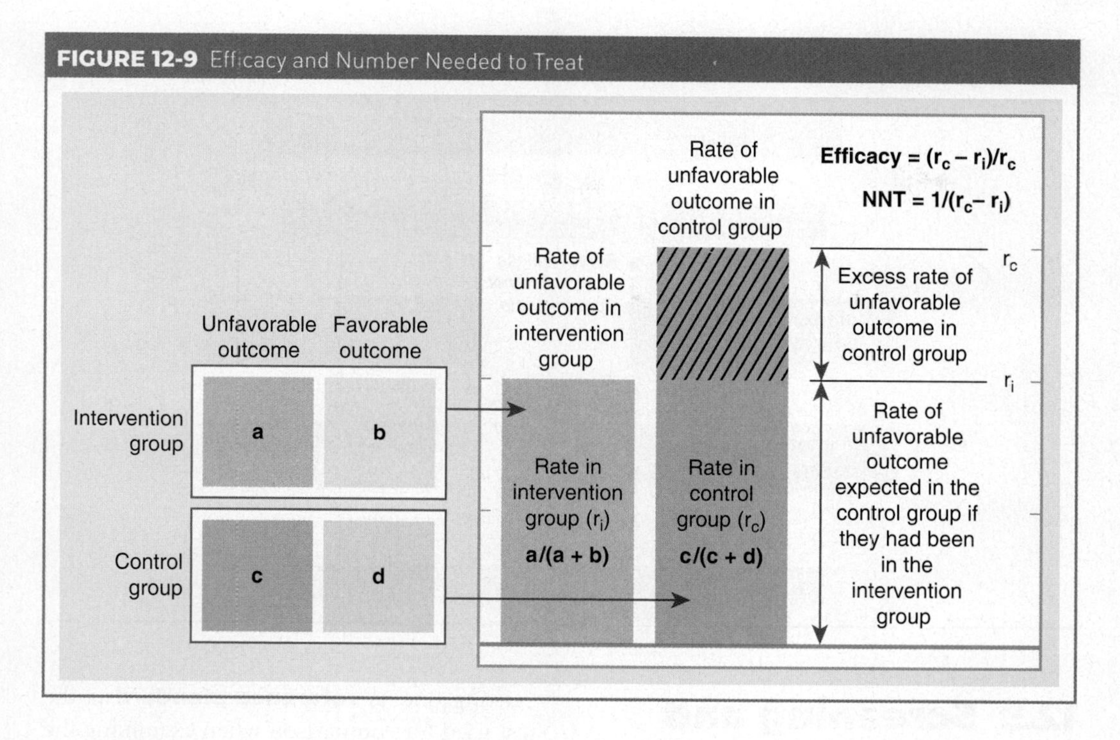

those people would have a particular adverse outcome. A large NNH indicates a safer intervention. NNT and NNH are often used for cost-effectiveness analysis.

Efficacy typically refers to results under ideal circumstances, such as when an experiment is conducted in a controlled laboratory setting and all participants are fully compliant with the protocol. **Effectiveness** is calculated with the same equation as efficacy but refers to results obtained under real-world, less-than-ideal conditions. For example, in a real-world setting, some participants might skip some doses of an experimental drug, or they might not take the doses at the exact specified times, or they might not store the pills at the ideal temperature. **Efficiency** is an evaluation of the cost-effectiveness of an intervention that is based on both its effectiveness and resource considerations.

Analysis for experimental studies typically uses either a treatment-received approach or a treatment-assigned approach. A **treatment-received analysis** of experimental data includes only the participants who were fully compliant with their assigned intervention or comparison protocol. Treatment-received analysis allows for the calculation of efficacy, because the only participants included in the analysis are those who never missed taking a pill at the prescribed time, never missed a scheduled clinical exam, and were otherwise exemplary study subjects. A **treatment-assigned analysis** (or **intention-to-treat analysis**) includes all participants, even if they were not fully compliant with their assigned protocol. Treatment-assigned analysis is better at measuring real-world (rather than ideal-world) effectiveness.

No matter which analytic approach is used, the research protocol should include specific plans for promoting compliance, minimizing dropouts, and ensuring the safety of participants. The flow of participants through the study, from the recruitment and enrollment stages through the analysis stage, should be included in reports of findings for any experimental study (**Figure 12-10**).

FIGURE 12-10 Flow of Participants in an Experimental Study

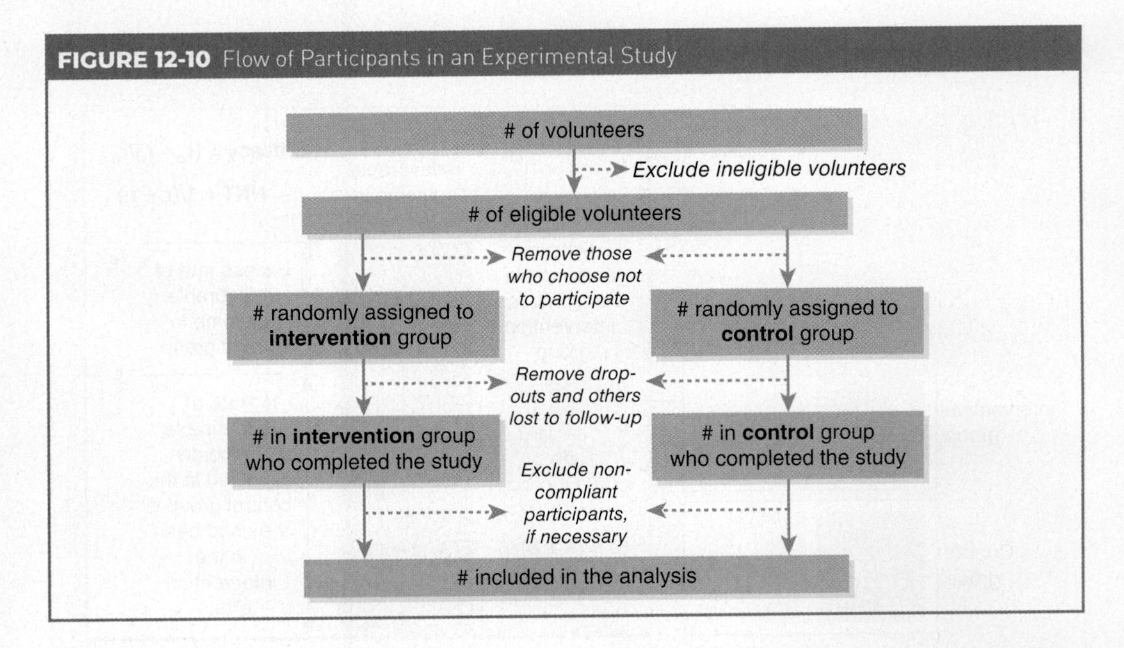

12.9 Screening and Diagnostic Tests

Screening is a type of secondary prevention in which all members of a well-defined group of people are encouraged to be tested for a disease based on evidence that members of the population are at risk for the disease and early intervention improves health outcomes. Screening tests are commonly used to detect hypertension, glaucoma, scoliosis, hearing loss, depression, and various types of cancer in previously undiagnosed people. Screening tests can also be used for infectious diseases, such as HIV, tuberculosis, intestinal parasites, and sexually transmitted infections. These programs often rely on the use of a **rapid diagnostic test (RDT)** that can detect the presence of a pathogen (or markers for a pathogen) in saliva, a drop of blood, or another body fluid within 15 to 30 minutes.

The goal of most studies of new or improved screening or diagnostic tests is to compare two assessments that are supposed to measure the same thing. In most situations, this goal involves comparing a new test to an existing one. A **reference standard** is the test used for comparison when examining the validity of a new screening or diagnostic test. For example, a new blood antigen test for a type of cancer might be compared to biopsy results. In this scenario, a biopsy is considered to be a **"gold standard"** because the test shows the actual presence of disease in affected people. Cancer cells will be visualized if they are present in the sample. Positive and negative biopsy results will serve as the reference standard against which the blood test will be evaluated. The hope is that the blood test will be cheaper, quicker, and less invasive than a biopsy but will yield results similar to those from the biopsy. For many other health conditions, there is no "gold standard" for visualizing the presence or absence of disease, so the current best test is used as the reference standard.

A **cutpoint** or **threshold** is a value that divides a numeric variable into separate categories. For example, a systolic blood pressure of 140 mm Hg could be selected as the value that defines the presence or absence of hypertension. People with a blood pressure of 140 mm Hg or greater would be coded as having hypertension, while people with a blood pressure of less than 140 mm Hg would be coded as not having hypertension. For a serological test, the threshold might define the concentration of antigens in the blood that will be considered to indicate a positive versus a negative test result. The threshold selected for a test must be scientifically justifiable. For example, it would be inaccurate to lower the cutoff for hypertension to a systolic blood pressure of 100 mm Hg, because there is no clinical evidence that this value is associated with the adverse clinical outcomes linked to elevated blood pressure. The threshold should be a value that can reasonably be said to correctly classify individuals as having or not having hypertension.

The **diagnostic accuracy** is the percentage of individuals who are correctly classified by the test as true positives or true negatives (that is, the percentage for which both the reference test and the new test yield the same result). An ideal test will have 100% diagnostic accuracy. The sensitivity, specificity, positive predictive value, and negative predictive value of a new screening or diagnostic test in comparison to a reference standard are other metrics of how well the test performs (**Figure 12-11**). The **sensitivity**, or **true positive rate**, is the proportion of people who actually have a disease (according to the reference standard) who test positive using the new test. The **specificity**, or **true negative rate**, is the proportion of people who do not have the disease who test negative with the new test. The **positive predictive value (PPV)** is the proportion of people who test positive with the new test who actually have the disease (according to the reference standard). The **negative predictive value (NPV)** is the proportion of people who test negative who actually do not have the disease. An ideal screening or diagnostic test would have 100% values for sensitivity, specificity, PPV, and NPV.

The **false positive rate** is the proportion of people who actually do not have a disease (according to the reference standard) who incorrectly test positive using the new test. The false positive rate can also be calculated as 1 – specificity (that is, the number one minus the specificity, which must fall between 0 and 1). The **false negative rate** is the proportion of people who actually have a disease who incorrectly test negative using the

FIGURE 12-11 Screening and Diagnostic Test Results

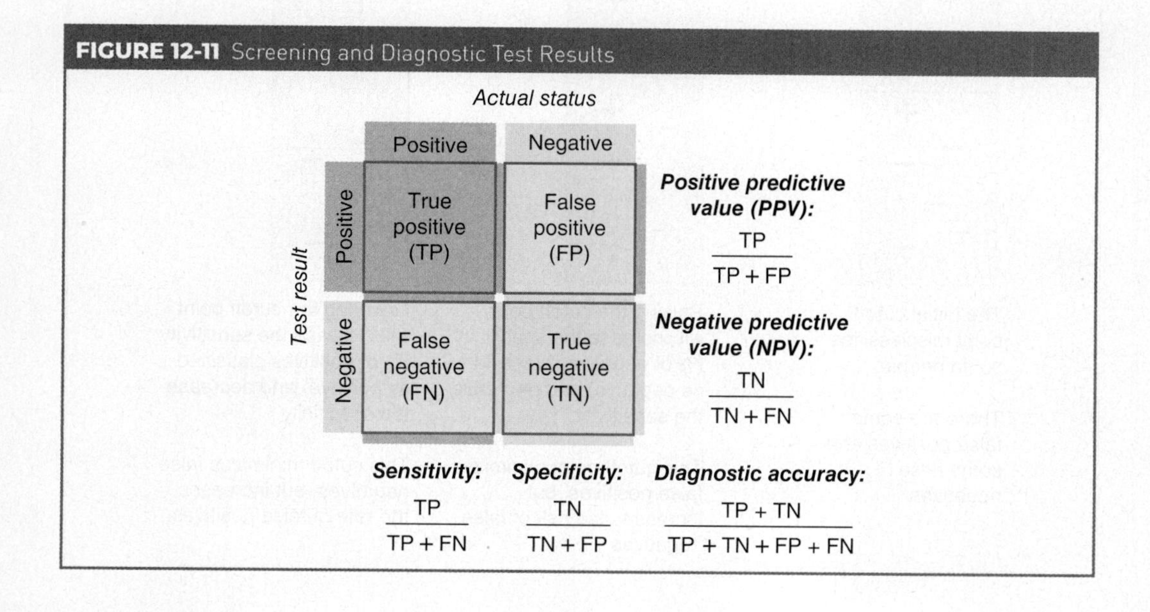

new test. An ideal screening or diagnostic test would have a 0% rate of false positive and false negative outcomes.

For tests with a flexible cutoff point for defining positive and negative test results, there is always a trade-off between sensitivity and specificity (**Figure 12-12**). Increasing the sensitivity decreases the specificity. Increasing the specificity decreases the sensitivity. Consider the use of systolic blood pressure as a sign of hypertension. Suppose that the cutoff for being classified as having clinically high blood pressure is reduced from 160 mm Hg to 140 mm Hg. The sensitivity of the test will increase, because a higher percentage of people with hypertension will be classified as hypertensive. The specificity will decrease, because a lower percentage of people without hypertension will be correctly classified as not being hypertensive.

A **receiver operating characteristics (ROC) curve**, or **ROC curve**, can be used to graphically examine the accuracy of a screening or diagnostic test by plotting the false positive rate (on the x-axis) versus the true positive rate (on the y-axis) for the different possible cutoff points of the test. In other words, the ROC curve plots 1 – specificity on the x-axis and sensitivity on the y-axis for a variety of cutoff points. The **area under the curve (AUC)** is an aggregate measure of how well a screening or diagnostic test performs across various cutoff points. AUC values can range from 0 to 1, with 0 indicating a test that is incorrect 100% of the time and 1 indicating a test that is correct 100% of the time.

Likelihood ratio tests are probability ratios used to evaluate the accuracy of screening and diagnostic tests. The **positive likelihood ratio (LR+) test** examines whether a new test is good at predicting the presence of disease. The LR+ is calculated as the probability that an individual with the disease has a positive test divided by the probability that an

FIGURE 12-12 Sensitivity and Specificity

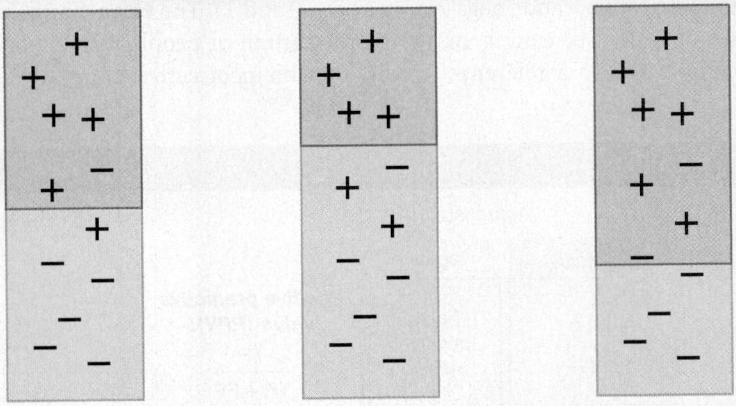

The initial cutoff point misclassifies some people.

There are some false positives and some false negatives.

Raising the cutoff point will increase the specificity (% of negatives classified as negative) and decrease the sensitivity.

This cutoff point minimizes false positives, but increases the risk of false negatives.

Lowering the cutoff point will increase the sensitivity (% of positives classified as positive) and decrease the specificity.

This cutoff minimizes false negatives, but increases the rate of false positives.

individual without the disease has a positive test. This is equivalent to:

$$LR+ = \frac{sensitivity}{1 - specificity}$$

A larger LR+ (such as LR > 10) indicates a good test. The **negative likelihood ratio (LR–) test** examines whether a new test is good at predicting the absence of disease. The LR– is calculated as the probability that an individual with the disease has a negative test divided by the probability that an individual without the disease has a negative test. This is equivalent to:

$$LR- = \frac{1 - sensitivity}{specificity}$$

A smaller LR– (such as LR < 0.1) indicates a good test.

Many comparative studies of laboratory-based tests can be considered observational because they do not require the researchers to do anything to the participants other than collect a biological specimen. However, some tests involve experimental procedures, such as biopsies in individuals who might otherwise forgo this kind of testing because they are almost certain to yield negative results. Those studies are appropriately classified as experimental.

Studies of new screening and diagnostic tests should have a clear set of eligibility criteria. The protocols may intentionally seek to recruit individuals known to have the disease of interest and individuals who are known not to have the disease. To reduce the risk of observer bias, a system should be put in place to blind the examiners—the clinicians or laboratory scientists conducting the assessments—to the health status of the participants that is indicated by the reference test.

CHAPTER 13

Qualitative Studies

A qualitative study looks for the meanings, themes, and theories that emerge from observing and interacting with study participants.

13.1 Overview

Quantitative research uses structured, hypothesis-driven approaches to gather data that can be statistically analyzed. **Qualitative research** uses in-depth interviews, focus group discussions, participant observation, and other unstructured or semi-structured methods to explore attitudes and perceptions, identify themes and patterns, and formulate new theories. Qualitative research seeks to answer questions like "why?" and "how?" that numbers-focused quantitative research cannot adequately answer (**Figure 13-1**).

FIGURE 13-1 Comparing Typical Qualitative and Quantitative Research Approaches

	Quantitative	Qualitative
Main types of questions answered	Who? Where? When? What?	Why? How?
Participants	A large randomly sampled population	A small purposefully recruited population
Data collection approach	Structured	Unstructured or semi-structured
Common data collection methods	Questionnaires	In-depth interviews, focus groups, observation
Types of questions asked	Closed-ended (fixed response options)	Open-ended (flexible response options)
Data collected	Numeric data	Textual data (words, images, and objects)
Goal of data analysis	Test existing hypotheses	Formulate new theories
Outcomes reported	Statistics	Themes and theories

The word **phenomenon** is often used to describe the central concept being studied during a qualitative research project. In the health sciences, many qualitative research projects aim to improve health promotion programs and clinical processes or to provide a foundation for social change. Other qualitative studies seek to understand how people experience health and illness as individuals and as members of communities, why they engage in or do not engage in various health-related behaviors, and how they make health-related decisions.

13.2 Ontology, Epistemology, and Axiology

Because qualitative research examines thoughts and beliefs that cannot be directly measured, researchers using qualitative methods must be aware of the ways their own thoughts, experiences, biases, and assumptions shape the design and interpretation of their research studies. The philosophical orientations that guide both quantitative and qualitative research are rooted in ontology, epistemology, and axiology (**Figure 13-2**).

Ontology is the study of the nature of reality and truth. Ontological perspectives extend from realism to relativism. **Realism** assumes that one reality exists, and it can be understood. **Objectivity** describes facts that can be evaluated without bias. A realist would say that meaning exists in an object independent of the subject investigating it. Quantitative research studies seek to make objective determinations about the world. **Relativism** assumes that there are multiple realities and they cannot be fully understood. **Subjectivity** describes claims and experiences that are interpreted based on an evaluator's beliefs, perceptions, and feelings. A relativist would say that a subject imposes meaning on an object. Qualitative research designs typically seek to understand subjective aspects of human existence.

Epistemology is the study of knowledge. Ontology explains how a researcher defines reality and truth, while epistemology explains how a researcher knows what is real and true. For example, a guiding principle for quantitative studies is **empiricism**, the assumption that the senses (such as seeing, hearing, and touching) are the best way to measure truth about the world. Quantitative research assumes that researchers are independent from their study subjects and that researchers can control for possible biases in order to make objective measurements. Qualitative research assumes that researchers and study participants are interdependent and create knowledge together as they interactively explore subjective topics.

FIGURE 13-2 Ontology, Epistemology, and Axiology

		Quantitative Studies	Qualitative Studies
Ontology	*What is reality?*	Realism (one reality exists)	Relativism (multiple realities exist)
		Objectivism	Subjectivism
Epistemology	*What is knowledge?*	Positivism (reality can be measured)	Constructivism/interpretivism (reality must be created and interpreted)
Axiology	*What are values?*	Inquiry should be value-free, context-free, and unbiased	Inquiry is value-laden, context-bound, and biased

Axiology, or value theory, is the study of values. Quantitative researchers usually apply an axiological perspective that assumes that rigorous procedures can eliminate the impact of values and biases on study outcomes. Qualitative researchers assume that a researcher's values affect the study. Researchers using qualitative approaches typically take time to examine their own backgrounds, beliefs, perspectives, and biases so that they can make informed decisions about the methods they use, their own role in the study, the ways in which they interact with study participants, and how they interpret data.

13.3 Theoretical Paradigms

Because the goal of qualitative research is to understand a complex phenomenon rather than to predict some observable event, qualitative researchers must select the theoretical perspectives that will guide their study design, data analysis, and interpretation. Four of the most prominent philosophical assumptions applied to health science research are positivism, constructivism/interpretivism, critical theory, and pragmatism (**Figure 13-3**).

Positivism is a paradigm in which researchers apply a realist perspective that assumes that reality is knowable and that inquiry should be logical and value-free. The word "positive" denotes definitiveness and certainty. Studies based on a positivist paradigm generally use empirical methods to test hypotheses and predict outcomes. They often employ quantitative experimental designs. **Post-positivism** is a related paradigm in which researchers aim to experimentally test theories about how the world works, but they acknowledge that the unpredictability of human behavior limits the validity of some empirical methods. Post-positivism applies a critical realist perspective to research. Studies using a post-positivist paradigm often employ quasi-experimental and mixed methods designs.

Constructivism is a paradigm in which researchers have a relativist perspective that considers each individual's reality to be a function of that person's lived experiences. The ontological assumption of constructivism is that there are many realities, not just one

FIGURE 13-3 Examples of Theoretical Paradigms

	Positivism	Constructivism/ Interpretivism	Critical Theory	Pragmatism
Ontology	There is a single reality (realism).	Reality is constructed by individuals in groups (relativism).	Realities are socially and historically constructed.	There is a single reality, but individuals interpret reality differently.
Epistemology	Reality can be measured (objectivism).	Reality needs to be interpreted (subjectivism).	Reality and knowledge are influenced by power relationships within societies.	The best method for understanding reality is the one that solves problems.
Common methodologies	Experimental research and other quantitative methods	Qualitative methods	Qualitative methods	Mixed methods

reality, and realities are created as researchers and participants interact. The epistemological assumption is that researchers and participants must work together to understand reality. The axiological assumption is that beliefs and values are social constructs. These philosophical stances make constructivism the opposite of positivism. **Interpretivism** is a related paradigm in which researchers consider the reality in the social world to be different from reality in the natural world. Interpretivism and constructivism are about understanding how various groups of people interpret reality. Studies based on constructivism and interpretivism use qualitative study designs.

Critical theory is a paradigm that considers reality to be dependent on social and historical constructs and assumes that reality can be uncovered by identifying and challenging power structures. The philosophical assumption of critical theory is that participants can actively construct realities that are shaped by beliefs and values. Research conducted under a critical theory framework is change-oriented. Studies based on critical theory often use participatory methods that empower participants and equip them to advocate for change. For example, studies based on critical theory may use an **action research** approach in which participants work together to solve a social problem. Action research conducted under a **transformative paradigm** assumes that reality can be changed when researchers collaborate with participants from marginalized populations to address a social justice issue.

Pragmatism is a paradigm in which researchers assume that reality is situational, and it is acceptable to use any and all research tools and frameworks to try to understand a particular problem so it can be solved. The goal of pragmatic research is to solve problems, so the focus is on the outcomes of the research project rather than the theories and processes that guide it. The ontological assumption of pragmatism is **symbolic realism**, which treats individuals' realities as being real to those individuals, because that orientation is the most useful and practical. The epistemological assumption is that reality can be known and understood using many different approaches. The axiological assumption is that beliefs and values are part of practical decisions. Pragmatists often use mixed methods in their study designs.

Theoretical **pluralism** occurs when a researcher draws on more than one theoretical framework to guide the design, analysis, and interpretation of a research project. Researchers using two or more theoretical perspectives must carefully explain why the theories were selected, how they were implemented, and how the use of multiple theories informed the insights gained from the study.

13.4 Qualitative Methodologies

The methodologic approaches selected for a qualitative research project must align with the goals of the study and the selected theoretical paradigm. Planning for a qualitative study often progresses from the identification of the underlying ontology and epistemology to the selection of a suitable paradigm and then the identification of methodology that aligns with the underlying theoretical perspective (**Figure 13-4**). In the health sciences, the most popular qualitative methodologies include phenomenology, grounded theory, ethnography, and case studies (**Figure 13-5**).

Phenomenology seeks to understand how individuals interpret and find meaning in their own unique life experiences and feelings. The researcher uses in-depth interviews to gather data from several people, then the transcripts of those interviews are examined so that meanings and themes can be identified and understood from the perspective of the participants. **Bracketing** is the process of a researcher intentionally setting aside any preconceived ideas about reality in order to be open to new meanings that might be expressed by participants.

FIGURE 13-4 Qualitative Research Planning Process

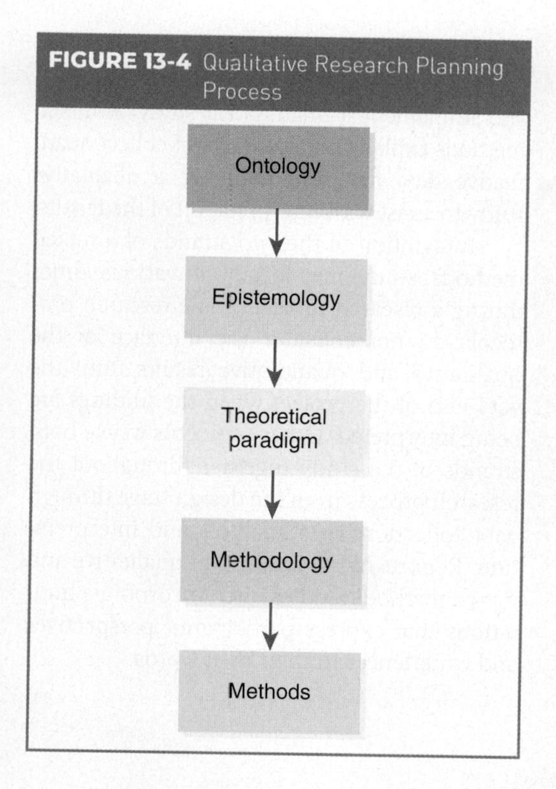

Grounded theory is an inductive reasoning process that uses observations to develop general theories that explain human behavior or other phenomena. Data collection and data analysis occur simultaneously, so that theories can be developed and refined. **Theoretical sampling** uses the emerging theory to guide the selection of new data sources. Data collection continues until data saturation has been reached.

Ethnography is the systematic study of people and cultures in their natural environments. Ethnography is an anthropological approach in which researchers aim to develop an insider's view (an **emic perspective**), rather than an outsider's view (an **etic perspective**), of how members of a particular sociocultural group understand their world. Ethnographers often use participant observation methods to understand a group's collective experiences, values, beliefs, and behaviors. It is typical for ethnographers to immerse themselves in the study community and to intentionally interact with the group for many months or years.

A **case study** is a qualitative research approach that uses multiple data sources to examine one person, group, event, or other situation in detail. A case study approach may be used when preparing a case report or a case series, or it may be applied to understanding an event, process, or program. The goal is not to develop generalizable knowledge, but to understand one event well. Multiple data sources may be used as part of a holistic examination of the case, including observations, interviews, and reviews of historic documents.

13.5 Mixed Methods Research

Qualitative research can be complete on its own, but sometimes the best option is to use qualitative methods in conjunction with

FIGURE 13-5 Examples of Qualitative Methodologies

	Phenomenology	Grounded Theory	Ethnography	Case Studies
Goal	Understand how individuals gain meaning from lived experiences	Construct a theoretical model to explain a phenomenon	Understand the collective experience of sociocultural groups	Understand an event, process, or program
Common methods	In-depth interviews	Interviews and observations	Participant observation	Observations, interviews, documents, and other reports

quantitative surveys. **Mixed methods** projects use both quantitative and qualitative methods in one research study. Some mixed methods projects use a convergent parallel design to collect quantitative and qualitative data concurrently and then compare the results and interpret them. Some studies collect the data sequentially, completing one type of study first and then designing and implementing the other type of study. Some qualitative studies are embedded within a quantitative study.

Social science research can often be classified as being exploratory, descriptive, or explanatory. **Exploratory research** aims to discover new ideas and develop hypotheses. Descriptive research seeks to understand key characteristics of a group. **Explanatory research** tests hypotheses about causal relationships. A mixed methods exploratory study might collect qualitative data first and then use the insights from that study to design and implement a quantitative study. A mixed methods explanatory study might collect quantitative data first and then use a qualitative study to assist with interpretation of the results.

Integration of the two strands of a mixed methods study may occur at various times during a research project. Some research protocols do not consider the interface of the qualitative and quantitative results until the very end of the project when the findings are being interpreted. Other protocols weave both strands of the study together throughout the research process from the design stage through data collection, data analysis, and interpretation. Reports of the findings of qualitative and mixed methods studies often incorporate quotations that express participants' perspectives and experiences in their own words.

CHAPTER 14

Correlational Studies

A correlational, ecological, or aggregate study uses population-level data to examine relationships between exposure rates and disease rates.

14.1 Overview

A **correlational study** uses population-level data to look for associations between two or more characteristics that have been measured in several groups (**Figure 14-1**). The variables included in correlational analyses are usually aggregate (grouped) statistics such as the proportion of a population with a particular characteristic or the average value of the variable in a population. A correlational study is sometimes called an **aggregate study** because

FIGURE 14-1 Key Characteristics of Correlational Studies

Objective	Compare average levels of exposure and disease in several populations.
Primary study question	Do populations with a higher rate of exposure have a higher rate of disease?
Population	Existing population-level data are used; there are no individual participants.
When to use this approach	The aim is to explore possible associations between an exposure and a disease using population-level data.
Requirement	The topic has not been previously explored using individual-level data.
First steps	1. Select the sources of data that will be used. 2. Decide on the variables to include in the analysis.
What to watch out for	The ecological fallacy Limited publication venues
Key statistical measure	Correlation

correlational studies look only at grouped population-level data and do not include any individual-level data.

For most correlational studies, at least one characteristic of the populations being examined is designated as an exposure and at least one is designated as an outcome or disease. Correlational studies then examine exposure–outcome pairs. For example, a correlational study could answer questions like:

- Does the percentage of adults with multiple sclerosis tend to be higher in countries farther from the equator?
- Does the rate of asthma tend to be higher in cities with higher levels of air pollution?
- Does the prevalence of diabetes tend to be higher in provinces with a higher prevalence of obesity?

Statistical methods can be used to control for interactions among related variables.

Some correlational studies examine links between the socio-demographic characteristics of populations and health outcomes. The key exposure for this type of study might be the percentage of adults in a state who have completed at least 12 years of education, the mean household income in the state, or the median age of the state's population. Alternatively, the key exposure for a correlational study might be an environmental one, such as a city's distance from the equator, its number of rainy days in a typical year, or the city's average ultraviolet radiation index during midday in the hottest month of the year. All of these environmental measures are likely to be experienced fairly consistently across the entire population of interest. It is unlikely that part of one city would experience many more sunny days than another part of the city. A correlational study that explores an environmental exposure may be called an **ecological study**.

14.2 Aggregate Data

Nearly all correlational studies are secondary analyses. Because existing data sources are almost always used for correlational studies, the key to success is identifying data sources that contain comparable information about the variables of interest. Information about all the variables of interest must be available for a suitable number of populations, which can be grouped by place or time. For example, place-based populations could consist of all member nations of the United Nations, all 50 states from the United States, the largest 20 metropolitan areas in the United Kingdom, all the counties in the U.S. state of Michigan, or a random sample of census tracts in Toronto. Time-based studies could use annual historical data for the past several decades from one or more place-based populations.

For any one variable in a correlational study, the best option is for all data to come from the same source. This helps to ensure that the data were all collected using similar methods and definitions. For example, data about a particular environmental exposure for all counties in one state might be downloadable from the state's environmental protection agency, and county-level data about the rate of a particular health issue might be accessible from the state's health department. Researchers should be aware that the definition of an exposure or a disease may change over time. A country might have used one definition for hypertension until 2015 and then used a different definition starting in 2016. If this happens, a plot of the variable's value over time will reveal a discontinuity in the data. This inconsistency will need to be addressed before the data can be analyzed.

When multiple data sources for one variable must be used because no single source of data is available, the measures from different sources might not be directly comparable. For example, countries might use very different definitions for what counts as access to clean drinking water or what constitutes literacy in an adult. Studies of adult health may present results for people ages 15 years and older, 18 and older, or 25 and older. The quality of the data may also vary when different data collection methods are used. Some methods may result in exposures being routinely undercounted or diseases being habitually overdiagnosed. When using multiple sources of data for one variable, the researcher should establish a scientifically justifiable set of inclusion criteria for the study and then exclude any data sources that do not meet all of those eligibility requirements.

Suppose that an ecological study is examining possible associations between weather and the prevalence of obesity among adults, and the populations selected for the study are the world's 100 most populous cities. The researcher can decide that only studies defining obesity as a body mass index (BMI) of 30 or greater (≥ 30) will be included. If one of the cities selected for the study reports the prevalence of BMI ≥ 35 rather than BMI ≥ 30, that statistic must be excluded from the analysis. The percentage of members of a population with BMI ≥ 35 will be much lower than the percentage with BMI ≥ 30, so including the study with an unusual definition for obesity may invalidate the results of the aggregate study. Or suppose that the only statistic about obesity in one of the cities is for adults between 18 and 35 years of age, rather than all adults ages 18 and older. That statistic would not be representative of the city's adults as a whole, so it must be excluded.

The eligibility criteria should also ensure that data points are not excluded for invalid reasons. It would be inappropriate to eliminate a city's statistic from the analysis just because the percentage of adults in that city with BMI ≥ 30 was higher or lower than most other cities' results or did not conform to the desired pattern related to that city's weather. If a city measured the prevalence of BMI ≥ 30 in a population that aligns with the correlational study's inclusion criteria, then that study must be included in the analysis even if its statistic appears to be an outlier. Strong case definitions and inclusion and exclusion criteria help minimize the risk of invalid data being included in an analysis and can help ensure that valid data are not inappropriately excluded.

Before conducting a statistical analysis of aggregate data, the data from each population must be entered into a spreadsheet. Each population should be assigned to its own row in the spreadsheet. Each exposure and each outcome should be assigned to its own column. The data should be filled into the cells in each column so that they line up with the correct population. **Figure 14-2** shows a sample data table. See Chapter 28 for more information about quantitative data management.

FIGURE 14-2 Sample Data Table

Population	Exposure 1	Outcome 1
A	48.2	14.1
B	65.1	17.0
C	37.8	14.9

14.3 Avoiding the Ecological Fallacy

Correlational studies compare groups rather than individuals. No individual-level data are included in the analysis, only population-level data. The **ecological fallacy** is the incorrect assumption that individuals follow the trends observed in population-level data. Even when a population with a higher rate of exposure has a higher rate of disease than populations with lower exposure rates, individuals in those populations who have a high level of exposure do not necessarily have a higher risk of the disease. The experience of individuals in a population may vary significantly from the population average. For example, it would be incorrect to assume that any one individual from a country with a high average BMI will be obese or that an individual from a country with a low average BMI will not be obese. However, it is appropriate to identify trends across populations and to use those observations to generate hypotheses for future individual-level studies that will test for relationships between the characteristics of interest in individuals. Correlational studies are a useful starting point for generating hypotheses about associations, but they are not the final word on risk factors for disease.

14.4 Correlation

Correlation is a statistical measure of the degree to which changes in the value of one variable predict changes in the value of another. On a scatterplot used to illustrate correlation, each point represents one population in the study. The exposure is plotted on the x-axis, and the outcome or disease is plotted on the y-axis (**Figure 14-3**). A trend line is fit to the data points, usually using a software program that calculates the line with the best fit.

- When all the points fall neatly along or very near to a sloped line, the correlation is strong. A positive upward slope shows that higher levels of exposure are associated with higher rates of disease, as shown in Figure 14-3A. A negative downward slope shows that higher levels of exposure are associated with lower rates of disease.
- When the points are not exactly linear but a line for trend can be drawn through them, the correlation is mild or moderate, as shown in Figure 14-3B.
- When the points appear to be randomly placed and no obvious line can be drawn through them, or when the best-fit line is horizontal, as shown in Figure 14-3C, the correlation is weak or nonexistent.

Different equations are used to calculate correlations between different types of variables. The **Pearson correlation coefficient**, often called "Pearson's *r*," is a statistical measure of the degree to which changes in the value of one numeric variable predict changes in the value of another numeric variable. Pearson's *r* is used for continuous variables and other variables with responses that can be plotted on a number line. The **Spearman rank-order correlation**, usually called "Spearman's rho (ρ)" or listed as r_s, is used when examining the correlation between variables that assign a rank to responses (like 1st place, 2nd place, 3rd place, and so on) or that have ordered categories (such as scales that range from 1 = strongly disagree to 5 = strongly agree). (Chapter 30 explains the difference between parametric tests like the Pearson correlation and nonparametric tests like the Spearman rank-order correlation and Kendall's rank correlation, which is often designated with the Greek letter tau [τ].)

For both of these types of correlation, the value of r (or ρ) ranges from −1, when all points lie perfectly on a line with a negative slope, to 1, when all points lie perfectly on a line with a positive slope. When $r = 0$, there is no association between the exposure and outcome. The association between two or more variables can also be reported as the **coefficient of determination**, r^2, which shows how strong a correlation is without

FIGURE 14-3 Correlation

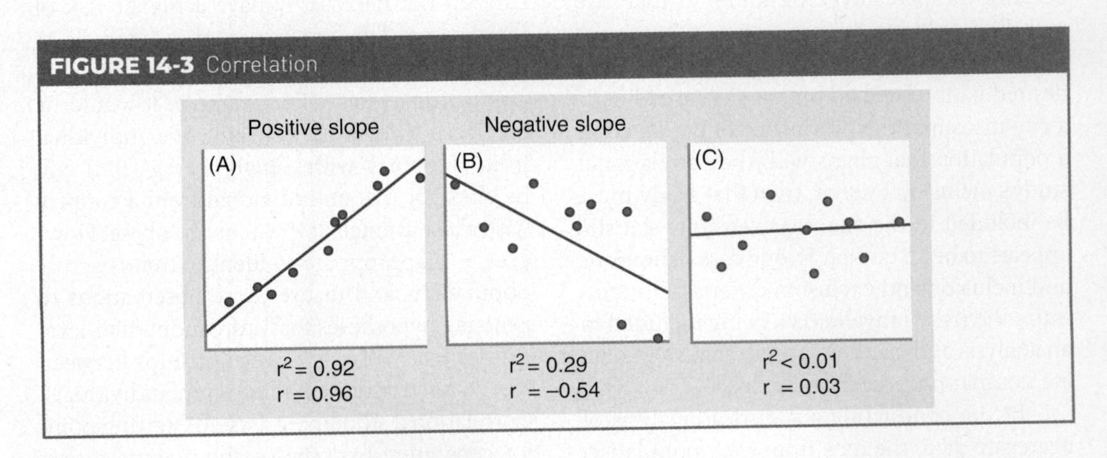

indicating the direction of the association. The value of r^2 ranges from 0 for no correlation to 1 for perfect correlation.

When more than two variables are being compared or the goal is to understand the relationship between two variables while controlling or adjusting for the effects of other variables, linear regression models can be used to assess the associations.

A correlation can be present even when the relationship is not causal. Correlation is a statistical measure of association, and tests of correlation cannot demonstrate that variables designated as exposures represent events that happened before the associated outcomes and caused the outcomes to occur. The term "correlation" should never be used as a synonym for causation.

14.5 Age Standardization

Sometimes populations that are being compared have very different age structures, with one or more populations considerably younger or older than the others. A younger population may have more favorable health statistics because fewer population members have developed the chronic diseases associated with aging. An older population may have less favorable health statistics simply because its members are more advanced in years. **Age adjustment** methods can improve the validity of comparisons of two or more populations with different age distributions. An **age-specific rate** is a rate for a particular age group, such as the prevalence of asthma in children who are 6 to 10 years old or the incidence rate of lung cancer among adults who are 60 to 74 years old. **Age standardization** applies age-specific rates from one or more study populations to a "standard population," or vice versa, to generate comparable rates for populations with different age structures. (Similar methods can statistically adjust for other demographic differences between populations that are being compared.)

The age standardization method used depends on how much is known about the age distributions in the study populations. **Direct age adjustment**, also called direct age standardization, applies age-specific rates in two or more study populations with different age structures to one standard population so that the rates in the study populations can be more fairly compared. For example, the rates from several cities can be standardized to the national population, or the rates in several countries can be standardized to the global population. Direct age standardization can be conducted only when age-specific data are available for all of the study populations being compared. This requires knowing the exposure and/or disease rates for each age group in each study population. To calculate an age-adjusted overall rate for each study population, each population's age-specific rates are applied, one at a time, to a standard population and then the expected numbers of outcomes for each age group are added together and the sum is used to calculate an all-ages rate in the standard population.

In **Figure 14-4**, two cities in Country C are compared. City A has a young population and an overall disease rate of 19 cases per 1000 adults. City B has an older population and an overall disease rate of 48.5 cases per 1000 adults. The raw data show that City B has more than twice the rate of disease as City A. However, the age-specific rates in these cities are similar, with City B having slightly lower—not higher—rates in some age groups. If the age-specific rates from City A were the rates in the national adult population in Country C, the overall rate of disease nationally would be 25.4 per 1000. If the age-specific rates from City B were the rates in the national population, the overall rate of disease nationally would be 22.9 per 1000. These age-standardized rates for the two cities show that the disease rates are similar after adjusting for differences in the age distributions of city residents, with City B having a slightly lower age-adjusted rate. These standardized rates are

FIGURE 14-4 Direct Age Adjustment

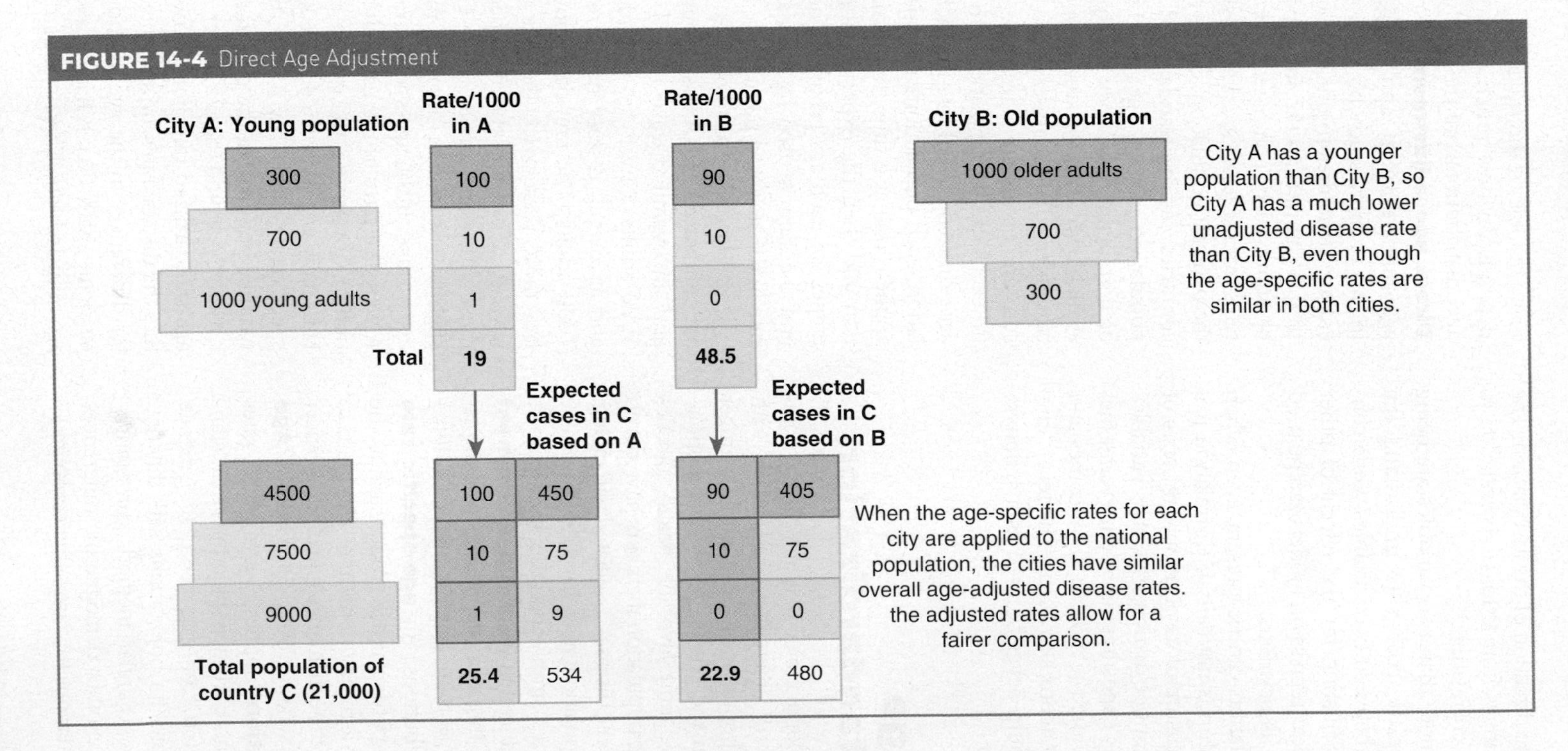

directly comparable because they are based on the age distribution of the same standard population, and they allow for a fairer comparison of disease status in the two cities.

Indirect age adjustment, also called indirect age standardization, applies age-specific rates in a standard population to a study population so that a determination can be made about whether the overall rate in the study population is greater or lesser than expected given the population's age distribution. Indirect age standardization methods can be used to compare study populations for which the age distributions are known but age-specific rates of exposure and/or disease are not known. A **standardized mortality ratio (SMR)** compares the number of deaths observed in the study population to the number of deaths expected in the study population based on the age-specific mortality rates in the standard population. Suppose that 51 people died in City D in 2020 and indirect age standardization methods based on a state reference population suggest that the city could have expected 55 people to die that year. The SMR would be SMR = Observed/Expected = 51/55 = 0.93 (**Figure 14-5**). This SMR is less than 1, which suggests that City D has a more favorable mortality rate than the state as a whole, after adjusting for City D having a slightly different age distribution than the state.

A **crude statistic** is a raw or unadjusted statistic. When only one population is being described, the crude statistic is usually the correct measure to report, because it accurately describes the true experience in the population. An **adjusted statistic** is a statistic that has been corrected to account for the effects of one or more other variables. Some adjusted statistics provide a more accurate depiction of a population's true status. For example, some adjusted statistics correct for discrepancies between the demographics of study participants and the source populations from which they were sampled. When an adjusted statistic corrects an error, it is a more accurate representation of the true value in the population than the crude statistic. Some adjusted statistics are intended to improve the comparison of two or more populations. In those situations, the adjusted values generally do not improve the accuracy of statistics. Instead, they improve the validity of comparisons across multiple populations.

An **age-standardized statistic** is a fictitious statistic for a study population that is created by applying age-specific rates to or from a standard population. Age standardization improves the validity of comparisons of populations with different age structures, but it does not improve the accuracy of the statistics themselves. The overall population-level rates calculated by age standardization procedures are artificial rates for the population. The standardized number is not an accurate portrayal of the true value of the rate in the study population. For example, in the

FIGURE 14-5 Indirect Age Adjustment

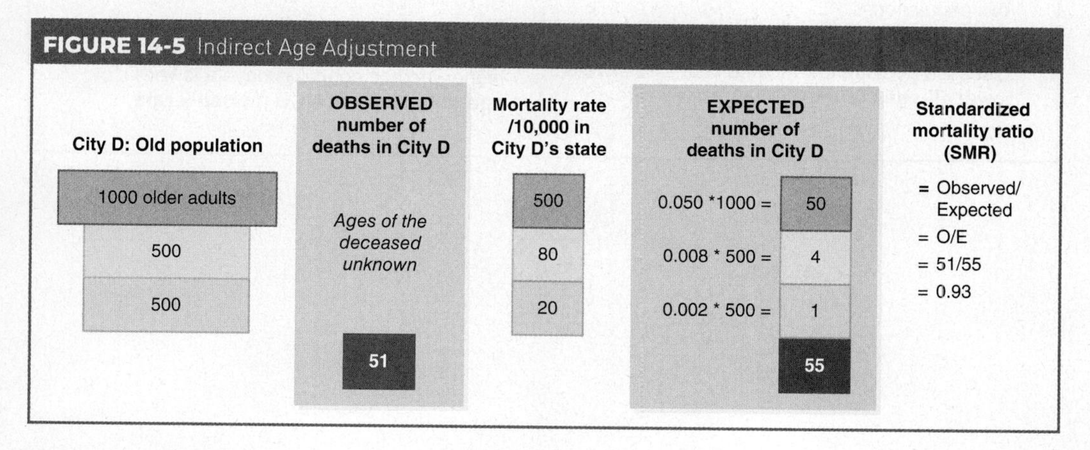

preceding indirect age-adjustment example, the true number of fatalities during the year in City D was 51. It would be incorrect to report that there were 55 deaths (the expected value based on state rates) rather than 51 deaths (the observed number of deaths in the city). The 51 represents the true number of deaths in the city, and the 55 is a way to fairly compare the city's mortality rate to the mortality rate in its state. When the value of a statistic for one population is being reported, the crude rate is usually the most appropriate number to report. When two populations are being compared, the standardized rates are the best numbers to use for the comparison (**Figure 14-6**).

FIGURE 14-6 Crude and Standardized Rates

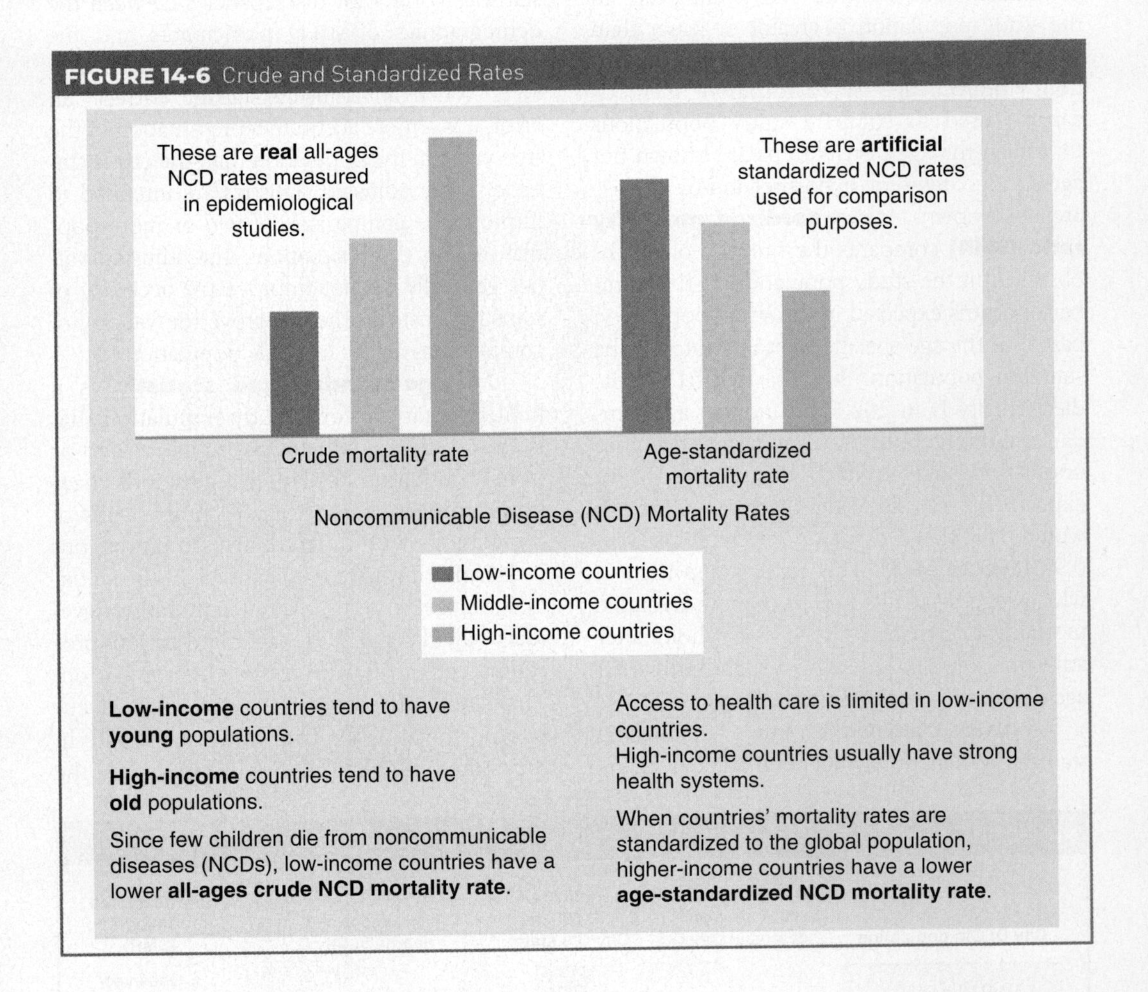

These are **real** all-ages NCD rates measured in epidemiological studies.
These are **artificial** standardized NCD rates used for comparison purposes.
Crude mortality rate
Age-standardized mortality rate
Noncommunicable Disease (NCD) Mortality Rates
Low-income countries
Middle-income countries
High-income countries
Low-income countries tend to have **young** populations.
High-income countries tend to have **old** populations.
Since few children die from noncommunicable diseases (NCDs), low-income countries have a lower **all-ages crude NCD mortality rate**.
Access to health care is limited in low-income countries.
High-income countries usually have strong health systems.
When countries' mortality rates are standardized to the global population, higher-income countries have a lower **age-standardized NCD mortality rate**.

© DmitriyRazinkov/Shutterstock

CHAPTER 15

Synthesis Research

Reviews and meta-analyses gather all prior publications on a specific topic and provide an integrated summary of them.

15.1 Overview

Most scientific research projects seek to identify new findings derived from a single study population, but the goal of tertiary analyses is to engage in the scholarship of integration. **Synthesis research** integrates existing knowledge from previous research projects. The common types of synthesis research in the health sciences include narrative reviews, systematic reviews, and meta-analyses (**Figure 15-1**). These types of research summarize what is already known about a topic, show the connections among previous studies, and offer new interpretations of previous studies' contributions to scientific knowledge. A review article in the health sciences requires:

- An extensive search of the literature
- The extraction of key information from relevant articles
- The clear and concise presentation of this information

Conducting synthesis research is one way to become an expert in the literature on a well-defined topic. This knowledge is a good outcome in and of itself, and a tertiary analysis can also be a helpful step in preparing for future primary or secondary analyses. Well-written and comprehensive review articles often become foundational for new research in the field because they summarize what is known about an area of inquiry. Because reviews provide a concise summary of the literature, published review articles may be cited more frequently than the typical article reporting the results of a primary or secondary analysis.

There also some limitations associated with synthesis research. Not all journals publish review articles, especially reviews that the editors do not solicit, so the likelihood of publication might be lower for tertiary studies than for other study approaches. Also, reviews are sometimes regarded as exhibiting less originality than other types of scholarship. A good review requires meticulous library work followed by the careful compilation and interpretation of scientific information, yet reviews are sometimes perceived to be a less rigorous form of research than projects collecting new data or involving statistical analysis.

FIGURE 15-1 Key Characteristics of Reviews and Meta-analyses

Approach	Narrative Review	Systematic Review	Meta-analysis
Objective	Synthesize existing knowledge	Synthesize existing knowledge	Synthesize existing knowledge
Primary study question	What conclusions about this topic are supported by previous studies?	When all previously published studies on this topic are examined, what conclusions can be drawn?	When the results of all previously published studies on this topic are merged, what is the summary statistic?
Population	Published literature	Published literature	Published literature
When to use the approach	The goal is to describe a new perspective on a topic that can be supported by the existing literature.	The goal is to compare the findings of previous studies on a well-defined topic.	The goal is to summarize previous findings using pooled statistics.
Requirements	The researcher has excellent library access.	The researcher has excellent library access.	The researcher has excellent library access.
	The researcher has a unique perspective on the topic.	The researcher can obtain every relevant article.	The researcher has strong quantitative skills.
First steps	Decide what key message the article will convey.	1. Decide on the specific objectives of the review. 2. Select the search methods that will be used to find potentially relevant articles. 3. Select inclusion and exclusion criteria for articles.	1. Decide on the specific objectives of the review. 2. Select the search methods that will be used to find potentially relevant articles. 3. Select the inclusion and exclusion criteria for the articles. 4. Decide how to assess the quality of the studies. 5. Decide how the results of the studies will be combined into one summary statistic.
What to watch out for	Limited publication venues	Publication bias	Studies that cannot be fairly compared
Key statistical measure	No statistics are required.	No statistics are required, but reporting some results from included studies may be helpful.	Summary measures for included studies must be reported.

15.2 Selecting a Topic

When starting a tertiary analysis, the most important decision is the selection of a topic that is narrow enough that all the relevant publications can be acquired. The topic may need to be modified if a preliminary literature search does not yield an appropriate number of articles. If an initial search of an abstract database yields only 8 possibly relevant articles, the topic probably needs to be expanded. If a search produces 352 articles, the topic needs to be narrowed to a more specific disease condition, a smaller geographic area, or a reduced scope. For example, a review of risk factors for cardiovascular disease would be cumbersome. A very long book would be required to cover all the identified risk factors, and an article-length summary would provide such a superficial level of information that it would not be useful. There is a greater likelihood of success for a review article with a narrower scope—one that limits the types of risk factors, the particular cardiovascular diseases, and the population groups included in the analysis.

15.3 Library Access

No review article can be written without excellent library access because *every* relevant article must be identified and obtained during a systematic review. The article acquisition process usually requires access to a university library that allows affiliates to make numerous interlibrary loan requests. Before starting a review project, a researcher should check with a university librarian about the library's journal access policies and the fees that users may have to pay to access articles that are not part of the library's collections or subscription services. The researcher must also prepare to maintain a meticulous system for tracking articles that have already been acquired, those that have been requested but not yet received, and those that need to be requested.

15.4 Narrative Reviews

A **narrative review** provides a unique perspective about a topic by using evidence from the literature to support the author's commentary. A narrative review might summarize important clinical aspects of a disease or summarize the epidemiological profile of a well-defined population. Because they are intended to convey a perspective and not merely compile facts, narrative reviews must be carefully organized by theme, methodology, chronology, or some other guiding principle. A narrative may also be appropriate when the researcher has developed a unique conceptual framework or theory that can be illustrated with examples from the literature. However, narrative reviews are becoming less common as editors and reviewers push for the use of systematic methods. Researchers must be prepared to justify their selection of a narrative approach. A narrative review works best when the researcher has a unique perspective on a topic and a particular expertise in the field that can be drawn on without using a systematic search strategy.

15.5 Systematic Reviews

A **systematic review** uses a predetermined and comprehensive searching and screening method to identify relevant articles. This process is designed to minimize the bias that might occur when researchers handpick the articles they want to highlight. After the identification of a focused study question, the most important decisions in a systematic review are the selection of keywords and inclusion criteria. The goal is to craft a search strategy that identifies all the articles ever published on the narrow, well-defined area covered by the review. Once potentially relevant articles have been identified from a search of one or several abstract databases, each candidate article is screened to see whether it meets all of the inclusion criteria. Relevant information is extracted from all eligible articles and presented in a summary table, then the trends and key observations are summarized. In sum, the systematic review process requires:

- Identification of an appropriately narrow study question
- Selection of a well-defined and valid search strategy
- Screening of all potentially relevant articles to determine whether they meet the predefined eligibility criteria
- Extraction of relevant information from all eligible articles
- Summarization of the findings of these articles

15.6 Meta-analysis

Meta-analysis is the calculation of a pooled statistic that combines the results of similar studies identified during a systematic review. The values included in the calculation of a summary statistic should come from high-quality quantitative studies that used similar methods to collect and analyze their data. The typical steps of a meta-analysis are:

- Define the study question and develop a study protocol
- Use a comprehensive systematic search strategy to identify every possibly eligible article
- Use predefined inclusion and exclusion criteria to identify candidates for inclusion in the meta-analysis
- Extract statistical results from each of the candidate studies
- Assess the quality and comparability of each candidate study, eliminating studies that do not meet predefined inclusion requirements
- Combine comparable statistical results into one summary statistic that adjusts for the sample sizes or confidence intervals of the contributing statistical measures

The inclusion criteria for meta-analyses are usually more restrictive than they are for general systematic reviews, because a summary statistic is meaningful only when every study included in the meta-analysis has very similar definitions for exposures and outcomes as well as similar populations, study designs, and research methods. Trying to combine dissimilar studies could hide real and meaningful differences among the study populations.

15.7 Meta-synthesis

A **meta-synthesis** is a tertiary analysis that integrates the results from several different qualitative studies. A meta-synthesis might take the form of a meta-ethnography, a meta-narrative, a thematic synthesis, or another type of process. The goal of a meta-synthesis is not to create some sort of aggregate measurement, but to enhance understanding of a particular phenomenon. A meta-synthesis might look for ways that the concepts in one study relate to the concepts developed from other studies, identify the ways that the concepts in different studies are different, or propose a new interpretation of a phenomenon that draws on several previous analyses. The methods for finding relevant studies and coding them, the key results of the analysis, and the interpretation of the meta-synthesis are included in the research report.

STEP 3

Designing the Study and Collecting Data

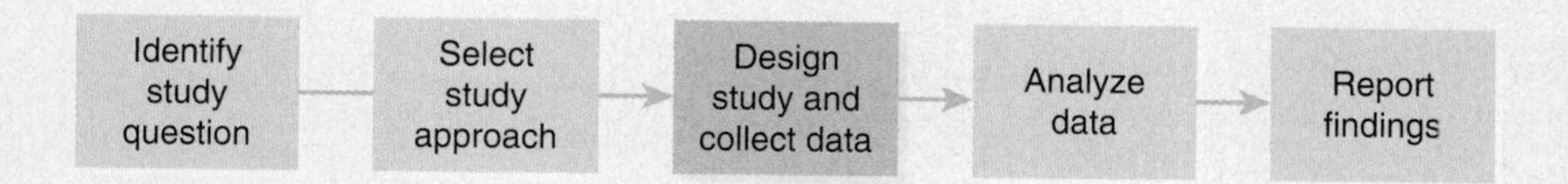

The third step in the research process is developing and implementing a detailed study plan. This section describes how to create a protocol and collect data for primary, secondary, and tertiary studies. The section also includes information about research ethics and about grant proposal writing.

- Research protocol development
- Ethical considerations, review, and approval
- Population sampling methods
- Sample size and power estimation
- Questionnaire development
- Quantitative data collection
- Qualitative data collection
- Additional assessments
- Secondary analyses of existing data
- Systematic reviews and meta-analyses
- Writing grant proposals

CHAPTER 16

Research Protocols

A research protocol is a detailed guide describing all the actions that will be taken during the implementation of a research project.

16.1 Overview of Research Plans by Study Approach

Once a study question and a study approach have been selected, the next step is to begin developing a research plan. A research **protocol** is a detailed written description of all the processes and procedures that will be used for participant recruitment (if relevant), data collection, and data analysis. For all types of studies—primary, secondary, and tertiary—it is helpful to create a research protocol that will guide each step of the data collection, management, and analysis process.

The components of the research protocol are specific to the selected study approach (**Figure 16-1**). For the collection of new data from individuals, the researcher needs to select appropriate methods for sampling and recruiting participants, using a questionnaire or other tools to collect data, recording participant responses, and analyzing the acquired data. All of these details must be included in an application submitted to a research ethics review committee. If existing data will be analyzed, an appropriate data source must be identified and the data file and supporting materials acquired. If a systematic literature review will be conducted, the search strategy must be defined, eligible articles identified, and relevant information from each article extracted into a database.

16.2 Writing a Research Protocol

A comprehensive research protocol describes the exact procedures that will be used for every step of the research process. For a primary study, the protocol will explain details about the study design, sampling, recruiting, research ethics, data collection, data management, and data analysis (**Figure 16-2**). For a systematic review, the protocol will have a different set of components but will be equally detailed. It will define the exact search criteria and databases that will be used, the specific eligibility criteria for inclusion in the review, and the processes that will be used to extract and compare information about the articles.

FIGURE 16-1 Research Plans for Primary, Secondary, and Tertiary Data Collection

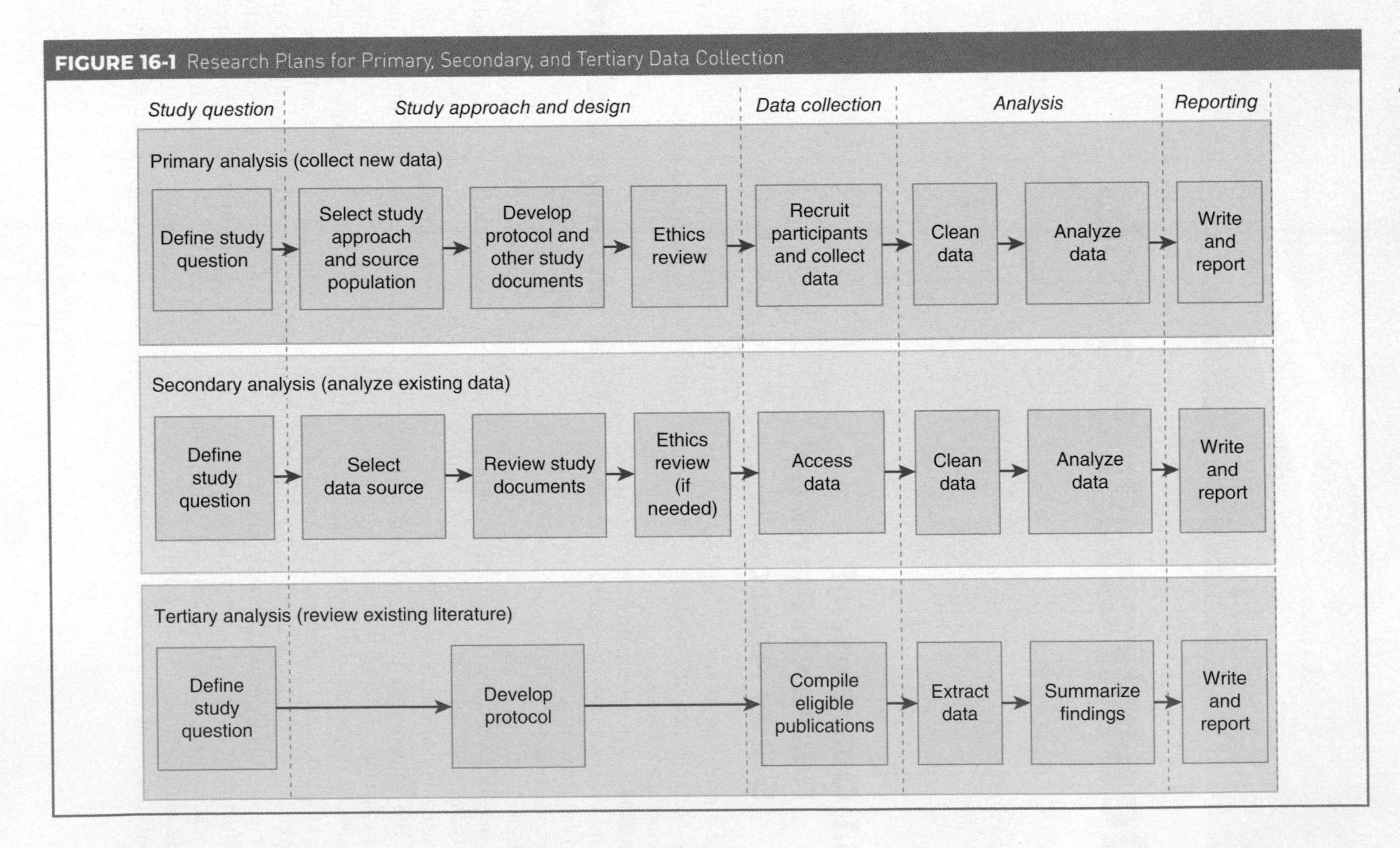

FIGURE 16-2 Examples of Items for a Primary Study Protocol

Goal	The main goal and specific aims of the study
Study design	The key features of the study design, including key variables
Sampling	Details about the source population and, if only a subset of the source population will be sampled for the study, a thorough description of the sampling process that will be used to generate the sample population
Sample size	The desired sample size, the calculations used to estimate the required number of participants, and the steps that will be taken to acquire an adequate number of participants
Recruiting	The specific processes that will be used for contacting and recruiting participants
Consent	The precise procedures that will be used to obtain and document informed consent (or to document refusal to participate)
Questionnaire	The exact questions that participants will be asked and, if interviews will be used to gather data from participants, details about how those questions will be posed
Additional assessments	If applicable, the precise laboratory procedures and other techniques that will be used
Data management	The exact ways that questionnaire responses and other data will be entered into a computer database, including the methods for recording missing responses, and the precise steps that will be taken to confirm the accuracy of the entered data
Confidentiality	The steps that will be taken to maintain the confidentiality of personal information that might be contained in the data set
Data analysis	The plans for data analysis, including the particular statistical tests that will be used to answer the study questions

For all study designs, the protocol should anticipate foreseeable challenges or areas of confusion and address them as completely as possible. For example, the protocol for a tertiary study should explain what the researchers will do if an exhaustive attempt to track down the full text of an article that appears likely to be eligible for inclusion is unsuccessful.

A complete protocol for a ready-to-launch project fully describes all the procedures that will be used for data collection and analysis, lists the anticipated dates of completion for each of the steps in the research process, provides details about the responsibilities of each member of the research team, and describes the mechanism for updating any part of the research plan if gaps become apparent during the study or if revisions to the protocol are deemed necessary (and are approved by all relevant research ethics committees before they are implemented). A strong protocol provides enough detail that another researcher could easily replicate the study in a new population. A completed protocol should be detailed enough that the entire methods section of any paper that will result from the project can be written before data collection begins.

16.3 Rigor and Reproducibility

Proposals submitted to the U.S. National Institutes of Health are expected to demonstrate both rigor and reproducibility. These two characteristics are valuable for any research protocol in the health sciences.

Rigor is the careful design, implementation, interpretation, and reporting of an exacting, unbiased, and ethical research protocol that answers a clearly defined scientific question. Rigor is a demanding standard. A rigorous sampling plan might rule out the use of a convenience sample. A rigorous data collection plan might necessitate the use of laboratory testing rather than relying on self-reported data from participants. A rigorous data analysis protocol might require a large number of participants to be recruited so that the statistical tests run on the resulting data set have sufficient power to answer the study question. A strong protocol also demonstrates that all data collection and analysis procedures closely align with the study goal and specific aims.

Reproducibility is the ability of an independent researcher to implement another researcher's data analysis protocol and generate the same results as the original researcher if given access to the original data set. Reproducibility requires a detailed set of explanations about how participants were sampled, how data were collected, which variables were included in the data set, and which statistical tests were used to analyze various types of data. All of these details should be included in a protocol. Reproducibility complements replicability. Reproducibility in health research is typically demonstrated when an independent researcher reanalyzes already-collected data according to the original research team's protocol and gets the same results. Replicability is demonstrated when the same protocol is used to collect and analyze new data from a new population, and similar conclusions are reached. Transparency about all methods is the foundation for both reproducibility and replicability.

16.4 Research Timelines

Most research protocols include a detailed schedule for the planned research project. It is therefore helpful to:

- Create a list of all the steps from planning the study through the dissemination of results.
- Create a calendar that shows when each of these steps is expected to be initiated and be completed.
- Identify fixed deadlines that must be met, such as grant application deadlines and abstract submission dates for conferences where the work might be presented.
- Agree with collaborators on intermediate deadlines for research tasks that will help the project stay on track toward timely completion and will ensure that no important fixed deadlines are missed.
- Set up regular meeting times for the research team (whether in person, online, or via other communication modes).

A **Gantt chart** is a type of bar chart that visually displays the research timeline and marks critical calendar dates and deadlines. A Gantt chart can be very helpful for visually displaying the research timeline (**Figure 16-3**).

The internal due dates set by the research team will need to be somewhat flexible because predicting how long some steps will take can be difficult. For example, waiting for ethics approval or for the disbursement of funds from a granting agency might take several months instead of several weeks. Data collection might be completed far more slowly or more quickly than expected. Data entry or data cleaning might take much more time than originally anticipated. Additionally, relying on collaborators to complete some aspects of the work may result in delays. This is especially likely when the lead researcher is not in a

FIGURE 16-3 Sample Gantt Chart for a Year-Long Secondary Analysis Project

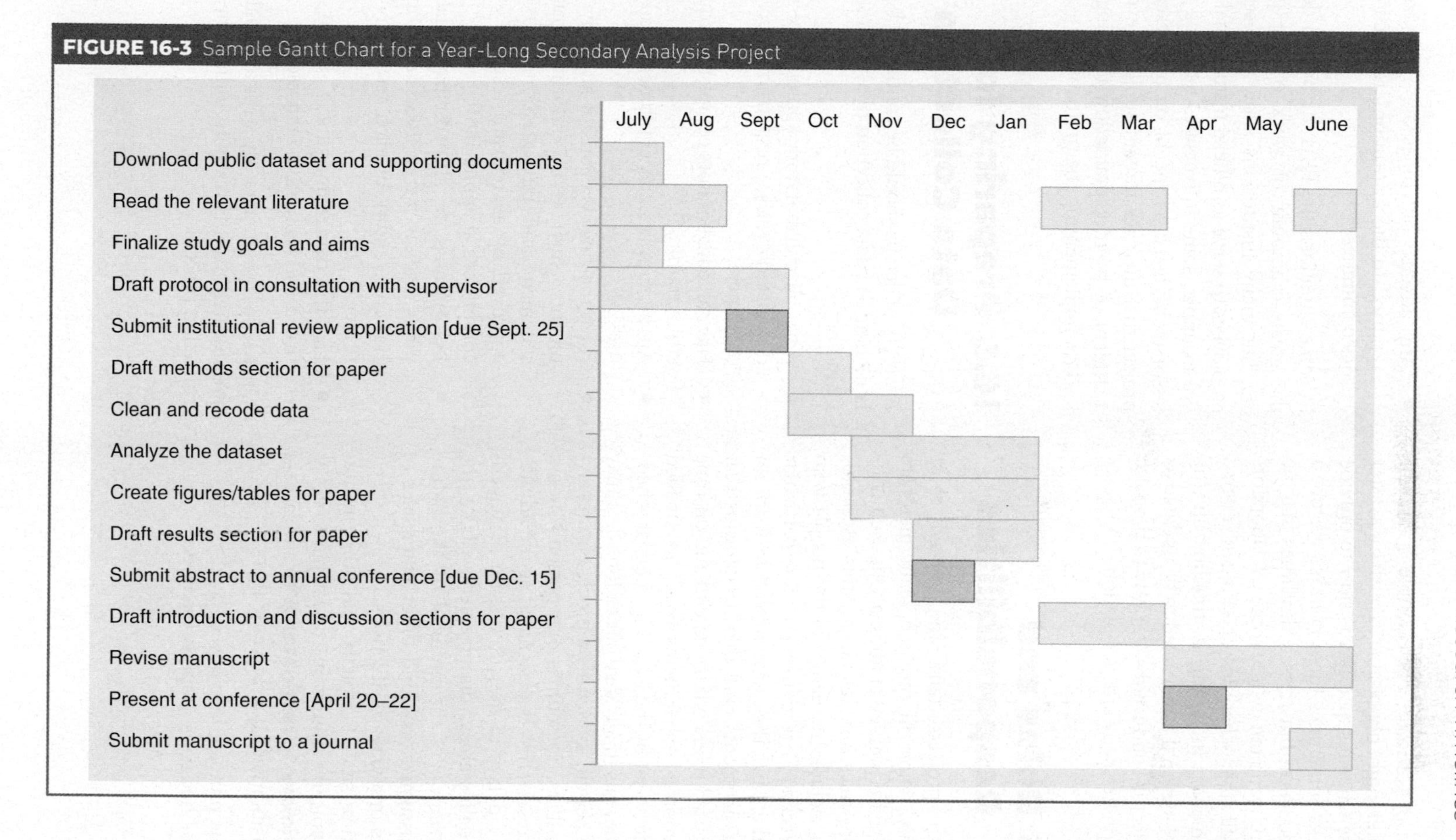

position of authority. For example, a student researcher might not be able to push for a faster response when a supervisor is slow to provide feedback. Sometimes these holdups are not a major concern, but missed deadlines may be a serious problem when some collaborators have inflexible schedules or stringent degree requirements. Including a timeline in the protocol—even if it may need to be updated as the project is conducted—is a way to promote steady progress toward completion.

16.5 Roles and Responsibilities

Research projects tend to proceed most smoothly when all research team members have a shared understanding about each contributor's roles and responsibilities. A protocol can include the names of the people who have accepted responsibility for particular tasks, the dates they have agreed to work on those assignments, and the mechanisms that will be used to encourage careful and on-time completion of those items (such as emailed reminders sent by the lead researcher at scheduled intervals in advance of deadlines). It may also be helpful to identify a process for resolving conflicts. Sometimes one person, often a senior researcher, is designated as the adjudicator of delays in the submission of agreed-upon deliverables, disagreements about the interpretation of the protocol or the nature of an assigned task, and other differences of opinion or awkward situations.

Universities, hospitals, and other institutions typically require one researcher to act as the **primary investigator (PI)** and accept responsibility for guaranteeing that:

- The protocol is followed.
- The budget is properly managed.
- Any adverse outcomes are immediately reported to the institution's research ethics committee.

In some situations the PI is the person doing the greatest amount of work on the project, but many institutions allow only senior employees to serve as official institutional PIs. For example, some universities require a professor to be listed as the PI on any research project that involves human subjects, even if a student or postdoctoral fellow is taking the lead on implementing the protocol.

16.6 Preparing for Data Collection

A variety of details should be confirmed before launching the data collection phase of a study, including:

- Have all collaborators approved of their designated roles, responsibilities, and deadlines?
- Have all collaborators completed required ethics training?
- Are all supplies and equipment ready for use?
- Have the final versions of all study documents (such as the informed consent statement and the questionnaire) been approved by all relevant research ethics committees, if applicable?
- If applicable, are all participating laboratories ready to begin processing samples? Has the validity of each test been confirmed?
- Has the data management system been tested and found to be reliable?

Once all of the preparatory steps are finalized, the research team is ready to initiate the data collection process.

CHAPTER 17

Ethical Considerations

Researchers have an ethical obligation to minimize the risks that research may pose to participants.

17.1 Foundations of Research Ethics

The ethical standards for human health research have evolved over the past several decades. One of the first sets of principles for ethical research was the **Nuremberg Code**, which in 1947 mandated voluntary consent for experimental studies of humans. The **Declaration of Helsinki** was written by the World Medical Association in 1964 to provide ethical guidelines for physicians conducting clinical trials. The **Belmont Report**, published by the U.S. National Commission for the Protection of Human Subjects of Biomedical and Behavioral Research in 1979, defined the key research principles of beneficence, respect for persons, and distributive justice. The *Belmont Report* is a foundational document for the current U.S. Federal Policy for the Protection of Human Subjects, which is often simply called the **Common Rule**. All of these documents have influenced research laws, policies, and regulations in countries across the globe.

Experimental studies have traditionally raised the greatest ethical concerns because the researcher assigns participants to try a new product, take a new drug or supplement, adopt a new behavior, or otherwise engage in an activity they would not normally do. Ethical principles require that adequate, but not coercive, benefits be offered to participants. An appropriate control for an experimental intervention must be selected. Safety must be monitored. The ability of participants to continue to have access to the new product or service after the conclusion of the research project must be considered prior to the study's implementation.

Observational studies have traditionally been considered less risky because the research team is not imposing changes on the participants. However, that does not mean that voluntariness and other research ethics principles do not apply. Under current standards, observational studies usually require that all participants grant their informed consent to be included in a study, all data collected by the researchers be kept confidential, and all physical, psychological, and other potential harms to participants be minimized. All patient protection regulations, such as the Health Insurance Portability and Accountability Act (HIPAA)

Privacy Rule in the United States, must be strictly adhered to for observational as well as experimental studies.

17.2 Respect, Beneficence, and Justice

The three core principles of biomedical research ethics are now usually considered to be respect for persons, beneficence, and distributive justice. Each protocol for a primary research project in the health sciences (and most secondary research projects) should be carefully inspected to ensure its compliance with these principles. This level of scrutiny is required by research ethics review committees and by professional standards.

Respect for persons is a research principle that emphasizes autonomy, informed consent, voluntariness, and protection of potentially vulnerable individuals. **Autonomy** is the ethical principle that only an individual (or his or her legal guardian) is authorized to decide whether to volunteer to participate in a research study. For almost all research projects that involve interaction with individual participants and/or their personally identifiable data, each potential participant must be fully informed about the benefits and burdens of the study, the procedures involved, and the plans for use of the data collected. Recruits must make an autonomous choice to participate or not to participate.

Respect for persons also involves many other considerations. Researchers should choose an appropriate source population for the research question or, if they are conducting community-based participatory research, should select an appropriate research question for the source population. The study protocol should be a scientifically valid and rigorous one that will answer the research question, because it would be disrespectful to ask participants to volunteer for a poorly designed study that will not yield meaningful results. The research procedures should be as minimally invasive as possible. A nondiscriminatory process should be used to sample and recruit participants. Researchers should recruit the correct number of participants to ensure adequate statistical power for the study. Recruiting too few participants would mean the study cannot generate significant results, and recruiting too many participants may burden volunteers unnecessarily. The researchers should confirm that all participants understand the informed consent materials and process. The confidentiality of all shared information must be maintained.

In research, **beneficence** is the ethical imperative for a study to maximize possible benefits and minimize possible harms. Beneficence is about doing good. To meet the requirement of beneficence, a research proposal must have a high likelihood of benefiting individual participants and/or the communities from which they are drawn. For most studies, the opportunity to contribute to scientific knowledge is considered an adequate benefit to participants. For some studies, more specific individual and community benefits are offered. Researchers conducting beneficent research should be able to justify the necessity for and the importance of the research project.

Beneficence is often paired with **nonmaleficence**, the ethical imperative for a research study to do no harm. Nonmaleficence requires the research team to minimize potential physical, psychological, financial, social, or other harms to participants, as well as to ensure an acceptable balance between risks and benefits. For example, experimental studies must identify ahead of time what events would lead to early termination of the study. Discontinuation might be appropriate when the intervention appears to be dangerous or when it appears to be so beneficial that it would be unethical not to immediately offer the intervention to the individuals assigned to the control group. For observational studies that ask participants to share about times of emotional distress, nonmaleficence requires disclosing the possibility of emotional harm in the informed consent statement and providing participants with contact details for local counseling services.

Distributive justice is a principle of research ethics that requires the benefits and burdens of research to be fairly allocated. Equitable burden means that vulnerable populations should not be preferentially selected as the source population for research studies targeting the general population, because that might unfairly burden a disadvantaged population group. At the same time, equitability means that members of under-studied populations who happen to be sampled at random for a study of the general population should not be excluded from participation unless there is a defensible reason for exclusion. Also, to be just and not exploitative, the source population must have access to the results of the research study. For example, if an experimental pharmaceutical therapy for a chronic disease proves to be effective and safe, participants in the clinical trial usually should have the continued opportunity to access the drug after the trial is over. Justice is about the long-term impact of the study, not just the immediate benefit to the individual participants and the communities from which they were drawn.

Figure 17-1 highlights some of the many questions that researchers should ask and answer about their own protocols prior to formal review by an ethics committee. **Figure 17-2** illustrates the kinds of questions that can be asked for community-based projects to complement key questions associated with individual-focused projects. International health research guidelines, such as those developed by the Council for International Organizations of Medical Sciences (CIOMS), and national and disciplinary research guidelines may identify additional areas of concern that need to be considered as protocols are developed.

17.3 Incentives and Coercion

Researchers need to consider the ethical implications of offering an inducement to potential participants to encourage them to enroll in a study, offering reimbursement for the direct or indirect costs of participation, or compensating participants for their efforts. The principle of beneficence does not require monetary compensation for participants. For many research projects in the health sciences, the innate reward of contributing to science is a sufficient benefit of participation. At the same time, the principle of nonmaleficence requires that participation in a research project not be an undue burden to participants. In some situations, reimbursing participants for their travel and other expenses and/or compensating them for their time may be appropriate.

Incentives are sometimes offered to research recruits and participants. To increase the participation rate, researchers may reasonably offer a small gift to all participants or enter everyone who completes a questionnaire into a drawing for a more substantial gift that one randomly selected participant will receive. It may also be appropriate to provide free treatment for some types of conditions examined by the study, such as iron pills for participants found to have anemia after a blood test or deworming medication for participants found to have intestinal parasites. These medical treatments must be provided with appropriate clinical supervision, and health education must also be provided to the recipients of these treatments to ensure that they complete the therapy correctly and safely. For some clinical trials, covering all medical expenses directly related to participation in the study may be expected and appropriate. The tests and procedures that will be covered and those that will not be provided and paid for by the research team must be fully disclosed to participants prior to their enrollment.

FIGURE 17-1 Eight Central Considerations ("8 Cs") in Research Ethics

Category	Examples of Questions to Ask
Contribution	■ Why is the proposed project important? ■ How will individuals and/or communities benefit from this study?
Compensation	■ Will individuals or communities participating in the study be offered any form of inducement, reimbursement, or compensation? If so, what will be offered, and is it appropriate? Is the offer so high that it could be seen as coercive or so low that the study could be seen as exploitative? ■ Are the risks of participation minimal? ■ How will study-related injuries be handled? ■ Are the risks and benefits balanced?
Consent	■ How will potential participants be informed about the study? ■ How will consent to participate be documented? ■ Will a test of comprehension of the informed consent statement be required? ■ If applicable, how will consent (and possibly assent) be acquired for children and other members of potentially vulnerable populations? ■ If applicable, will community meetings be held prior to beginning the study?
Confidentiality	■ How will the privacy and confidentiality of participants and their personal information be maintained?
Community	■ Is the source population appropriate for the goals of the research study? ■ Will the selection process be fair? ■ Will the sample size be adequate? ■ Are potentially vulnerable individuals and communities adequately protected? ■ Has the protocol been adapted to address the cultural expectations of the source population? ■ If applicable, has the community agreed to participate in this project?
Conflicts of interest	■ Who is contributing to the project's finances and/or logistics? ■ Might potential conflicts of interest inhibit the ability of a researcher to conduct ethical and unbiased research?
Collaborators	■ Are all members of the research team adequately trained to conduct ethical research? ■ What steps will be taken during data collection and analysis to ensure that the protocol and all ethical standards are adhered to by all members of the research team?
Committees	■ Which research ethics committee(s) need to review the project? ■ If applicable, what community organizations have been consulted about the proposed project?

The desire to reward or thank participants with gifts must be balanced with the need for participation in any research project to be voluntary. **Voluntariness** describes a decision made of an individual's own free will without undue outside influence. When an individual feels coerced into participation, the principle of voluntariness is violated. **Coercion** involves compelling an individual to participate in a research study in violation of the principles of

FIGURE 17-2 Sample Ethical Considerations for Individual- and Community-Based Research Projects

	Individual Participants	Community Participants
Respect	■ What steps have been taken to protect individual rights? ■ Has the risk of coercion in recruitment been considered and minimized? ■ Is the informed consent process more than just signing a piece of paper? ■ Do participants in sensitive studies have privacy? Will their participation be kept secret? ■ Will data shared with the researchers be kept confidential? Will files be protected and not shared unless individually identifiable information is removed?	■ What steps have been taken to ensure that a community's values are respected? ■ Are appropriate community-based research methods being used? ■ Have community representatives and a local oversight committee been consulted about the project?
Beneficence	■ How will individuals benefit from participation? Free services, supplies, or medicines? Free health education? Gifts or money? Contribution to knowledge?	■ How will a participating community benefit from the research project?
Nonmaleficence	■ What steps have been taken to minimize physical, psychological, financial, social, and other risks to participants? ■ Is counseling available for participants in sensitive studies? ■ Is appropriate reimbursement for travel costs and other expenses being offered?	■ What steps have been taken to ensure that a community is not burdened by research participation?
Justice	■ What are the long-term benefits for individual participants? For example, will they gain increased knowledge about their health status? ■ What will happen to participants after the study is completed? Will the results of the study be shared with them?	■ What are the long-term benefits of participation to the community? ■ Will the researchers have an ongoing relationship with the community?

autonomy and respect for persons. Coercion could include social pressure or requests from authority figures that make it difficult for an individual not to agree to enroll in a study. For example:

- Employees asked by their supervisors to enroll in an occupational health study may fear losing their jobs if they do not agree to participate.
- Patients asked by their own physicians to register for a clinical trial may fear that their medical care will suffer if they do not comply with the request.
- People in jail or prison may believe that participation in a research study is mandated or will yield unspecified rewards, even if they are told that participation is voluntary and there will be no direct benefits.

Coercion can also be instigated by generous incentives, such as free medical care and monetary compensation, which could significantly impair an individual's ability to make an informed decision about the risks and benefits of participation. To minimize the risk of coercion, researchers must be transparent about what participants will gain from participation in a research study and what they will not gain.

17.4 Informed Consent Statements

Informed consent is an individual's voluntary decision to participate in a research study after reviewing essential information about the project. Informed consent statements provide crucial information about research projects to potential research participants so that they can make a thoughtful decision about whether to enroll in a study. The key components of an informed consent statement are summarized in **Figure 17-3**. The statement must use clear, simple language to describe the study aims, the procedures, the expectations of participants, and the benefits and the possible risks of participation. The statements should emphasize that participation is voluntary and that any participant can withdraw from the study at any time without penalty. Many research institutes provide templates for informed consent statements that use language approved by ethics experts and the institution's legal advisors. The template might need to be modified to

FIGURE 17-3 Content for the Informed Consent Statement

Content Area	Description
Research	A definition of "research" and a statement that the study involves research
Purpose	An explanation of the purpose and aims of the research process (except in rare situations in which this explanation interferes with the research goals)
Participants	A description of how and why certain individuals and/or communities were invited to participate in the research project and an estimate of the total number of individuals who will be recruited
Procedures	A description of the study procedures (including any physical exams, collection of biological specimens, randomization or blinding processes, interventions, and other procedures that are part of the study protocol) and the expected duration of the individual participant's involvement in the study
Benefits	A description of benefits to participants and/or to society, including a clear explanation of the compensation to be offered or a clear statement that the participant will receive no direct benefits
Risks	A description of the possible risks, discomforts, and costs associated with participation, a statement that involvement in the project might involve unforeseeable risks, and a description of how study-related injuries will be handled
Confidentiality	A description of the steps that will be taken to maintain confidentiality
Voluntariness	A statement that participation is voluntary and that the participant may withdraw from the study at any time with no penalty, along with a description for the process of withdrawing from the study
Contact details	Contact details for the researchers
Signature	Space for the participant's signature

ensure that the language in the consent statement can be understood by the members of a particular study's source population.

17.5 Informed Consent Process

Informed consent is intended to be a process, not merely a piece of paper. The principle of autonomy dictates that potential participants in a research study have the right to make their own decisions about whether to participate and that they must be provided with information that will allow them to make informed choices. The goal of the consent documentation process is not to acquire signatures from potential participants, but to ensure that participants truly understand the research process.

The process of documenting informed consent for an experimental study typically consists of reading the informed consent statement aloud to a potential participant or allowing the individual to read a copy of the statement, allowing adequate time for the potential participant to consider whether he or she wants to participate, answering any questions, and then asking whether the individual wants to participate in the study and is willing to sign an informed consent form. Acquiring a signature is not the end of the process. The lines of communication between researchers and participants must remain open during and even after the data collection process. All participants must be given a copy of the informed consent statement that includes contact details so that they can call or email the researchers if they have concerns about the study or desire to withdraw.

The researcher should ensure that participants understand the research process and the consent document. A brief test of comprehension may be helpful. For example, recruits for an intervention study may be asked to say in their own words what "randomization" means. A correct answer will demonstrate an understanding that each participant may be assigned to a control group rather than to the active intervention group and that participants do not have a choice in the matter. An incorrect or incomplete answer may require additional explanation of the research process prior to acquisition of a signature on a consent document. The goal is not merely informed consent, but understood consent. **Understood consent** requires evidence that a potential study participant comprehends the study benefits, risks, and procedures and knows his or her rights as a study participant prior to agreeing to participate.

17.6 Informed Consent Documentation

For most experimental and observational projects that involve in-person contact with participants or their identifiable data, the expectation of research ethics committees is that each individual participant will sign his or her name on a printed informed consent statement. This written record provides legal protection for the institution sponsoring the research project because it shows that participants agreed to the terms of the study. For telephone interviews, informed consent documents may be mailed to potential participants, signed, and mailed back to researchers prior to the interview. For computer-based surveys, an electronic signature can be provided.

When the source population has a low literacy rate, signatures may not be the best way to document consent. Participants who are not able to read or write might provide a thumbprint or some other mark to indicate their consent Alternatively, if it is deemed inappropriate to ask people who cannot read a document to sign it, oral consent may be preferable. **Oral consent**, also called **verbal consent**, is informed consent for participation in a study that is spoken and witnessed rather than requiring a participant's signature. Oral consent must usually be witnessed by an independent third person (someone other than the researcher or the participant). In some cases, a declaration of consent is also audio-recorded.

For surveys that are not collecting identifiable information about participants, individuals may be asked to read an informed consent statement but not asked to sign it. Instead, the consent statement may state that completing the questionnaire will be considered proof of voluntary consent to participate in the study. A consent process that does not require a signature may be granted when:

- The responses cannot be linked to individuals.
- The survey instrument does not ask sensitive questions.
- The researchers will not physically examine individuals or collect biological specimens.
- The questionnaire is so short that describing the study would take longer than completing the questionnaire form.
- There are no foreseeable risks to participants.

In some exceptional situations, an institutional review board (IRB) may grant a complete **waiver of consent**, which gives the researchers permission not to provide an informed consent statement to the individuals who will have their data included in an analysis as well as permission not to give those individuals the opportunity to opt into or out of a study. Waivers of consent are granted only for low-risk studies for which it would not be practicable to contact participants. For example, if researchers will observe groups of individuals in public places, where participants have no reasonable expectation of privacy and will not interact with the researchers, consent is not required. Any request not to require the full consent process for a primary data collection process must be approved by a research ethics committee.

Some research designs call for the use of **deception**, which is the intentional misleading of research participants about the true purpose and procedures of a study. For example, suppose that a psychologist is testing how people react to unfairness by having them play a game against someone who is cheating. If participants are told that the study is about how people respond to cheating and they are informed that they will be playing a game against a member of the research team who will be blatantly cheating, the study will be worthless. Because deception is critical to the study, the IRB may approve an informed consent statement that does not state the study's true goals and procedures. Researchers who want to use deception in a study must carefully describe the research protocol to the IRB, explain why deception is necessary, provide evidence that the harms associated with deception do not outweigh the benefits, and indicate whether they will debrief the participants about the true goals of the study after participation or do not plan to disclose the deception to participants. The IRB has the option to deny the application or to approve deceptive informed consent statements or issue a waiver of consent.

17.7 Confidentiality and Privacy

Privacy is the assurance that individuals get to choose what information they reveal about themselves. The right to privacy means that:

- Individuals have the right to refuse to allow their personal information to be shared with researchers.
- Individuals who agree to participate in a study involving face-to-face interviews should have the option of meeting with researchers in a place where no one outside the research team will be able to observe or overhear the interview.
- The identities of participants in a research study should not be disclosed to unauthorized people.

Confidentiality is the protection of personal information provided to researchers. One way to guarantee confidentiality is not to collect any personally identifiable information, such as names, addresses, government-issued identification numbers, telephone numbers, birthdates, or other data that can easily be linked to an individual. This approach is often an option for cross-sectional studies, but it is not possible for prospective or longitudinal studies in which baseline data about individuals must be linked to their own follow-up data. When individually identifying information must be collected, many steps throughout the research process can be taken to protect the information.

- All paper records should be stored in a locked file box in a locked room, and all computerized data files should be password-protected.
- Names and other personal identifiers should not be included in data files that contain sensitive personal information. Instead, two separate files should be created, one for identifying information and one for all other data. These should be linked only by a unique study identification number.
- Only essential research personnel should have access to the file containing personally identifying information.
- At some point after the end of the study, and in compliance with the rules of the relevant research ethics committees about how long documentation of informed consent must be stored, individually identifying records should be destroyed.

17.8 Sensitive Issues

Researchers asking questions about sensitive issues must decide ahead of time how to handle disclosures. Sensitive issues may include questions concerning:

- Drug or alcohol abuse
- Sexual practices and preferences
- Psychiatric illnesses
- Immigration status
- Participation in illegal activities
- Genetic disorders
- Other information that could materially damage a participant if it were made known to the public

When written documentation of consent is not acceptable because participants could be harmed by being linked to the study, a waiver of the need to document consent can be requested. **Waiver of consent documentation** is permission from an IRB not to collect signed consent forms from participants because they could be harmed by being able to be linked to a study on a sensitive topic. This is not a waiver of the consent process. It is just permission not to collect written documentation of consent. The research team can also apply for a **certificate of confidentiality** (or the equivalent in the study country), which is a legal document that protects the identity of participants in a study of sensitive topics from being subject to court orders and other legal demands for information. When working with vulnerable populations or sensitive information, the relevant research ethics committees should be consulted about what alternative methods for documenting consent they will consider acceptable.

In some situations, guaranteeing confidentiality may be impossible because withholding critical information from authorities would violate the law. For example, legal mandates may obligate researchers to alert the police about child abuse, intimate partner violence, or suicidal ideation and planning, and to inform public health authorities about diagnoses of infections designated as notifiable conditions. The decision about whether or how to collect data related to these issues may require consultation with a legal expert and local authorities.

Participants in studies of serious genetic diseases should be offered genetic counseling and given the opportunity to decide whether they want to know the results of tests. A qualified genetic counselor can assist with development of appropriate protocols.

17.9 Cultural Considerations

Culture is a way of living, believing, behaving, communicating, and understanding the world that is shared by members of a social unit. **Cultural competency** is the ability to communicate effectively with people from different cultures and backgrounds. A research protocol must be appropriate to the culture or cultures of the expected study participants.

For example, culturally appropriate recruiting may take different forms. In some cultures, a small gift may be expected as a token of goodwill before an individual is asked to participate in a study. In other parts of the world, this would be considered coercive because it would create a perceived debt owed to the researcher. In some cultures, participants may expect a small gift upon completion of the study. In other cultures, volunteers might feel that a gift devalues their donation to science. Participants from some cultures expect to share tea or coffee or a light meal with researchers before any questions are asked. People from some cultures may expect that all health research will be conducted in an impersonal clinical setting. In some parts of the world, prospective participants might need to know that community leaders, such as government officials, religious leaders, or tribal leaders, have approved of the project and are involved in making sure it is conducted well. In other cultures, the association of authorities with a research project may raise concerns about voluntariness, confidentiality, and the potential misuse of data.

The informed consent process may need to be adapted to local custom. Individual adults invited to participate in a study must consent to their own participation, and no one else (except a legal guardian) can consent for them. However, potential participants may need time to consult with their spouses, parents, adult children, or other family members prior to making a final decision about participation. For some community-based studies, a meeting of the whole community should be held so that everyone is confident that they are all hearing congruent messages from the research team. It may be helpful to have a local advisory board facilitate communication between the community and the research team.

The survey instruments and data collection processes should be culturally appropriate, and researchers should be trained in culturally respectful interview techniques. Topics that are openly discussed in one culture may be sensitive in another. Tests that are only mildly uncomfortable in one culture may be distressing in another. For example, although people in some cultures are sensitive about the measurement of body weight, other cultures may not care about weight but may be uncomfortable with the measurement of height. There may be formal or informal restrictions on who can conduct an interview or a physical examination. Female participants may be unwilling to be examined by a male, or older participants may be uncomfortable being interviewed by a much younger person. Some participants may expect to be alone with just a researcher, and others will expect to have a family member present for the entire process. The informed consent statement and study materials may need to be available in multiple languages, and multilingual interviewers may need to be hired.

If the research team does not include members of the population being recruited for the study, it is important to consult with representatives of the source community when developing and revising the protocol. Those community members may represent a cultural or language group, patients with the disease being studied, or other stakeholders.

Additionally, some research ethics committees require a cultural expert to examine the protocol as part of the review process.

17.10 Vulnerable Populations

Although most members of vulnerable populations can make their own choices about whether to participate in a research project, children and some adults with cognitive impairments may not be considered competent to make an informed decision. In this situation, a legally approved guardian is allowed to grant consent on behalf of the study participant. **Assent** is the expressed willingness to participate in a study by a child or another person who is deemed not legally competent to provide his or her own consent. Whenever possible, in addition to having the legal representative's consent, potential participants should assent to their own participation.

People in prison, comatose patients, and some other groups of people who have restricted autonomy or might be at elevated risk of harm from research participation may be considered to be vulnerable research populations. Special considerations may also apply to research involving females who are pregnant or may become pregnant, adolescents, people with cognitive impairments, traumatized patients with altered mental status, terminally ill people, older adults, members of some racial and ethnic minority groups, students, employees, healthy volunteers who cannot therapeutically benefit from clinical research projects but may be at risk of harm, and international populations. A protocol for a study focused on a potentially vulnerable population must justify the necessity of the project and ensure that the sampling process is fair, participation is voluntary, and volunteers (and/or their legal representatives) understand the requirements of participation and the possible benefits and risks of the study.

17.11 Ethics Training and Certification

Research ethics committees usually require everyone who will be in direct contact with research participants and/or their identifiable data to complete formal research ethics training. Many institutions offer their own courses, either in person or online, and several funding agencies and nonprofit organizations also offer training programs. After completing modules on various aspects of research ethics and passing an exam, the investigator receives a training certificate that is typically valid for 1 to 3 years. Copies of these certificates should be saved, because research ethics committees often require proof of ethics training for all members of the research team.

CHAPTER 18

Ethical Review and Approval

Research ethics committees protect study participants, researchers, and host institutions by carefully reviewing research protocols prior to their implementation.

18.1 Ethics Committee Responsibilities

An **institutional review board (IRB)**, sometimes called a **research ethics committee**, is a group responsible for protecting people who participate in research studies. The three primary goals of IRBs are to:

- Protect the "human subjects" who will participate in observational or experimental studies or whose personal information will be examined by researchers.
- Protect researchers by preventing them from engaging in activities that could cause harm.
- Protect the researcher's institution from the liability that could occur as a result of research activities.

An **Institutional Animal Care and Use Committee** (IACUC) oversees research with animals and operates separately from an IRB.

The major functions of ethics review boards are to:

- Review new and revised research protocols.
- Approve or disapprove of those protocols.
- Ensure that informed consent is documented (if required).
- Conduct continuing review of long-term research projects.

To verify the achievement of these goals, IRBs maintain careful records of their procedures and membership; all proposals, consent statements, and supporting documents; all correspondence; and minutes of all meetings that chronicle the decisions made to exempt, approve, or disapprove proposals and the justifications for these decisions. Researchers must provide all documents requested by the review committee, including all requested status reports.

Federalwide assurance (FWA) is a status that applies to IRBs that are registered with the U.S. federal government. IRBs overseeing research funded by the federal government must be certified as having FWA. Institutions in the United States and in other countries can apply for FWA status and FWA renewal. IRBs with FWA agree to adhere to U.S. research laws, regulations, and policies (that is, the "Common Rule") and to release their written operational procedures to the government if requested to do so.

18.2 Ethics Committee Composition

Research ethics committees are usually composed of at least five members, preferably from diverse backgrounds, including both scientists and nonscientists. Each member reviews the proposal and then meets with the others to discuss it and to determine whether it meets the requirements of the institution. An outside scientific expert and/or community representative may also be consulted about the research plan.

Because of the number of individuals involved in protocol review, even the most efficient ethics review committees may need a month or longer to issue an exemption or an approval or to make a request for a revision to be made to the protocol, which must then be reconsidered by the committee before final approval. The review process may take several months for complicated proposals, such as studies involving:

- Invasive procedures
- Sensitive questions
- Potentially harmful interventions
- Waiver of written informed consent
- Deception about the study aims
- Multiple sites
- International research

A research timeline should assume a lengthy review period. The application should be submitted to the ethics committee as early as possible in the planning process in order to minimize the risk that delays in the approval process will complicate the timing of data collection and other research activities.

18.3 Application Materials

Research ethics committees examine many aspects of a research protocol during the review process (**Figure 18-1**). Some research ethics committees ask applicants to provide a narrative statement that addresses a list of possible ethical concerns. The research protocol or narrative statement about a planned project should address each of the topics required by the committees evaluating the proposal. To ensure the completeness of the review, some committees mandate the completion of many pages of forms that require answers to a long series of questions about the project. All questions must be answered, even when most questions warrant an answer of "not applicable."

FIGURE 18-1 Examples of Information Requested and Examined by Ethics Review Committees

Category	Considerations
Participants	■ What is the anticipated composition and size of the study population? ■ Is the source population appropriate for the study question? ■ How will participants be recruited? Does the recruitment method raise any concerns about coercion? ■ What are the inclusion and exclusion criteria? Will the exclusion criteria screen out participants with a higher-than-typical risk of harm? Will the criteria generate a study population that is reasonably representative of the source population? (For example, if the study question applies to all adults, are there any restrictions on participation by reproductive-age women that are not directly related to safety?) ■ If applicable, are potentially vulnerable subjects protected?

Category	Considerations
Risks and benefits	■ Why is the study important and necessary? How will the proposed study benefit participants and/or their communities? ■ How will data be collected? Will existing data, documents, records, or specimens be used? Will individuals or groups be examined using surveys, interviews, focus groups, or other methods? Will interviews be audio- or video-recorded? Will noninvasive clinical measures be used? Will participants be asked to engage in exercise or tests of endurance, strength, or flexibility? What machines will be used to collect data, and will collection involve radiation exposure? Will blood, hair, nail clippings, sweat, saliva, sputum, skin cells, or other biological specimens be collected noninvasively? Will drugs or devices be tested? ■ What are the potential physical, psychological, financial, or other risks to participants? ■ Are the risks minimal (or at least minimized)? ■ Are the risks reasonable compared to the anticipated benefits?
Informed consent	■ Does the informed consent statement adhere to institutional guidelines? ■ How will informed consent be sought? ■ How will informed consent be documented? ■ Is any modification to the usual methods of documenting informed consent being requested? If so, is the request reasonable? (For example, is a waiver of a signed consent form being requested because the source population has a low literacy rate? Or is a request being made to have no documentation of consent because the existence of a form linking an individual to the study could harm the participant?)
Privacy and confidentiality	■ How will privacy and confidentiality be maintained? ■ What are the plans for the protection of computerized and noncomputerized data?
Safety monitoring	■ What constitutes an adverse event? How will such events be handled? ■ Does the informed consent statement clearly state how research participants can contact the research team and the ethics review board if they have concerns?
Conflicts of interest	■ How is the project being funded? ■ Do any financial or personal conflicts of interest need to be disclosed to participants and/or addressed in other ways?
Researcher training	■ Are the investigators prepared to conduct ethical research?
Documentation	■ Are copies of all recruitment materials attached? ■ Are copies of the questionnaire and other assessment tools attached? ■ Is a copy of the informed consent statement attached? ■ If applicable, are letters of approval from study sites and collaborating institutions attached? ■ If applicable, is a copy of the grant proposal attached? ■ Are copies of research ethics training certificates for all members of the research team attached?

Proposals for the analysis of existing data may be significantly shorter than proposals for new data collection, but both primary and secondary analysis proposals need to:

- Describe the expected study participants.
- Explain the sample size estimate, the inclusion and exclusion criteria, and the recruitment plans (if applicable).
- Discuss the risks and benefits of the study.
- Describe the plans for seeking and documenting informed consent and monitoring safety (if applicable).
- Explain how confidentiality will be maintained.
- Disclose potential conflicts of interest.
- Provide proof of ethics training.
- Supply all relevant documentation.

For primary studies, the required documentation typically includes a protocol along with any recruiting materials, the informed consent statement, and the questionnaire. For secondary analyses, the application typically includes evidence that the data are in the public domain or that appropriate individuals or organizations have granted the researcher permission to analyze the data.

18.4 Exemption from Review

Once all application materials have been submitted to a research ethics committee, there are three possible next steps: (1) exemption, (2) expedited review, and (3) full review. The ethics review board decides which action is appropriate. **Exemption from review** is a determination by an IRB that a research protocol does not require full IRB review because it does not meet the IRB's definition of human subjects research.

Exemption can be granted—but does not have to be granted—only after the IRB professionals review a protocol and determine that it meets their criteria for exemption. Each IRB has its own rules about what kinds of studies are eligible for exemption. Many IRBs will exempt analyses of existing data and biological specimens that cannot be linked to individuals. They may require that the data sources be publicly available or anonymized so that study subjects cannot be identified. Some IRBs have a generous definition of the types of projects that can be exempted. Some rarely exempt any research protocols.

Some funding agencies require review to be conducted unless their rules for exemption are met. For example, the U.S. Department of Health and Human Services allows some research protocols to be exempted from review when:

- The study procedures do not involve an intervention.
- Researchers will not interact with participants.
- The participants will not be prisoners, young children, and other individuals in protected populations.
- Individually identifiable information will not be collected.
- Data will be collected using commonly accepted educational practices or tests, established survey or interview procedures, or observation of public behavior, or the study will rely on existing data or biological specimens, data on public benefits or public service programs, or evaluations of taste, food quality, or consumer acceptance.

An exemption can also be granted for data collected as part of routine professional practice that is not intended to contribute to generalizable knowledge. Practice activities include teachers assessing their students' knowledge of class material, clinicians examining their patients, community health organizations initiating monitoring and evaluation projects, and public health officials collecting surveillance data and conducting outbreak investigations. None of these activities requires review by a research ethics committee, because all are considered to be within the accepted scope of practice.

However, ethics review is required if these practitioners or organizations choose to engage in research activities, such as:

- An educator plans to have students take special pre- and posttests to assess a new pedagogical approach and hopes to publish the results in a teaching journal.
- A clinician reviews patient records so that they can be presented as a case series at a professional conference.
- The results of a survey of clients of a community organization might later be published in a professional journal.

When a researcher is considering transitioning from a practice-based inquiry to an intentional research project, the IRB should be consulted about what application materials are required. The decision about whether a practice-based project is exempt from review is up to the IRB, not the researcher.

18.5 Review Process

Expedited review is a determination by an IRB that a proposal requires review but a review by the full committee is not required. An expedited review may be possible when a minor change to a previously approved protocol is requested. Sometimes expedited review is also possible for new studies in which the risk to participants is no greater than what is encountered in ordinary daily life or, in the case of clinical work, during routine examinations or procedures. Exemption from review is not allowed for research focused on vulnerable populations. Expedited review may allow the chair of the ethics committee to approve the protocol without a full meeting of the committee. However, all members must be notified of the decision and given an opportunity to express concerns.

Full review is a determination by an IRB that the full committee must discuss a study protocol in order to ensure that the requirements for the protection of human subjects are met. Full review of a research proposal is usually required when an intervention will be tested in individuals or a community, data will be collected through interaction with individuals, identifiable private information will be collected, or other criteria for expedited review are not met.

The ethics review committee has the right to approve each proposal or to deny approval. If a protocol is not satisfactory at initial review, the committee usually informs the investigators of the protocol changes that are necessary to make the proposal acceptable. Some requests may be easy to accommodate, and researchers should simply comply with them. At other times, the requested changes would significantly alter the nature of the project or would not be feasible with the resources available for the study. In this situation, the researchers need to present their concerns to the IRB and to try to work with them to find an acceptable resolution. However, the committee does not have to acquiesce to the desires of the researchers. The ethics review board has the right to deny approval of any protocol that does not meet its standards. The board can also demand proof that certain standards (for example, standards for data storage or investigator training) are met prior to approving the protocol.

18.6 Review by Multiple Committees

Multiple research ethics committees may be required to review studies that involve researchers from multiple institutions and/or participants from multiple countries or multiple study sites. Additionally, funding agencies may require review by their own ethics boards. For example, a student planning to conduct thesis research in another country must typically have the research protocol reviewed by both the student's own university and an ethics committee in the study country (often a local university or a teaching hospital).

At least three matters must be resolved prior to submission of a research proposal to multiple committees: the application documents that will be required, the wording of the informed consent statement, and the order of review.

First, each review board must be consulted about the application materials it wants to receive. It is sometimes possible to submit the same paperwork to all committees. However, the more likely scenario is that each board will require its own unique application materials, perhaps in addition to copies of all documents submitted to other ethics committees. The researchers are responsible for ensuring that each application packet describes the study objectives and protocol in the same way.

Second, many institutions have their own preferred wording for informed consent statements. The informed consent statement is seen as a legal document, and institutions want to be sure that the wording protects them. However, the preferred wording may differ for each participating institution. A resolution must be reached about how to merge consent statement requirements while making sure the study participants will understand the language.

Third, the order of review must be established. Sometimes, all the committees independently review the proposal at the same time. At other times, the reviews are conducted "domino" style, with the proposal being independently reviewed and approved by one committee, then passed to the next committee, and so on. Committees commonly stipulate that approval by their institution will be contingent on approval from all other participating institutions. If a modification of the protocol or informed consent document is mandated by one committee, then all other committees must re-review the proposal. A significant amount of extra time for ethics review should be built into the project timeline when multiple research ethics committees will be involved.

18.7 Ongoing Review

All ongoing research protocols must be re-reviewed annually (or more often, at the discretion of the ethics review committee) until the completion of data collection or, in some cases, until the completion of data analysis. Institutions typically require several documents to be included in progress reports:

- Current versions of the protocol, informed consent statement, questionnaire, and other study documents (even if these documents have not changed since the start of the study)
- A report on the study population, including the number of participants who have enrolled in the study, the number who have dropped out of the study, the demographic characteristics of the study population, and basic details about the number of participants who are members of vulnerable populations
- A report of any adverse events, complaints, or unanticipated problems, including details about any issues reported to the ethics committee since the last annual review
- A list of any amendments to the protocol or study materials that are being requested
- A summary of findings (which are especially important for experimental studies that might need to be stopped early if the intervention appears to be harmful or very beneficial)

All adverse events must immediately be reported to the IRB. Any desired changes to recruiting materials, the informed consent statement, the questionnaire, or other study documents must receive approval prior to being implemented. At the end of a study, most committees require a final report to be submitted that at minimum states the number of participants, affirms that no adverse events occurred, and declares that the project is concluded.

18.8 Conflicts of Interest

Most ethics review committees and an increasing number of journals require researchers to disclose potential conflicts of interest (or competing interests) related to the study. A **conflict of interest** (COI) is a financial or other relationship that could influence the design, conduct, analysis, or reporting of the study, or could appear to have caused bias. A potential conflict of interest is most likely to occur when a new product is being tested, such as a new medication or medical device, and a member of the research team earns a salary (or a consulting fee or honorarium) from or holds equity interests (like stocks or ownership) in the company that produced, developed, or will market the product. Similar concerns arise when intellectual property rights (such as the ownership of a patent or copyright) may result in earnings for a researcher or a close family member. However, these are not the only potential competing interests.

When a financial or other relationship could bias the design, conduct, or reporting of the study—or could merely appear to have the possibility of biasing the study—the potential conflict of interest must be disclosed. Several types of relationships might be required to be disclosed to employers, funders, and publishers:

- Personal fees paid as salaries by employers, as honoraria, or as compensation for consulting, lecturing, giving expert testimony, or providing other services
- Financial relationships such as ownership of stocks, shares, or equity
- Income from patents, copyrights, and other intellectual property related to the project, or pending patents that might result in future income
- Nonfinancial support such as donated equipment or supplies, travel support, or writing assistance
- Service on the board of directors of a company doing work related to the research project
- Personal relationships with individuals or organizations that could influence the work, such as having a spouse, parent, child, sibling, or other family member who works for a company with a direct interest in the research project

The disclosure of a potential conflict of interest is not a confession that bias has occurred or an admission that bias will occur. It is, however, an important assurance of transparency. Most universities, hospitals, and other institutions involved in scientific research have policies about what constitutes a conflict of interest and about when and how potential conflicts need to be disclosed. For example, some universities require only interests exceeding $5000 to be disclosed to the university, but others have a lower threshold for reporting.

18.9 Is Ethics Review Required?

Ethics review is required for almost every proposal that will involve living human subjects, whether those people will be directly contacted by the research team (in person, by telephone, by mail, by Internet, or via any other method) or their existing personal information will be analyzed. A small subset of projects might be exempted from review, but the decision to exempt a project from review can be made only by the relevant ethics committees. Most institutions do not allow researchers simply to declare that their projects do not need to be reviewed. Exemption usually involves a formal process of having an appropriate IRB confirm that a project meets set criteria for exemption from review.

Many incentives are in place to encourage participation in the formal review process. First, institutional approval provides a degree of legal protection to the researcher.

An approval letter is evidence that the research plan was deemed reasonably safe by a committee of experts prior to the initiation of data collection and analysis. Another incentive is that many research sponsors will not release grant or contract funds until a research plan has been approved by a research ethics committee. Also, an increasing number of journals require authors to provide details about which research ethics committee(s) reviewed the project, even if it was subsequently exempted from review. Some even require copies of the official approval letters. Research protocols cannot be retroactively approved, so researchers must take the time to undergo a formal review prior to collecting any data or analyzing any data files.

Researchers should also remember that approval is not the end of the involvement of an IRB with a research project. Any problem or potential problem associated with the implementation of an approved or exempted research protocol should be reported to the IRB immediately. Just like informed consent is a process and not a signed document, IRB oversight is an ongoing process and not something that ends with the issuance of an initial approval letter.

CHAPTER 19

Population Sampling

Primary studies require an accessible and appropriate source of study participants.

19.1 Types of Research Populations

At least four types of populations must be considered when preparing to collect data (**Figure 19-1**). Several different names are used to describe each of these four entities, but the general categories are the same for all health science fields.

- The broadest group is the **target population**, the broad population to which the results of a study should be applicable.
- The **source population**, sometimes called a **sampling frame**, is a well-defined subset of individuals from the target population from which potential study participants will be sampled.

FIGURE 19-1 Types of Populations

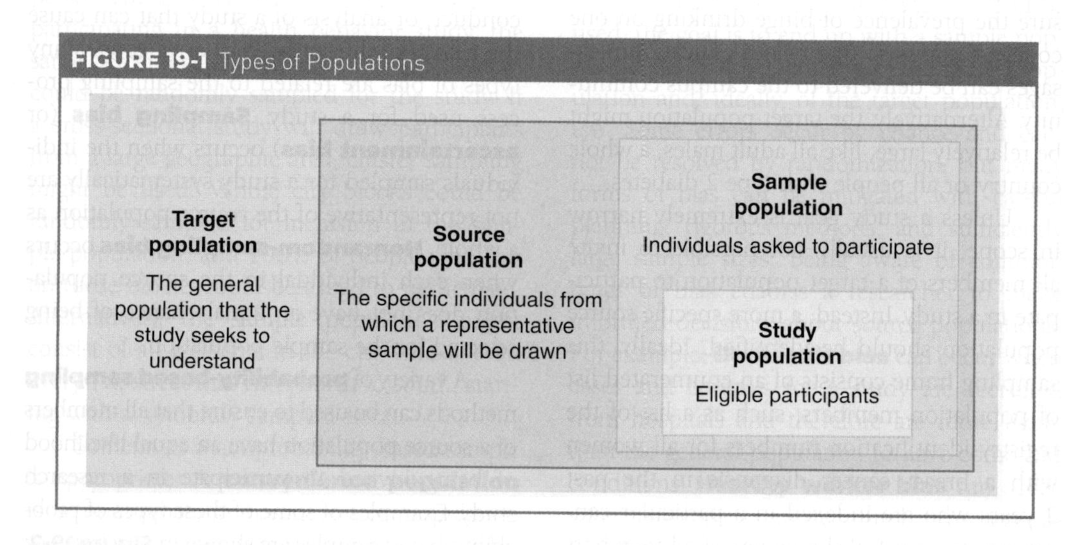
Target population
The general population that the study seeks to understand
Source population
The specific individuals from which a representative sample will be drawn
Sample population
Individuals asked to participate
Study population
Eligible participants

related to an exposure are excluded but cases with those comorbidities are not excluded. These study design problems can be avoided by selecting more appropriate source populations and applying inclusion and exclusion criteria consistently.

19.4 Study Populations

The individuals identified as the sample population will later be asked to participate in the study. The study population will consist of the members of the sample population who can be located, consent to participation, and meet all eligibility criteria. The **participation rate** is the percentage of members of a sample population who are included in the study population.

In ideal situations, all of the sampled individuals agree to participate in the study. This helps prevent the **selection bias** that occurs when the members of the study population are not representative of the source population from which they were drawn. However, a 100% participation rate is extremely rare. At least some of the individuals in the sample population will ignore an invitation to participate. Some who respond to the invitation will choose not to participate. Others will turn out to be ineligible because they do not meet the inclusion criteria. A low response rate may result in **nonresponse bias** if the members of a sample population who agree to participate in a study are systematically different from nonparticipants. However, a less than 100% participation rate is usually not a problem as long as the researcher:

- Uses acceptable and carefully explained sampling methods
- Takes appropriate steps to maximize the participation rate
- Recruits an adequately large sample size
- Reports the number of potential participants at each stage

Even so, researchers should develop a recruiting strategy that will encourage a high rate of participation.

19.5 Populations for Cross-Sectional Studies

The results of cross-sectional studies are often used to make important resource and policy decisions, so the study population for a cross-sectional study should adequately represent the target population. A **population-based study** uses a random sampling method to generate a sample population that is representative of a well-defined larger population. The strongest cross-sectional surveys are population-based studies that use probability-based sampling rather than a convenience population. The goal is to select a source population that is reasonably representative of the target population and then to sample and recruit a set of study participants who are reasonably representative of the source population.

The most rigorous population-based studies use probability-based sampling methods to generate sample populations and then confirm that study populations are reasonably representative of the source populations from which they were drawn. A **census** is a complete enumeration of a population, such as a count of every resident of a country, the number of inpatients at a particular hospital at noon on a selected day, or the number of employees of a large company. When a census of the source or target population has been recently conducted, the sample and study populations for a cross-sectional study should reflect the demographics of that census.

For example, suppose the goal of a study is to quantify the prevalence of tobacco use among high school students in a county. The county's 14 high schools together serve as the target population and the source population. Selecting only 1 high school as the sample population is probably not sufficient. Working intensely with 1 school might maximize participation rates. However, the selected school might enroll students who are different from county students as a whole—more rural or urban, more or less diverse, or from more or less wealthy households. In such a situation, the results from that one high school would not be an accurate reflection of adolescent health across the county. A better option is to sample some students from each of the 14 schools. For example, 20% of the classes in each high school that meet during the first period of the school day could be randomly sampled, and 100% of the students in those sampled classes could be invited to participate (**Figure 19-3**). After the data are collected, the researcher can validate the representativeness of the study by confirming that the proportion of students by grade, age, and sex in the study population is similar to the distribution of these characteristics among the county's total high school student population.

When census data are not available for a source or target population, careful planning can help a researcher recruit a study population that is reasonably representative of the target population. Recruiting participants for a general population survey from among the spectators at a football game, the shoppers in a particular grocery store, or the donors at a volunteer blood drive will likely result in a sample population that does

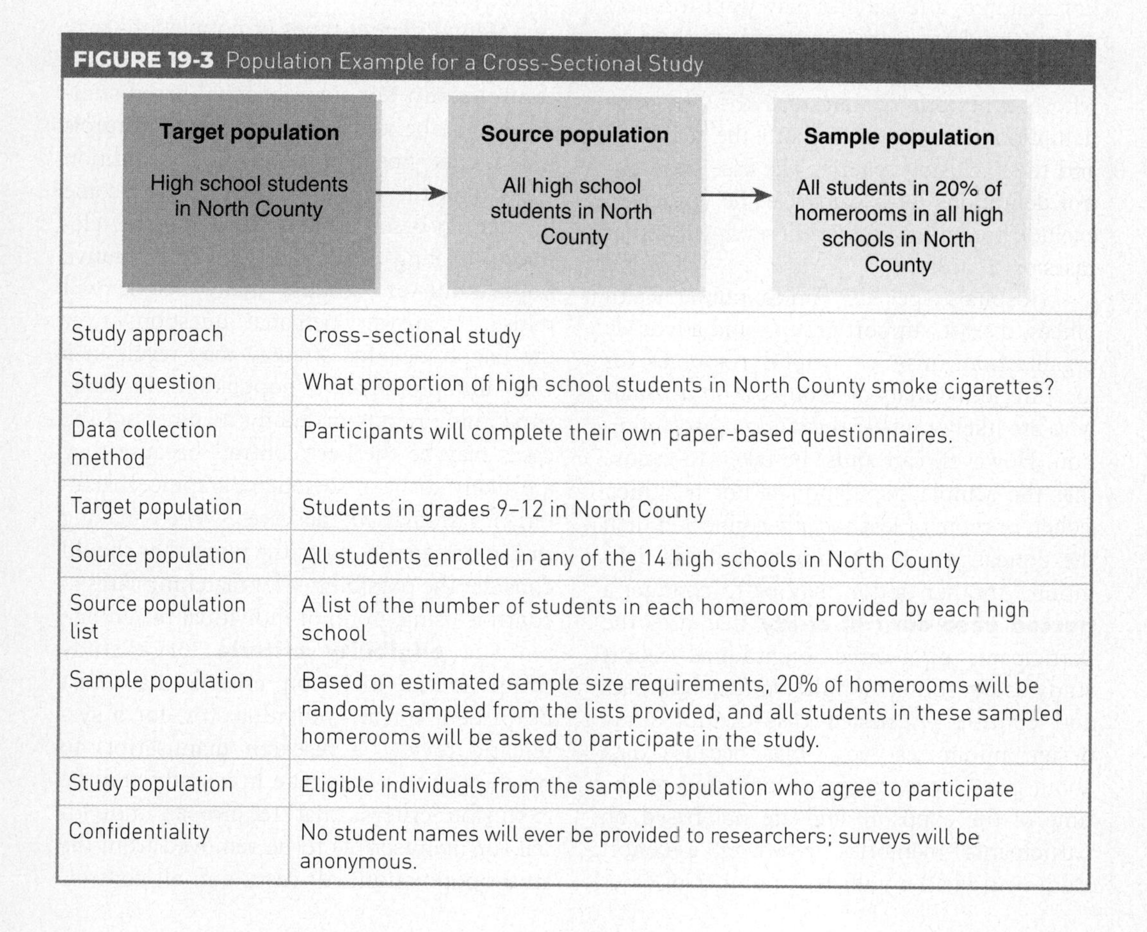
Target population
High school students in North County
Source population
All high school students in North County
Sample population
All students in 20% of homerooms in all high schools in North County

FIGURE 19-3 Population Example for a Cross-Sectional Study

Study approach	Cross-sectional study
Study question	What proportion of high school students in North County smoke cigarettes?
Data collection method	Participants will complete their own paper-based questionnaires.
Target population	Students in grades 9–12 in North County
Source population	All students enrolled in any of the 14 high schools in North County
Source population list	A list of the number of students in each homeroom provided by each high school
Sample population	Based on estimated sample size requirements, 20% of homerooms will be randomly sampled from the lists provided, and all students in these sampled homerooms will be asked to participate in the study.
Study population	Eligible individuals from the sample population who agree to participate
Confidentiality	No student names will ever be provided to researchers; surveys will be anonymous.

not represent the target population. Convenience populations are not suitable for most cross-sectional studies because they are not population-based.

19.6 Populations for Case–Control Studies

When identifying possible participants for a case–control study, the first step is to find an appropriate and available source of cases. All cases must have the same disease, disability, or other health-related condition. The study's case definition should be very clear about the characteristics that must be present and absent for an individual to be categorized as a case. For example, a researcher may want to select only candidates with advanced disease or, alternatively, may prefer to study only cases whose symptoms began recently. The case definition should specify both the inclusion and the exclusion criteria. The case and control definitions for a study can be written to exclude borderline cases from serving as either cases or controls.

Hospitals, specialty clinics, public health offices, disease support groups, and advocacy organizations may be helpful resources for locating individuals or groups of individuals who are likely to meet the study's case definition. However, care must be taken to ensure that the sample population is not healthier, sicker, or more or less socially connected than the typical person who meets the case definition. Another option may be to conduct a **nested case–control study** that uses the participants of a large longitudinal cohort study as the source population for both cases and controls. A nested case–control study design minimizes recall bias because data about past exposures were collected at the time of the exposure and are not based on participants' memories. However, a cohort study will yield a sufficient number of cases only when the disease being studied is relatively common.

Once a source of cases is identified, a valid control group must be selected. The controls must be similar to the cases in every way except for their disease status. For example, it would be inappropriate to compare cases with heart disease to controls who are marathon runners. A study examining a chronic disease should choose a control population representative of the general public, not a population that is unusually physically active. Similarly, it would be inappropriate to compare older adult women to teenage boys, or to compare big-city businessmen to men who are subsistence farmers in remote areas. All cases and controls for any one study should come from similar sociodemographic and geographic source populations.

Many different types of populations may be suitable sources of controls for a case–control study. For some hospital-based studies, it may be appropriate to use as controls individuals hospitalized with a condition other than the one being studied. For some population-based studies, random-digit telephone dialing may yield a representative population—or, because many people will refuse to answer personal questions over the telephone, this strategy may result in a very unrepresentative population. In some situations, friends or family members of the cases may be the best controls because they are likely to have sociodemographic characteristics similar to the cases. When making this important decision, the researcher should consider the possibilities for matching cases to controls using group or individual matching.

The **eligibility criteria** for a study comprise the inclusion criteria that must be present for an individual (or, for a systematic review, a research manuscript) to be allowed to participate in a study and the exclusion criteria that require an individual (or manuscript) to be removed from the study population. All cases and all controls in a case–control study must meet the same eligibility criteria, except for the ones relating to disease status. For example, a study targeting septuagenarian women should require both cases and controls to be women in their 70s who live in the same general area (**Figure 19-4**).

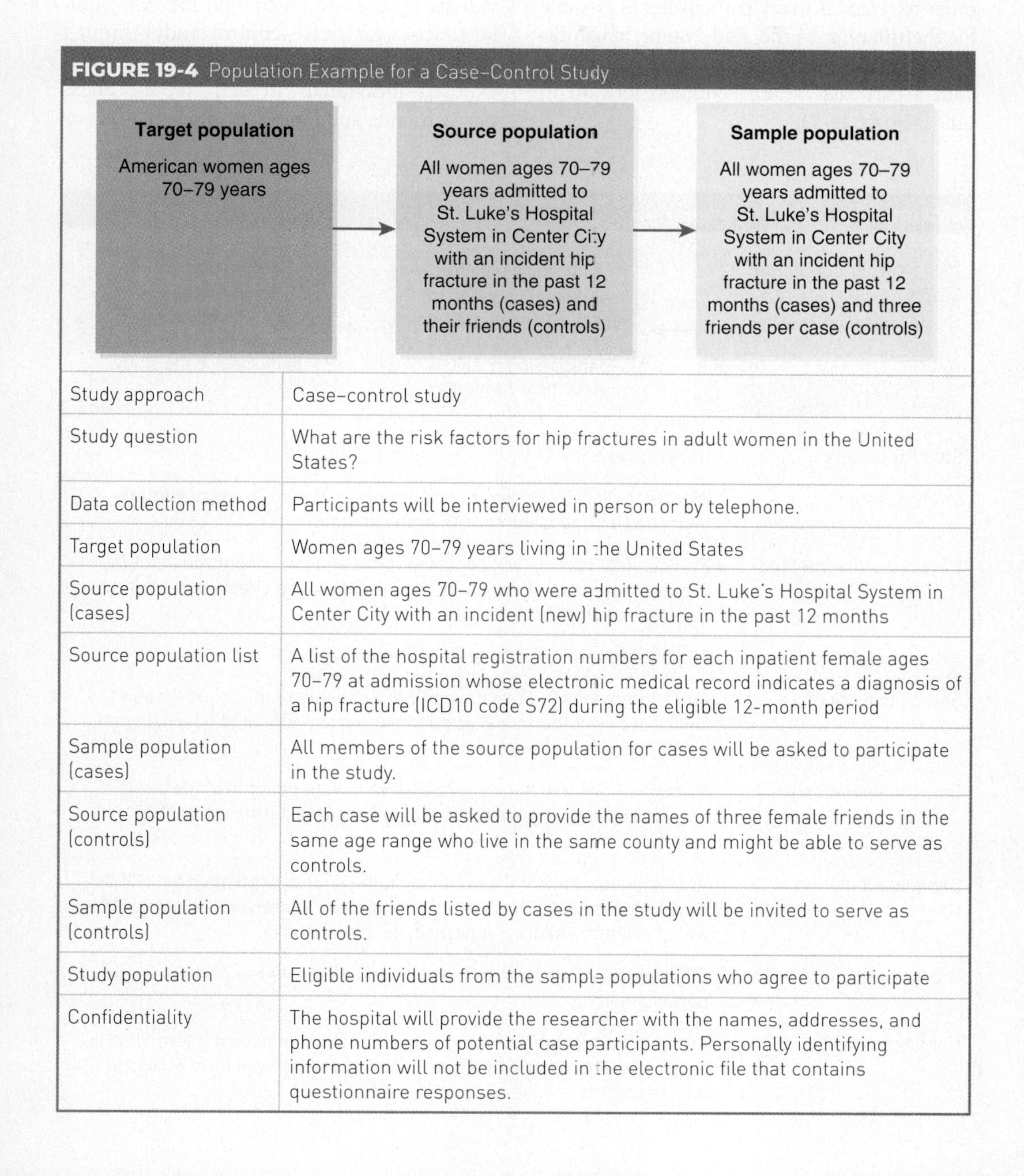
Target population
American women ages 70–79 years
Source population
All women ages 70–79 years admitted to St. Luke's Hospital System in Center City with an incident hip fracture in the past 12 months (cases) and their friends (controls)
Sample population
All women ages 70–79 years admitted to St. Luke's Hospital System in Center City with an incident hip fracture in the past 12 months (cases) and three friends per case (controls)

FIGURE 19-4 Population Example for a Case-Control Study

Study approach	Case–control study
Study question	What are the risk factors for hip fractures in adult women in the United States?
Data collection method	Participants will be interviewed in person or by telephone.
Target population	Women ages 70–79 years living in the United States
Source population (cases)	All women ages 70–79 who were admitted to St. Luke's Hospital System in Center City with an incident (new) hip fracture in the past 12 months
Source population list	A list of the hospital registration numbers for each inpatient female ages 70–79 at admission whose electronic medical record indicates a diagnosis of a hip fracture (ICD10 code S72) during the eligible 12-month period
Sample population (cases)	All members of the source population for cases will be asked to participate in the study.
Source population (controls)	Each case will be asked to provide the names of three female friends in the same age range who live in the same county and might be able to serve as controls.
Sample population (controls)	All of the friends listed by cases in the study will be invited to serve as controls.
Study population	Eligible individuals from the sample populations who agree to participate
Confidentiality	The hospital will provide the researcher with the names, addresses, and phone numbers of potential case participants. Personally identifying information will not be included in the electronic file that contains questionnaire responses.

19.7 Populations for Cohort Studies

The sampling methods used for a cohort study must align with the particular type of study design that will be applied. Some longitudinal studies follow a population-based sample of individuals forward in time. For longitudinal cohort studies, the process of identifying representative source and sample populations is similar to the process for identifying these populations for a cross-sectional study. The researchers should try to recruit a stable study population in order to retain as many participants as possible for the duration of the study. Some longitudinal studies are case series that follow individuals with a particular health condition forward in time (**Figure 19-5**).

For prospective cohort studies that seek to compare exposed and unexposed populations, identifying exposed and unexposed participants is similar to the steps for identifying cases and controls for a case–control study. For example, studies of the aftereffects of occupational exposures often recruit individuals exposed to on-the-job hazards (similar to cases for a case–control study) through employers, and the researchers then ask those exposed individuals to help recruit unexposed friends and family members to serve as

FIGURE 19-5 Population Example for a Cohort Study

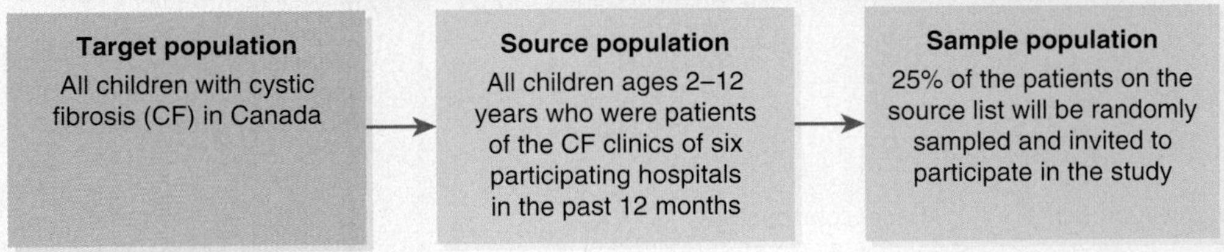

Target population
All children with cystic fibrosis (CF) in Canada

→

Source population
All children ages 2–12 years who were patients of the CF clinics of six participating hospitals in the past 12 months

→

Sample population
25% of the patients on the source list will be randomly sampled and invited to participate in the study

Study approach	Cohort study
Study question	What is the incidence rate of pneumonia and other lower respiratory infections among children with cystic fibrosis?
Data collection method	Participants' parents will be asked to log all infections throughout the 5-year prospective study period, and these will be checked against the patients' medical records.
Target population	All children with cystic fibrosis in Canada
Source population	All children ages 2–12 years who were patients of the cystic fibrosis clinics at any of the six participating university-affiliated hospitals in the past 12 months
Source population list	A list of all children ages 2–12 years who were patients of the cystic fibrosis clinics of the six participating hospitals during the past 12 months
Sample population	Approximately 25% of the patients on the source population list will be randomly sampled. The parents of sampled children will be asked if they will allow their children to participate in the study.
Study population	Eligible individuals from the sample population whose parents consent to participation
Confidentiality	All guidelines and regulations for the protection of patient information will be strictly adhered to, and only essential personnel will have access to patient records.

members of the comparison population (similar to the controls for a case–control study).

19.8 Populations for Experimental Studies

Sampling methods for experimental studies focus on the validity of the study and the safety of participants. Validity usually requires a sample population that is reasonably representative of the target population. For example, suppose that the goal of an experimental study is to test whether nutritional counseling during the first semester at a residential college prevents weight gain during the first year of college. For this intervention study to be valid, the researcher needs to recruit a reasonable cross-section of the first-year student population (**Figure 19-6**). If the researcher recruited students majoring in nutrition, asked for volunteers, or sampled from among student athletes, the study population would likely be much more concerned about weight than the average first-year student. Members of all three of these populations—nutrition majors, volunteers for a wellness study, and varsity athletes—are likely more attuned to their nutritional status than is typical for a first-year college student. These individuals are therefore less likely than the typical student to

FIGURE 19-6 Population Example for an Experimental Study

Target population	→	Source population	→	Sample population
First-year students attending primarily residential colleges	→	All students enrolled in mandatory first-year seminar courses at East State College	→	A randomly selected sample of students from the list of enrolled first-year students

Study approach	Experimental study
Study question	Does nutritional counseling during the first semester of college prevent weight gain?
Data collection method	Half of the participants will be assigned to meet weekly with a nutritionist during their first semester, and half will have no intervention. All participants will complete nutritional assessments during the first and last weeks of the fall and spring semesters of their first year at college.
Target population	First-year students at primarily residential colleges
Source population	All first-year students at East State College
Source population list	A list of all students enrolled in the mandatory first-year seminar class at East State College
Sample population	A randomly selected sample of students from the source population
Study population	Eligible individuals from the sample population who agree to participate
Confidentiality	Nutritional counseling and assessment sessions will be conducted in a private setting, and only essential personnel will have access to participants' records. Participation in the study will be voluntary, and professors teaching first-year seminars will not know which students have enrolled in the study.

gain significant weight during the observation period. Minimal changes in weight would be observed during the year both in the intervention group receiving nutritional counseling and in the control group not being offered dietary advice as part of the study. Because there would be no significant difference in weight gain in the intervention and control groups, the intervention would be deemed unsuccessful. If the researchers recruited a more representative study population, the intervention might be more likely to be deemed a success.

Some experimental studies require participants to be exposed to potentially risky substances or activities. In such studies, the risk of harm can be reduced by selecting an appropriate source population and defining strict inclusion and exclusion criteria. For example, studies that involve exercise must target potential participants who are likely to be healthy enough to engage in physical activity. Studies of new drugs for advanced forms of cancer are often open only to extremely ill patients for whom standard therapies have not been effective. Safety should always be the top priority when designing and implementing an experimental study. Precautions must be taken to protect participants from injury. For example, if all volunteers for an experiment will be injected with a solution, the researchers must ensure that potential participants have no known allergies to any of the ingredients in either the experimental substance or the placebo. An allergy to any ingredient must be listed as one of the exclusion criteria.

19.9 Sampling for Qualitative Studies

Qualitative data collection is not a detached, structured process based on a random sample of individuals. Instead, researchers typically have intense contact with a selected group of informants. **Key informants** are individuals selected to participate in a qualitative study because they have expertise relevant to the study question. **Purposive sampling** is a nonprobability-based sampling method that recruits participants for a qualitative study based on the special insights they can provide. For example, focus group participants are typically recruited through purposive sampling because they are able to provide valuable perceptions about the study question. Research projects using focus groups often hold discussions with several different sets of informants. Membership in each group is designed to minimize power differentials and allow for the open sharing of perspectives. A study of workplace safety issues might include separate groups for factory workers who use heavy equipment, their supervisors, and safety engineers at the factory. If supervisors and safety engineers were included in the discussions with the floor workers, it might be difficult for the people operating the machinery to be honest about their concerns. They might feel pressured to provide the answers they thought the company wanted to hear rather than acknowledging the need to address hazards.

Qualitative studies often do not start with a set number of participants who are supposed to be recruited. Instead the goal is to reach **data saturation**, a time in the research process in which no new information about a particular theory is emerging from additional data collection because variations across population members have already been captured. Some studies with homogeneous populations might reach saturation after 15 or 20 interviews. Some studies might require larger numbers of participants, especially if they include members with a diversity of perspectives.

19.10 Vulnerable Populations

Vulnerable populations are populations whose members might have limited ability to make an autonomous decision about volunteering to participate in a research study. Potentially vulnerable populations in health research include young children, some individuals with serious health issues, people in prison and some other socially marginalized populations, and others who might have limited ability to make an independent decision about volunteering to participate in a research study. These populations should not be selected as the source population for studies that do not require their participation.

At the same time, it is problematic when members of vulnerable populations are systematically excluded from research, since the only way to study health issues of special importance to potentially vulnerable populations is to allow members of those populations to participate in relevant research studies. Pregnant women and children must be included in tests of the safety of pharmaceutical agents before those drugs can be approved for wider use. New therapies for life-threatening diseases must be tested in people with advanced illnesses. Individuals with severe mental health disorders must be included in studies of psychiatric diseases. The critical health concerns of prisoners can be identified only by conducting research in prisons. The impact of interpersonal violence on health can be understood only by asking survivors about their experiences.

Research studies including members of vulnerable populations require extra consideration of the potential risks of research to participants. The study must be sufficiently important to justify gathering new data from members of a vulnerable population. The study must allow every participant (or, for young children and those with significantly diminished cognitive abilities, a legally recognized representative) to provide informed consent free from coercion. Further, it must address concerns about the increased risks of adverse effects from study participation. For example, people with fragile health may have an elevated risk of injury from physical tests, and people with histories of abuse or mental illness may have a heightened risk of psychological damage from answering questions about sensitive topics. Institutional review boards can assist researchers with the design of ethical and culturally appropriate research protocols.

19.11 Community Involvement

Some studies benefit from or require the participation and support of geographic, cultural, educational, religious, or social communities and their leaders. A cross-sectional survey that will collect data from students may require the permission of school authorities in addition to the consent of the students and/or their parents and the approval of a research ethics committee. A clinical study that seeks to enroll participants with a rare disease may benefit greatly by partnering with an active disease support and advocacy network. A longitudinal study that intends to recruit and monitor whole villages will be most successful if formal and informal community leaders and other local representatives are actively involved in planning, recruitment, and retention. These connections with community representatives should be established early in the research planning process and maintained throughout the data collection and dissemination period. The approval of community leaders does not negate the requirement to obtain individual informed consent from participants whose individual data will be collected. However, community buy-in for a project often facilitates access to source populations and improves participation rates in addition to enhancing the cultural appropriateness of the research protocol and ensuring that the study's outcomes are valuable to the community.

CHAPTER 20

Sample Size and Power

An adequate number of study participants is required to achieve valid and significant results.

20.1 Importance of Sample Size

When determining how many participants are needed for a quantitative or qualitative study to be meaningful, the goal is to recruit just the right number of participants—not too many and not too few. Resources are wasted when a study recruits more participants than it needs. When study materials, tests, or other implementation activities have a per-participant cost, researchers do not want to spend money on excess participants. They also do not want to waste the time of the research team or study participants by continuing to collect new data after they already have a sufficient number of participants. Resources are also wasted when a study recruits too few participants. If a primary research project does not collect sufficient data to answer the study question, then all of the time and money invested in the project will have been wasted. Most researchers do not have the luxury of worrying about a surplus of participants, but many struggle to recruit a sufficient study population. A shortage can make getting statistically significant results almost impossible.

20.2 Sample Size and Certainty Levels

In statistics, **sample size** is the number of observations in a data set. In the health sciences, the sample size is usually the number of individual humans in the study population. The desired sample size for a quantitative study is based on statistical estimations about how many data points are required in order to answer the study question with a specified level of certainty.

Large samples from a population are usually better than small ones at yielding a sample mean close to the true population value. For example, suppose that the mean age in a population consisting of 20 people is 39 years (**Figure 20-1**). If a sample of only 3 people is taken (15% of the total population), the sample mean might be close to the population mean of 39 years but there is some possibility that the sample mean will be quite distant from the population mean. If a larger sample of 8 people (40% of the total population) is selected from the population, then the sample mean is likely to be fairly close to the population mean.

FIGURE 20-1 Sample Size and Means

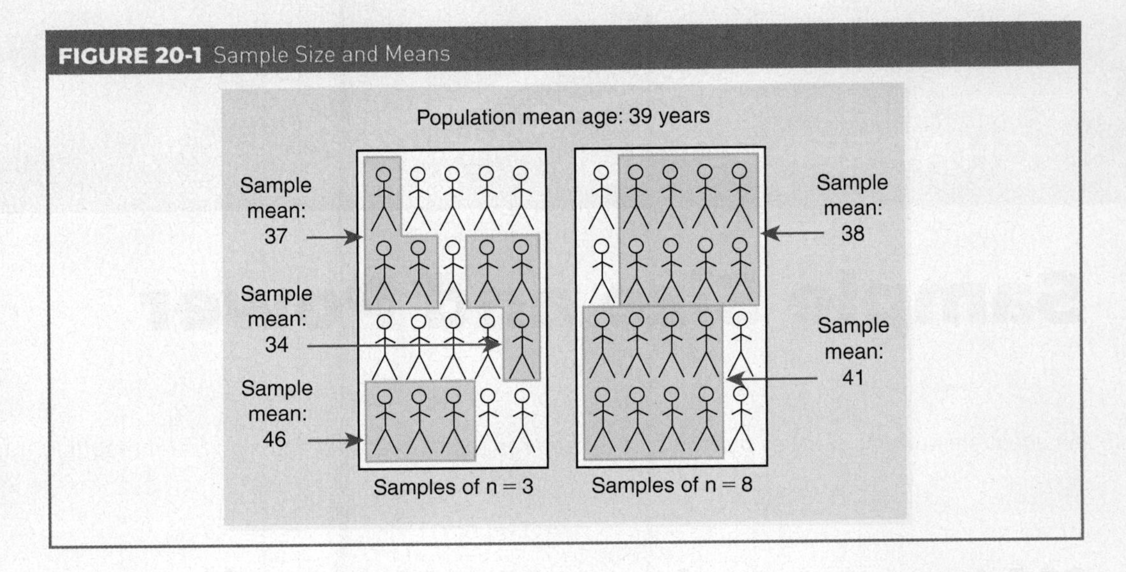

A **confidence interval** (CI) is a statistical estimate of the range of likely values of a parameter in a source population based on the value of that statistic in a study population. A narrow CI indicates more certainty about the value of the statistic than a wide CI. When the sample size is small, the sample mean might be quite far from the mean in the total population. This is represented by a wide CI that reaches far to the left and right of the sample mean. When a greater proportion of the total population is sampled for inclusion in the estimation of the mean age, the CI for the mean age will be narrower because there is greater certainty about the sample mean being close to the population mean.

Figure 20-2 shows an alternative display of the sample means from Figure 20-1 that combines each sample's mean age with its 95% CI. The dots at the center of each of

FIGURE 20-2 Sample Size and Confidence Intervals

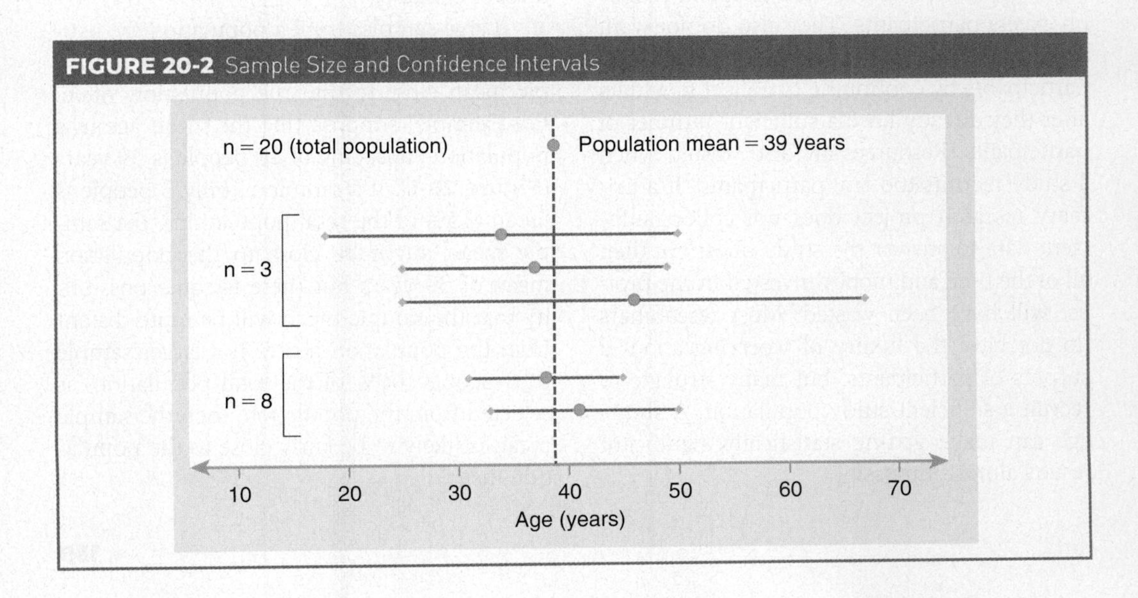

the five lines in Figure 20-2 represent the five sample means from Figure 20-1. Those are the point estimates of the mean that were calculated for each sample. The 95% CI for each sample population is represented by the lines extending from the sample means. The 95% CI is calculated for each sample based on the number of individuals included in the sample, the mean age of those individuals, and the standard deviation of their ages, which is a measure of how far apart the ages of the individuals in the sample are. If all 20 people in the total population are included in the analysis, no CI is required because the population mean age will be known exactly.

For the top line in Figure 20-2, the sample mean is 34 years, and the CI stretches from 18 to 50 years. Based on this sample, a researcher can be 95% confident that the mean age in the total population is somewhere between 18 and 50 years. If hundreds of random samples of 3 individuals are drawn from the total population of 20, about 95% of those samples will have a 95% CI that includes the true population mean of 39 years. About 5% of the time, the random sample of 3 individuals will include an unusually young or unusually old set of individuals, and the CI will not overlap the population mean. The incorrect conclusion that would be made about the population as a whole if the sample happens to be extremely young or old is a random error.

Larger sample sizes generally result in sample means that are close to the population mean. Large data sets have CIs that are narrower than the CIs generated by smaller sample sizes. If a sample of 18 of the 20 members of the population illustrated in Figure 20-2 is drawn, the sample mean age will be very close to 39 years and the CI will be so narrow that it will hardly extend beyond the dot representing the sample mean. More generally, larger sample sizes make it more likely that a study will yield statistically significant results.

20.3 Sample Size Estimation

A **sample size calculator** is a tool used to identify an appropriate number of participants to recruit for a quantitative study. A sample size calculator is more accurately called a sample size estimator, because the range of suggested sample sizes is based on a series of assumptions about the expected characteristics of the sample population. Sample size calculators are available online, often at no cost, and are bundled with most statistical software programs.

Figure 20-3 shows examples of the kinds of inputs that must be provided for various study approaches in order to get a rough estimate of the required number of participants. A best guess must be used for most of these inputs because accurate information will not be available until after the study has been completed. The values for inputs can be informed by previous studies, but those prior publications will likely not allow a researcher to be certain about whether, for example, the proportion of cases with the exposure of interest will be the same 25% mentioned in the article or will instead be 20% or 30%. Trying a variety of values for the input variables in a sample size calculator will show that even slight changes in input values may result in a considerable difference in the estimated sample size required (**Figure 20-4**). When the level of certainty about inputs is low, it is better to err on the side of a larger sample size.

20.4 Type 1 and Type 2 Errors

Population-based studies aim to have study populations that are representative of their source populations. Sometimes the study population does not capture the true experience of the source population, either because of random chance or because of a study design flaw, such as a sample size that is too small.

FIGURE 20-3 Examples of Sample Size Calculation

Characteristic	Cross-Sectional Study	Case–Control Study	Cohort Study	Experimental Study
Study question	What proportion of the population has the exposure or disease?	Are cases more likely than controls to have the exposure?	Are exposed people more likely than unexposed people to develop the outcome?	Are exposed people more likely than unexposed people to have a favorable outcome?
Population size	5000	—	—	—
Anticipated percentage with exposure or disease	15%	—	—	—
Confidence for anticipated exposure percentage	±3%	—	—	—
Ratio of controls to cases	—	2	—	—
Ratio of unexposed to exposed	—	—	1	1
Anticipated percentage of controls exposed	—	25%	—	—
Anticipated percentage of unexposed with disease or outcome	—	—	10%	70%
Odds ratio worth detecting	—	1.5	—	—
Rate ratio worth detecting	—	—	1.3	1.25
Confidence level (1 – α)	95%	95%	95%	95%
Power (1 – β)	—	80%	80%	80%
Estimated sample size	~500	~350 cases and 700 controls	~1850 exposed and 1850 unexposed	~90 exposed and 90 unexposed

In statistics, an **error** is a difference between the value obtained from a study population and the true value in the larger population from which the study participants were drawn that occurs by chance rather than as a result of systematic bias. Bias is a flaw in the way a study was designed or conducted that leads to an inaccurate result. Errors, by contrast, occur randomly. Even the most rigorous study protocol might by chance generate a study population that has an extreme distribution, such as one that is extremely young or extremely old.

A **type 1 error** (also written as a type I error) occurs when a study population yields a statistically significant test result even though a significant difference or association does not actually exist in the source population (**Figure 20-5**). For example, sometimes a study

FIGURE 20-4 Sample Size Estimates for a Case-Control Study

Situation	Ratio of Controls to Cases	Anticipated Percentage of Controls Exposed	Anticipated Percentage of Cases Exposed	Odds Ratio (OR) Worth Detecting	Estimated Sample Size Required	Estimated Number of Cases Required	Estimated Number of Controls Required
Base case (Figure 20-3)	2	25%	33%	1.5	1050	350	700
1:1 ratio	1	25%	33%	1.5	950	475	475
10:1 ratio	10	25%	33%	1.5	2750	250	2500
Lower % exposed	2	10%	14%	1.5	2025	675	1350
Higher % exposed	2	30%	39%	1.5	976	325	650
Lower OR	2	25%	29%	1.2	5400	1800	3600
Higher OR	2	25%	40%	2	345	115	230

population sampled from a larger source population in which there is no association between sex and a particular adverse health outcome happens, by chance, to have an unusually large difference in the proportion of males and females who have that disease of interest. The probability of a type 1 error is often noted by the Greek letter **alpha** (α). Most studies use α= 5% as the value for statistical significance, which corresponds to statistical tests using a 95% CI and a *p* value of *p* = .05. While most statistical tests conducted with sample data can be assumed to represent true characteristics of the source population, some test results will be inaccurate due to random sampling error. When α is set at 5%, about 1 in 20 statistical tests will result in a type 1 error.

FIGURE 20-5 Type 1 and Type 2 Errors

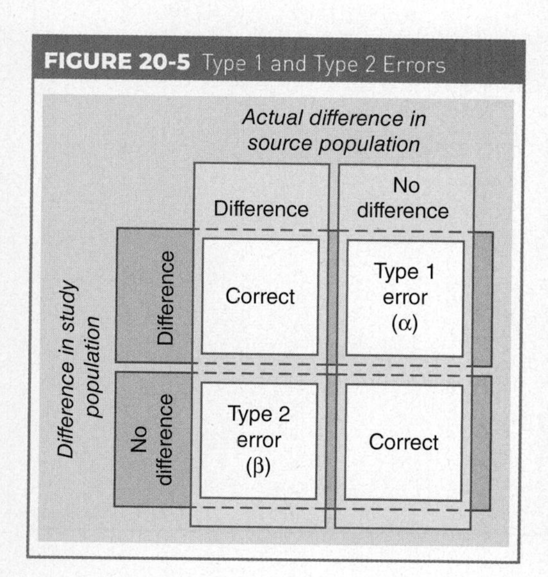

A **type 2 error** (or type II error) occurs when a statistical test of data from a study population finds no significant result even though a significant difference or association actually exists in the source population. The probability of a type 2 error is often referred to using the Greek letter **beta** (β). The best way to minimize the likelihood of type 2 errors is to have a large sample size.

20.5 Power Estimation

In statistics, **power** is the ability of a test to detect significant differences in a population when differences really do exist. Power is defined as 1 – β, so a 20% likelihood of a type 2 error (that is, β = 20%) corresponds to a power of 80%. The standard expectation is that the statistical tests that will be used to answer a study's main research question should have a power of 80% or greater.

The power of statistical tests is increased when the number of participants included in the analysis is large. When two means or proportions are close to one another, very large sample sizes are required to detect a significant difference between those proportions. When an incidence rate ratio or an odds ratio has a point estimate close to 1, a very large sample size will be required to yield a CI that is statistically significant and does not overlap 1.

Studies with too few participants lack adequate power to detect meaningful differences or associations in source populations. A study that recruits a small number of participants will not generate statistically significant test results even when there really is a substantial difference in the mean survival duration of males and females in a case series or when an exposure in a cohort study is truly a risk factor for the disease outcome of interest in the source population.

One of the ways to check whether a particular number of participants will be sufficient for a research study is to work backward from the number of participants likely to be recruited to see whether a study population of that size will provide adequate statistical power for the study design. Examples of power estimation for various study approaches are shown in **Figure 20-6**. Like sample size estimates, power estimates require best guesses about the expected findings of the study. If the power estimates for a study are lower than desired, the easiest way to improve the power is to increase the sample size.

FIGURE 20-6 Examples of Power Calculation

Characteristic	Cross-Sectional Study	Case-Control Study	Cohort Study	Experimental Study
Number of exposed	100	—	2500	70
Number of unexposed	250	—	1500	70
Number of cases	—	250	—	—
Number of controls	—	490	—	—
Percentage of exposed with disease or outcome	40%	—	13%	85.70%
Percentage of unexposed with disease or outcome	26%	—	9%	64.30%
Percentage of cases with exposure	—	32%	—	—
Percentage of controls with exposure	—	25.50%	—	—
Confidence level (1 – α)	95%	95%	95%	95%
Estimated power (1 –β)	~70%	~45%	~97%	~80%

20.6 Refining the Study Approach

Initial sample size estimates should be generated early in the study design process so that researchers can determine whether they are likely to be able to recruit a sufficient number of participants. Researchers must be prepared to rethink the study approach if their estimated number of available participants will not yield sufficient power. For example, the intended study design will not work if a researcher expects to be able to recruit about 300 participants but the sample size estimates suggest that 870 participants will be required. In this situation, the study question, the study approach, and/or the source population may need to be modified. A new plan must be crafted that is suitable for the number of participants that the researcher can reasonably expect to recruit.

The sample size estimates generated by sample size and power calculators refer to the study population (the actual number of participants), not the sample population (the number of individuals invited to participate in the study). The number of people sampled for a study needs to be larger than the required number of participants, because the participation rate is unlikely to be 100%. Suppose that a specialty clinic that has hosted several case–control studies has determined that about half of the patients invited to participate in research studies actually volunteer for them. If a protocol for a new case–control study requires 220 cases to participate in order to have adequate statistical power, the researchers need to be sure that at least 440 cases can be contacted through the clinic. If the clinic does not have an adequately high volume of patients, the study design needs to be revised or other clinics need to be invited to help with recruiting for the study.

© DmitriyRaznkov/Shutterstock

CHAPTER 21

Questionnaire Development

A questionnaire is a tool for systematically gathering data from study participants.

21.1 Questionnaire Design Overview

A **questionnaire**, or **survey instrument**, is a series of questions used as a tool for systematically gathering data from study participants. A good questionnaire is carefully crafted for a specific purpose. Questionnaire design usually works best when it starts with the identification of the general and specific content to be covered by the survey instrument and progresses to choosing the types of questions and responses that will be most appropriate for each topic (**Figure 21-1**). Most research studies must design new data collection instruments, but for some research topics, validated question banks are available and questions can be selected from them. Once a survey instrument is drafted, the wording of each question and its associated response items, if applicable, should be checked carefully. The sections and the questions within each section should be in logical order. The formatting of the document or the computerized data entry form should be visually appealing and easy to read. After initial design, the survey instrument should be pretested and revised to improve content and ease of use prior to launching the data collection phase of the study. This chapter provides details about each of these steps in questionnaire development. Specialty references can provide additional information about designing valid and useful questionnaires for particular research questions and study types.

FIGURE 21-1 Questionnaire Design Plan

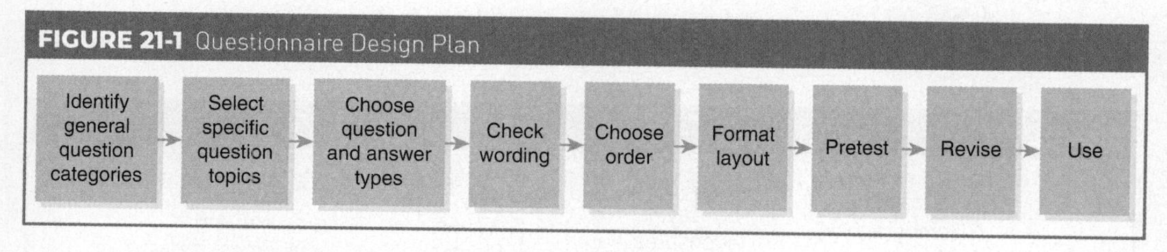

21.2 Questionnaire Content

The first step in designing a questionnaire is to list the topics that the survey instrument must cover. The first set of questionnaire items typically enables the researcher to confirm that participants meet the eligibility criteria for the study. For example, if only currently registered students are eligible to participate in a university-based cross-sectional survey, one of the first questions should be about school enrollment status. Participants who report that they are not students must be excluded from analysis based on that answer. The remaining questions ask about all of the exposure, disease, and population (demographic) areas that are the focus of the study question (**Figure 21-2**).

Several questions may be required in order to accurately assign participants to key exposure and disease categories. For example, in case–control studies, researchers need to ask questions that allow them to confirm that all cases meet the case definition and that all controls meet the control definition. When the case or control definition lists multiple inclusion and exclusion criteria, each component of the definition may be evaluated using a separate question. Prospective cohort studies examining rates of incident disease require a series of questions about both exposure status and disease status. The answers to these questions must allow each participant to be classified as exposed or unexposed, and they must provide evidence that no participant had the disease of interest at the start of the observation period. This may also require a series of questions about various aspects of exposure and disease histories.

A thorough search of the literature for studies related to the research question will help with the identification of the range of additional topics that should be included in the questionnaire. For example, a study about possible risk factors for breast cancer might have sections on:

- Sociodemographics (such as age, ethnicity, education level, and income)
- Family health history
- Personal health history (such as previous diagnoses of benign breast diseases and the date of the last screening mammogram)
- Reproductive history (including questions about menstrual cycle characteristics, pregnancies, and use of hormones)
- Lifestyle factors (such as alcohol use, exercise history, and working night shifts)

Systems thinking is the process of identifying the underlying causes of complex problems so that sustainable solutions can be developed and implemented. The survey instrument for a study examining complicated causal pathways should include questions about the various factors that might influence the relationships between key exposures and outcomes. For example, adults who smoke tobacco products may be more likely than other adults to consume large volumes of alcohol. In a study of the relationship between smoking and liver disease, alcohol consumption could be a potential confounder. Because smokers are more likely than nonsmokers to drink, tobacco users may appear to be at a greater risk of liver disease than nonsmokers, even if the

FIGURE 21-2 Question Areas

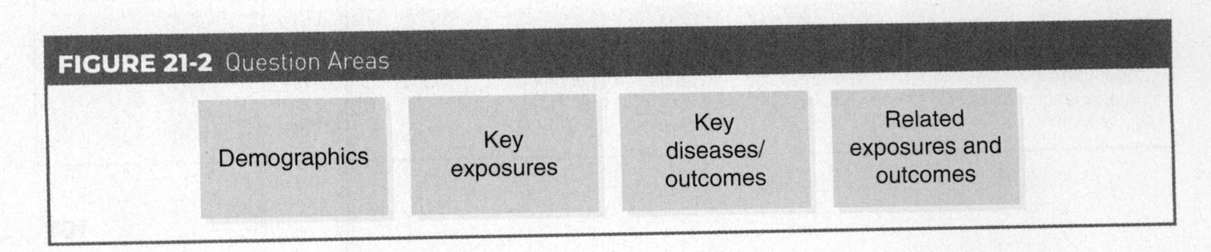

rate of liver disease is the same in smokers who drink as it is in nonsmokers who drink. Asking questions about both tobacco use and alcohol use enables the researcher to statistically adjust for different levels of alcohol use by smokers and nonsmokers and thus to more accurately examine the possible relationship between smoking and liver disease.

Theoretical frameworks can inform the diversity of questions that may be relevant to include in the survey instrument. For example, the infectious disease epidemiology triad provides a framework for exploring the *a*gent, *h*ost, and *e*nvironment characteristics (the "AHEs") that contribute to the spread of infectious diseases through populations. (Studies of infections transmitted by insects or arachnids add the *v*ector as a fourth component.) An **agent** is a pathogen or a chemical or physical cause of disease or injury. Infectious agents differ in how contagious, pathogenic, or virulent they are and whether they are drug resistant. Drugs and other chemical agents vary in how toxic or addictive they are. A **host** is a human (or animal) who is susceptible to an infection or another type of disease or injury. Host factors describe the intrinsic characteristics that influence an individual's vulnerability to an agent. People may have different risks of exposure, illness, and adverse outcomes as a result of age, genetics, immunology, comorbidities, nutritional status, health behaviors, psychological factors, and other characteristics. The **environment** includes the external factors that facilitate or inhibit health. This construct may refer to physical characteristics of the natural and built environment, such as water quality, air quality, climate, extreme weather events, noisiness, occupational hazards, or proximity to recreational areas, major roads, and various types of healthcare services. The environment can also refer to the social and political context in which health and disease occur. Researchers seeking to understand the causes of an epidemic should gather data about all of the AHE components they think might have contributed to the event. While some agent characteristics must be determined through laboratory testing, data about many of the host and environmental factors can be collected from patients and controls using a survey instrument.

After creating a list of all possible questions, it may be necessary to remove some of the less critical items so that the length of the survey will be manageable. A survey that is too long may yield a low response rate. A survey that is too short will miss potentially crucial data. It is usually better to err on the side of asking more questions rather than fewer questions.

21.3 Types of Questions

After determining the broad categories of questions and the specific topics to be addressed in each section, the next step is to decide which types of questions are most appropriate. Each survey item should be assigned a specific question type, such as a date question or a yes/no question. Examples of various question types are shown in **Figure 21-3**. The questions that are asked will determine the types of statistical tests that can be used to analyze the collected data. For example, the tests used with numeric data (such as *t* tests, ANOVA, and linear regression) are different from the tests used with categorical data (such as chi-square tests and logistic regression). A consultation with a statistician early in the planning phase of a study can ensure that the data analysis plan is strong and valid.

Closed-ended questions (also called closed questions or close-ended questions) allow a limited number of possible responses. **Open-ended questions**, also called **free-response questions**, allow an unlimited number of possible responses. Closed-ended questions are usually easier to statistically

FIGURE 21-3 Examples of Types of Questions

Type	Sample Question	Sample Response Option for the Sample Question
Date	What is your birth date?	__ __ - __ __ - __ __ __ __ d d - m m - y y y y
Numeric	What is your height without shoes (rounded to the nearest half inch)?	__ __ . __ inches
Yes/no	During your lifetime, have you smoked more than 100 cigarettes?	❐ Yes ❐ No
Categorical/ multiple choice: nominal (no rank)	What is your sex?	❐ Female ❐ Male
	What is your favorite type of film?	❐ Action / Drama ❐ Comedy / Romance ❐ Documentary ❐ Other: ______________
Categorical/ multiple choice: ordinal (ranked)	What is the highest level of education you have completed?	❐ Less than high school ❐ High school ❐ Some college but no degree ❐ College/university degree or advanced degree
	How much do you agree with this statement?: "No matter how much I exercise, I will not be able to lose weight."	❐ Strongly disagree ❐ Disagree ❐ Neutral ❐ Agree ❐ Strongly agree
	On a scale of 1 to 5, with 1 meaning poor and 5 meaning excellent, how would you rate your hearing (without the use of a hearing aid)?	Poor Excellent 1 2 3 4 5 ❐ ❐ ❐ ❐ ❐
Paired comparisons	Do you prefer to drink coffee or tea?	❐ I prefer coffee. ❐ I prefer tea. ❐ I like coffee and tea equally. ❐ I do not drink coffee or tea.
Rank ordering	List the following four political issues in order from most important to you (1) to least important to you (4).	Number from 1 (most important) to 4 (least important): ___ Crime/safety ___ Environment/energy ___ Foreign policy/defense ___ Taxes/revenue
Open-ended/ free response	What is your biggest personal health concern at present?	______________ ______________

analyze than open-ended questions. The main limitation of closed-ended questions is that they may force respondents to select answers that do not truly express their status or opinions. If a question asks whether someone prefers coffee or tea, and "coffee" and "tea" are the only two responses options, there is no way for study participants to report liking them equally, disliking them equally, or having no opinion. Open-ended questions allow participants to explain their selections and qualify their responses, to give multiple answers, and to provide responses not anticipated by the researchers. However, open-ended questions take longer to ask and answer, and they may result in irrelevant responses. Coding free-response answers into objective and meaningful categories for analysis is often a time-consuming and imprecise process. Open-ended questions on a survey form are often most useful when they capture initial impressions or clarify responses to closed-ended questions.

Closed-ended questions come in a variety of formats, including date and time questions (which can be used to calculate the duration of time between events), numeric questions, and categorical questions. Categorical questions can have as few as two response options (called dichotomous responses), like just having "yes" and "no" responses, or they can have dozens of possible answers. Categorical variables can also be ranked (ordinal) or unordered (nominal). Ordinal responses have an inherent order, and nominal responses do not have any built-in order. For example, a question about educational level is ordinal because some levels of education involve more years of school than other levels do. A question about occupational category is nominal because there is no obvious way to rank occupations as diverse as plumbing, farming, teaching, nursing, sales, and law. Less commonly used question types include paired comparisons and rank ordering. (Section 29.2 provides more information about variable types.)

21.4 Types of Responses

For closed-ended questionnaire items, the researcher must decide what types of response options are appropriate. The responses should be ones that participants can record accurately and completely.

For numeric responses, the question should state exactly how specific the answers should be. Should height be reported to the nearest inch, to the nearest half-inch, or to the nearest centimeter? Should height be reported in feet and inches (like "5 feet 6 inches"), or should it be reported only in total inches (as "66 inches")? This decision should accommodate the likely preference of participants. If participants may struggle to convert their heights to total inches, then the response option should be the "feet and inches" one rather than "total inches."

For categorical questions, researchers should consider all possible responses and determine how many response options are needed to allow all participants to answer each question. Four response options might be enough to capture all possible levels of education, but ten response options might be insufficient for capturing occupational areas. After drafting a list of response options for a question, the researcher should confirm that every possible answer can be coded. For example, an ordinal question should not allow response of only "less than half" and "more than half," because that would not allow a response of "exactly half" to be recorded. For nominal questions, it may be helpful to include an "Other" category that allows respondents to fill in their own answers if none of the listed responses is applicable. (Typically this means asking two questions rather than one. If the response to the main closed-ended question is "Other," the participant is asked to record a response to a second question that is open-ended.) When multiple responses to a single categorical question might be true for a participant, it may be better to ask a series of yes/no questions rather than asking respondents to "check all that apply."

For ranked questions, the researcher chooses how many entries to include on the scale and whether there will be a neutral option. A **Likert scale** presents ordered responses to a questionnaire item that asks participants to rank preferences numerically, such as by using a scale for which 1 indicates strong disagreement and 5 indicates strong agreement. Most scales with a neutral option list 5 to 7 categories. Most scales without a neutral option list 4 or 6 categories. Sample response scales are shown in **Figure 21-4**.

For self-report survey items, the researcher must decide whether to add a category for "not applicable" or "I do not know." Some questions must be answered by all participants because they are essential for determining eligibility for the study. For example, all participants in a study comparing adults with and without tattoos must answer a question about their tattoo history. Most people can answer that question very easily, so an "I do not know" response is not required. Anyone who skips that question when completing the survey form will be deemed ineligible for inclusion in the analysis.

However, for a question like "When was the last time your blood sugar levels were tested?" it may be important to know whether the respondent is uncertain about the answer. Uncertainty about if or when blood sugar testing has occurred might be a valid and important response. Neglecting to list "I do not know" or "I do not remember" as possible responses would preclude respondents from revealing important information. If participants record an answer they are not sure about because they feel they must choose one of the response options, those unreliable responses might cause systematic inaccuracies in the data. For example, information bias may occur if participants who do not know the answer to this question systematically default to providing the answer they assume the researcher wants to hear, such as reporting that they have had their blood sugar checked within the past year even if they are not sure that this test occurred at their last annual checkup. Allowing participants to record responses like "no opinion," "not sure," "hard to say," "no answer," "I prefer not to answer," "I do not understand," and "I forget" may increase the percentage of participants who complete a questionnaire rather than quitting mid-survey. If interviewers will read survey items to participants and record participants' responses, a "refused to answer" category is needed.

FIGURE 21-4 Examples of Responses for Ranked Questions

Strongly Disagree	Disagree	Neutral	Agree	Strongly Agree
Dissatisfied	Somewhat dissatisfied	Neutral	Somewhat satisfied	Satisfied
Very negative	Somewhat negative	Neither negative nor positive	Somewhat positive	Very positive
Poor	Fair	Good	Very good	Excellent
None	Few	Some	Many	Very many
Never	Rarely	Sometimes	Often	Always
Not important	Slightly important	Somewhat important	Very important	Extremely important

21.5 Anonymity

For some study topics, researchers must ensure that the responses given by participants do not reveal their identities to researchers or to others who might access study materials. Clients of a community organization serving people with addictions should not be asked to complete a questionnaire about illegal activities if there is any risk that their identities could be ascertained from their response forms. Community members sampled for participation in a telephone-based opinion survey might not feel comfortable expressing candid perspectives about healthcare policies when they are aware that the researchers know their telephone numbers and addresses and can therefore determine their names and identities. Patients completing a satisfaction survey might not report genuine concerns about their physical therapists if they fear that their comments will allow their identities to be unmasked and their comments reported to their care providers. **Anonymity** is the inability of a participant's identity to be discerned from his or her responses to a survey instrument or records in a database. Anonymity protects participants and allows them to provide honest answers to sensitive questions.

Many questions can be asked in more than one valid way. The researcher must decide which question type and which level of precision for responses is most appropriate for the study goal and the study population. For example, participants' ages can be ascertained in several ways. The researcher could ask for the date of birth so that the number of years between the birthdate and the interview or survey date can be calculated. However, asking for such specific personal information may raise concerns about anonymity because birthdates could be personal identifiers in a small population. Additionally, the fear of providing that individually identifying information may mean that many participants will skip the question about age entirely. Some may even drop out of the study rather than provide a birthdate. To protect participants and reduce fears about anonymity, the researcher could ask for each participant's current age in years rather than asking for a birthdate. Even then, age in years could reveal the identity of some individuals in a small study population. In surveys of residential college students, for example, most participants will fall into a narrow age range, so younger and older students might be identifiable if they provide their age as a whole number. In such situations, a categorical question with responses of < 21 years and ≥ 21 years will allow all participants to remain anonymous.

Similar decisions must be made about each component of the questionnaire that could allow for the ascertainment of the identities of participants in a survey intended to be anonymous. If participants could be harmed by being identified as a member of a study population or by having their responses linked to their identities, or if identifiability might limit the truthfulness of survey responses, then there must be a rigorous and transparent plan in place for protecting the privacy of all participants and the confidentiality of the information they share.

21.6 Wording of Questions

Figure 21-5 lists examples of problems with the wording of questions that should be avoided, including ones related to language, content, and responses. After drafting the questionnaire, check each question for clarity and confirm that the responses are also carefully worded. For example:

- Does each question ask what it is intended to ask?
- Is the language of each question clear and neutral?
- Will members of the study population understand the language?
- Do questions about sensitive topics use language acceptable to the source population?
- Are the response options clearly presented?

FIGURE 21-5 Questionnaire Problems to Avoid

Problem	Example	Problem with the Example
Big words/jargon	Have you ever had a myocardial infarction?	Participants may not know that a "myocardial infarction" is a technical term for a heart attack.
Undefined abbreviations	Have you ever been told that you have high BP?	Participants may not know that the researchers are asking about elevated blood pressure.
Ambiguous meanings	What kind of house do you live in?	Without seeing a list of appropriate responses, it is not clear if the answer should be "an apartment," "a rental," "a split-level duplex," or "a single-family home."
Vagueness	Do you exercise regularly?	"Regularly" is not defined. A person who exercises most days of each week might assume that "regularly" means daily and say "no." Another person who exercises one day per month may consider that to be regular. It would be better to ask "In a typical week, how many days do you exercise for at least 30 minutes?" or "Last week, on how many days did you exercise for at least 30 minutes?"
Double negatives	I did not find this visit with my doctor to be unpleasant. ❒ Disagree ❒ Neutral ❒ Agree	The wording of this question makes it hard to figure out whether a person who was satisfied with a visit should agree or disagree.
Faulty assumptions	Do your gums bleed during regular dental cleanings? ❒ Yes ❒ No	The question assumes that everyone has routine dental cleanings. If "I do not visit the dentist" is not an answer option, a person who does not have dental cleanings is forced to answer no.
Two-in-one	Do you exercise at least 3 times a week and eat a healthy diet? ❒ Yes ❒ No	Two separate questions are combined: one for exercise and one for diet.
Impossible to recall accurately	How many apples did you eat in a typical week when you were a child?	Adults will not be able to remember this level of detail about their childhood diets, especially since consumption levels may have changed by season and across years.
Too much detail	List any prescription medications you have taken for 1 month or longer in the past 10 years.	Unless the respondent has had very few prescriptions, answering this question accurately may be impossible. A better option would be to ask for permission to access medical or pharmacy records.
Sensitive questions	Have you ever hit, scratched, bruised, or otherwise physically injured an intimate partner?	This question is unlikely to be answered truthfully if the response should be yes, and it may raise concerns about confidentiality and potential legal requirements for reporting abuse.

Problem	Example	Problem with the Example
Hypothetical questions	Have you ever thought that you would like to lose 10 or more pounds?	Anyone could have felt this at some point in time, but the question does not clarify whether this is a long-term longing or a thought that crossed the respondent's mind for the first time upon reading the question.
Leading questions	What is your impression of the work done by the dedicated public servants who serve our community every day through their tireless work at the county health department? ❒ Fair ❒ Good ❒ Great ❒ Excellent	The phrasing of this question clearly intends to lead respondents toward a positive answer (and may unintentionally have the opposite effect).
Leading answers	How would you rate the availability of visitor and patient parking at Center City Hospital? ❒ Good ❒ Great ❒ Excellent	This question's response options allow only positive evaluations. There is no "poor" option.
Answers with a poor scale	How many hours of television do you watch each week? ❒ 0 ❒ 1–3 ❒ 4–7 ❒ 8 or more	Even though most people watch more than 1 hour of television daily, which would put them in the "8 or more" response category, participants may not want to choose an "extreme" answer. These response options may also cause respondents to misread the question as how many hours they watch television each day, which will also result in inaccurate responses.
Lack of specificity	What is your income?	It is not clear if income refers to earnings per hour, week, month, or year, or whether it refers to pre- or post-tax income.
Missing answer options	What color are your eyes? ❒ Brown ❒ Blue	Many possible eye colors are missing.
Overlapping answer options	In a typical week, how many days do you eat fish? ❒ 0 ❒ 1–3 ❒ 3–5 ❒ 5–7	Participants who eat fish 3 days a week or 5 days a week will not know which response to select.

- For scaled questions, is the rank order clear? (For example, is it obvious that 1 is "strongly disagree" and 5 is "strongly agree"? Or, alternatively, that 1 is "excellent" and 7 is "poor"?)
- For questions with unranked categories, is the order of possible responses alphabetical or otherwise neutral?

21.7 Order of Questions

Many questionnaires start with easy or at least general questions before moving to more difficult or sensitive questions. The questions should be in an order that flows naturally from one topic to another.

It is often best to group similar questions with similar response types, so that they are asked consecutively. However, sometimes it is better to mix up such questions to prevent habituation. **Habituation** is an error that occurs when participants completing a questionnaire or interview become so accustomed to giving a particular response (like "agree . . . agree . . . agree . . .") that they continue to reply with the same response even when that does not match their true perspectives.

Survey developers must carefully consider how previous questions could taint the answers to later ones. For example, once a participant has considered a variety of opinions about a topic, he or she can no longer provide an unbiased first impression. Thus, the researcher may want to order questions about impressions this way:

- First, an open-ended question to garner an initial impression from participants: "What do you do most often when ______?"
- Second, a series of yes/no questions to clarify beliefs, perceptions, and practices: "Do you ever ______?"
- Last, a concluding open-ended question to allow participants to express final impressions: "Now that you have considered the possibilities, what would you say you do most often when ______?"

21.8 Layout and Formatting

The next step in questionnaire design is formatting the data-collection form so that it is organized, easy to read, and easy to record answers on. Both paper-based and electronic survey pages should be carefully checked for grammatical errors, misspellings, missing questions, gaps in logic, unclear instructions, formatting errors, layout, spacing, and other organizational issues. If the questionnaire appears to be too long, it may be helpful to identify less important questions and remove them at this stage.

When a paper-based survey (sometimes called a paper-and-pencil survey) is used, the pages should not be too crowded. **White space** is the blank areas between printed content on a page. White space is helpful for separating sections and making the pages visually appealing. When an Internet-based or computer-based survey is used, there must be a high level of contrast between the colors selected for backgrounds and text. For both paper-based and electronic data entry platforms, the selected typefaces and fonts must be readable and sufficiently large.

When designing electronic data collection forms, researchers need to decide whether to show all of the questions on one page or to present questions over several pages. Using multiple pages is advantageous when there are many questions in the survey instrument, when some questions will be irrelevant to some participants, and when the researcher wants to require participants to provide an answer to one or more questions before the next set of questions is revealed. Computer-based data entry programs allow a researcher to force participants to provide an answer to selected questions. However, this option should be used judiciously, because required fields force a person who does not want to provide an answer to one required question to quit the survey and leave all subsequent items unanswered.

A **filter question** or **contingency question** is one that determines whether the respondent is eligible to answer a subsequent question or set of questions. For example, participants who indicate that they have never used tobacco products can be prompted to skip a series of questions about smoking habits. For paper-based forms, instructions for skips must be carefully described in words. In a computer-based survey, **skip logic** codes can automatically hide irrelevant questions from participants based on their responses to filter questions. (In some data collection programs, skip logic is called routing or branching.)

The layout of the data collection form will vary depending on the mechanism of data collection used. For self-response surveys in which participants record their own answers, a cover letter or list of instructions should explain how answers should be recorded, such as:

- "Fill in the oval in front of your answer completely using blue or black ink."
- "Write your answer in block capital letters, as shown in the example below."
- "Select the one answer that best describes you."
- "Circle all options that apply to you."
- "If you answered 'NO' to Question 4, then skip to Question 8. If you answered 'YES' to Question 4, then please answer Questions 5, 6, and 7 before moving on to Question 8."

The response form should clearly indicate where and how responses should be marked (**Figure 21-6**).

When survey data are collected through interviews, the interviewer reads the questions aloud and records the respondent's spoken answers. The script for the interview must include an opening statement, the survey questions, sentences to read during the transitions between sections of the survey, and closing sentences. The questionnaire should clearly indicate the sections to read aloud, the procedures for recording responses on paper or in a computer file, and other instructions (**Figure 21-7**). If applicable, very clear instructions for skipping irrelevant sets of questions should be included within the script.

21.9 Reliability and Validity

A reliable and valid questionnaire (or other assessment tool) measures what it was intended to measure in the population being assessed. **Reliability**, or **precision**, is demonstrated when consistent answers are given to similar questions and when an assessment yields the same outcome when repeated several times. The **validity**, or **accuracy**, of a survey instrument (or a diagnostic test or other assessment tool) is established when the responses or measurements are shown to be correct.

FIGURE 21-6 Example of a Self-Response Questionnaire

Basic Details			
1. What is today's date?	__ __ - __ __ - __ __ __ __ m m d d y y y y		
2. What is your date of birth?	__ __ - __ __ - __ __ __ __ m m d d y y y y		
3. What is your sex?	❒ Female	❒ Male	❒ ______
Health History (*Check one answer box for each question.*)			
4. Have you ever been diagnosed with breast cancer?	❒ Yes	❒ No	→ *If No, then skip to question 8.*
5. Have you had a mastectomy (either partial or complete)?	❒ Yes	❒ No	

FIGURE 21-7 Example of a Telephone Interview Script

	→ Fill in today's date.	__ __ - __ __ - __ __ __ __ m m d d y y y y		
Read:	*Thank you for agreeing to participate in this health study.* *I'm going to start by asking you some basic questions.*			
1	What is your date of birth?	__ __ - __ __ - __ __ __ __ m m d d y y y y		❒ Refused to answer
2	What is your sex?	❒ Female	❒ Male	❒ ________
Read:	*Now I'm going to ask you a few questions about your medical history.*			
3	Have you ever been diagnosed with breast cancer?	❒ Yes	❒ No	→ *If* ***No****, then* ***skip*** *to question 7.*
4	Have you had a mastectomy (either partial or complete)?	❒ Yes	❒ No	❒ Refused to answer

Figure 21-8 illustrates the differences between these terms. A dart thrower who hits the same spot on a target consistently is reliable, but if that cluster is not centered at the bullseye, the thrower lacks accuracy.

One aspect of reliability is **internal consistency**, which is present when the items in a survey instrument measure various aspects of the same concept. Some survey instruments ask the same question several different ways, or ask a series of similar questions, in order to confirm the stability of participants' responses. For example, a questionnaire might include two questions that are opposites of one another, like "I enjoy eating most fruits" and "In general, I do not like to eat fruit." The expectation is that all respondents who say the first item is true will say that the second item is false, and vice versa. If a very high proportion of respondents' answers meet this expectation, then that is evidence that the responses are reliable.

Various statistical tests can be used to examine concordance within a data set, assess the validity and consistency of assessment tools and procedures, and conduct other aspects of quality control. For example, internal consistency can be confirmed with tests of intercorrelation such as Cronbach's alpha and the KR-20.

FIGURE 21-8 Reliability and Validity

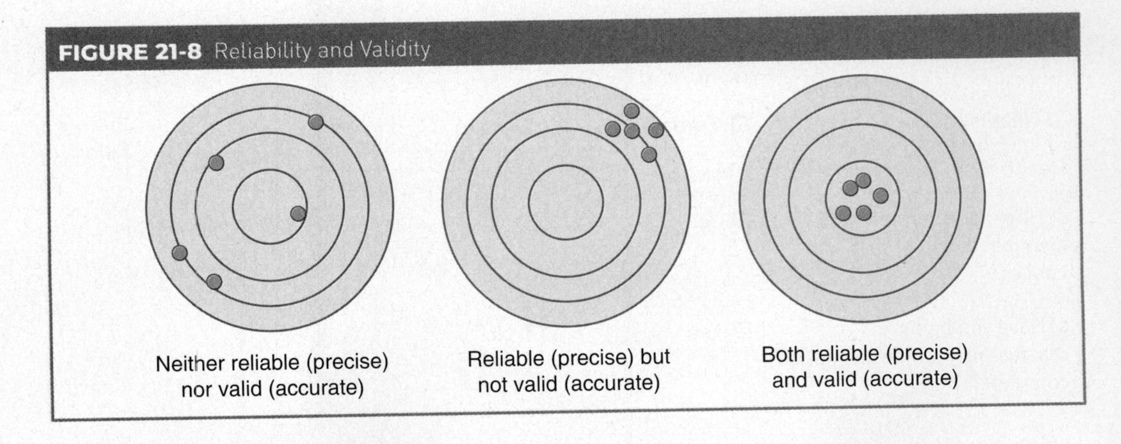

Intercorrelation is present when two or more related items in a survey instrument measure various aspects of the same concept.

Cronbach's alpha is a measure of internal consistency that is used with variables that have ordered responses. The **Kuder-Richardson Formula 20 (KR-20)** is a measure of internal consistency that is used with binary variables. Both of these statistics are expressed as a number between 0 and 1. Scores near 1 indicate an assessment tool with minimal random error and high reliability. (Tests of intercorrelation that examine the reliability of survey instruments are not the same as tests of correlation that compare two or more independent variables.)

Agreement between two or more evaluations is another facet of reliability. **Test–retest reliability** is demonstrated when people who take a baseline assessment and then retake the test later have about the same scores each time they are tested. **Interobserver agreement**, or **inter-rater agreement**, describes the degree of concordance among independent raters assessing the same study participants. The **kappa statistic**, also called **Cohen's kappa** and represented with the Greek letter κ, determines whether two assessors who evaluated the same study participants agreed more often than is expected by chance (**Figure 21-9**). For example, the value of the kappa statistic can indicate whether two radiologists examining the same set of x-ray images reach the same conclusion about the presence or absence of a fracture more or less often than expected. If the two radiologists agree as often as expected, $\kappa = 0$. If they agree on the interpretation of 100% of the x-ray images shown to both of

FIGURE 21-9 Inter-rater Agreement

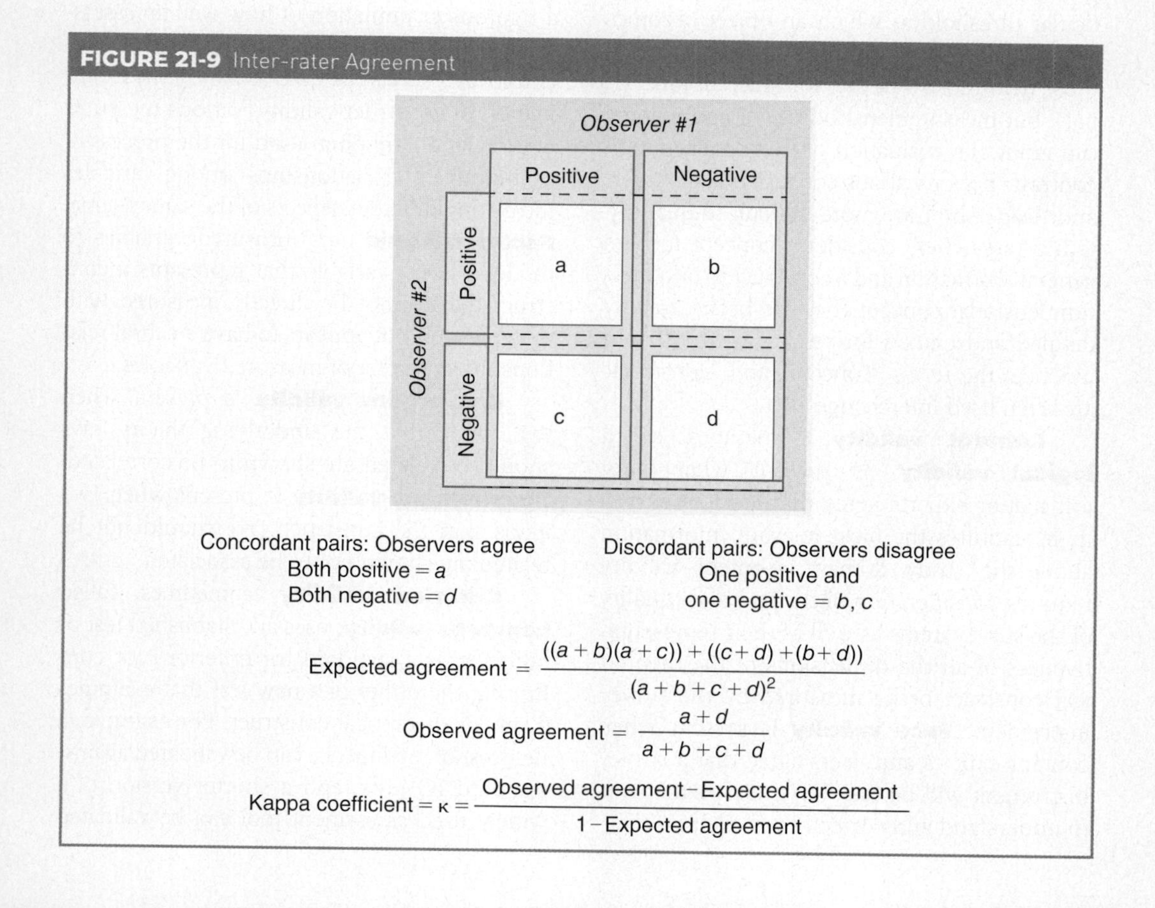

them, $\kappa = 1$. If they agree more often than expected, kappa will have a positive value somewhere between 0 and 1. Although complete agreement is rare, a valid study will have a value of kappa that is close to 1.

Several approaches are used to evaluate the accuracy of assessment tools that rely on self-reporting, such as psychometric tests and surveys about attitudes and perceptions. These tools are often considered to be proxy measures for an underlying theoretical construct that cannot be directly measured. For example, happiness and intelligence cannot be directly measured with physical or chemical tests, but survey instruments can be designed to measure them indirectly. Some researchers use the word **concept** to describe a theory informed by observations and use the term **construct** to describe a theory informed from more complex abstractions. For example, there is no particular threshold at which an object becomes "heavy" or a person becomes "rich," because those definitions will vary for different individuals, but measurements of weight and income can guide the evaluation of these concepts. In contrast, notions about constructs like "trust" and "leadership" are more difficult to quantify. Other researchers consider a concept to be a general abstraction and a construct to be a multidimensional concept that has been carefully defined and crafted for research purposes. In practice, the terms "concept" and "construct" are often used interchangeably.

Content validity, sometimes called **logical validity**, is present when subject-matter experts agree that a set of survey items captures the most relevant information about the study domain. Content validity requires consideration of the technical quality of the survey items as well as their representativeness of all the dimensions of the theoretical construct being measured by the survey instrument. **Face validity** is present when content experts and users agree that a survey instrument will be easy for study participants to understand and correctly complete.

Some statistical methods can provide information about which items in an assessment tool might be redundant or unnecessary and therefore can be removed without compromising the validity of the survey instrument. For example, **principal component analysis (PCA)** creates one or more index variables (called components) from a larger set of measured variables. The index variable is generated from the linear combination of measured variables, so it is a weighted average of the contributing variables. The PCA process determines the optimal number of components, the best measured variables to combine, and the best weights to use for the calculation.

Construct validity is present when a set of questions in a survey instrument measures the theoretical construct the tool is intended to assess. Construct validity requires the development of an explicit theoretical construct and a rigorous examination of how well an assessment tool represents that construct. Ideally, empirical tests can be used as part of the examination of construct validity. Various measures of correlation are often used for the process of examining interrelationships among variables measuring different aspects of the same theme. **Factor analysis** uses measured variables to model a latent variable that represents a construct that cannot be directly measured with one question but appears to have a causal relationship with a set of measured variables.

Convergent validity is present when two items that the underlying theory says should be related are shown to be correlated. **Discriminant validity** is present when two items that the construct says should not be related are shown not to be associated.

Criterion validity, sometimes called **concrete validity**, uses an established test or outcome as a standard (or criterion) for confirming the utility of a new test that examines a similar theoretical construct. For example, a new test of intelligence can be validated against standard IQ tests, and a shorter version of a widely used assessment tool can be validated against the longer original version. There are two main approaches to examining criterion validity. **Concurrent validity** is evaluated when participants in a pilot study complete both the existing and new tests and the correlation between the test results is calculated. A strong correlation between the tests is evidence that the new test is valid. **Predictive validity** is appraised when the new test is correlated with subsequent measures of performance in related domains. Suppose researchers create a new test intended to predict success in medical school. Concurrent validity could be demonstrated by comparing scores on the new test with scores on the MCAT, which is the current standard test for medical college admission in the United States. Predictive validity could be demonstrated by administering the test to incoming medical students and comparing their results on the new test with their performance on the initial licensing exam (USMLE Step 1) taken 2 years later.

21.10 Commercial Research Tools

One way to improve validity is to include survey questions or modules that are identical to the ones used in previous research projects. Unfortunately, access to survey questions is often not possible in the health sciences. Copies of questionnaires from some disciplines are almost never included with published papers and are only rarely posted on researchers' websites. As a result, new research projects usually require the development and testing of a completely new survey instrument. However, several widely used and validated tests are available to researchers, such as:

- The Beck Depression Inventory and the General Health Questionnaire (GHQ), which assesses psychological status
- The Mini-Mental State Examination (MMSE), which evaluates cognitive function
- The SF-36 and SF-12, which both measure **health-related quality of life**, a multidimensional construct that captures an individual's perceived physical, mental, emotional, and social well-being and the perceived impact of health status on the quality of daily life

The Buros Center for Testing's *Mental Measurements Yearbook* provides reviews of thousands of available tools used in psychological and educational assessment. Some of these tools are free of charge, but most are commercial products that require payment for use. For some instruments provided at no charge, researchers must pay to have the results scored and validated against previous users of the survey instrument.

21.11 Translation

Translation of the survey instrument into one or more additional languages may be necessary if the source population contains speakers of more than one language. (Translation may also be required when an ethics review committee requires materials to be presented in the committee's preferred language as well as in the language of the source population.) Researchers using multiple languages must be certain that the translated version expresses the same meaning as the original survey. Accuracy may require the rephrasing of whole sentences, not just direct word-for-word translations.

One way to ensure that the correct meaning is being conveyed is to use **back translation**, or **double translation**, in which one person translates the questionnaire from the original language to a new language and then a second person translates the survey instrument in the new language back into the original language. A comparison of the original version of the survey with the back-translated version will reveal where the second-language translation does not match the intended meaning of the original version. A second approach is to have two translators independently translate the survey instrument from the original to the new language. The two translations are compared, and a consensus process is used to decide which words and phrases best convey the precise meaning and complexity of the original questionnaire.

21.12 Pilot Testing

A **pilot test**, or **pretest**, is a small-scale preliminary study conducted to evaluate the feasibility of a full-scale research project. A pilot test of a questionnaire is helpful for checking, among other issues:

- The wording and clarity of the questions
- The order of the questions
- The ability and willingness of participants to answer the questions
- The responses given, and whether the responses match the intended types of responses
- The amount of time it takes to complete the survey

The researcher should ask several volunteers to help with the pilot test. These volunteers should be from the target population and meet the eligibility criteria for the study (in terms of age, exposure and disease status, and other key factors), but they should not be members of the sample population. They should be asked to complete the preliminary survey and then provide feedback about content, clarity, layout, timing, and other factors. Feedback may be provided individually or as part of a focus group. The survey instrument is revised based on these observations. Several rounds of pilot testing may be required to develop a sound survey instrument.

© Dmitriy Razinkov/Shutterstock

CHAPTER 22

Collecting Quantitative Data

Most primary studies collect data from individual participants using interviews or self-administered questionnaires.

22.1 Interviews Versus Self-Administered Surveys

The first decision to make about collection of quantitative data is whether to have a member of the research team interview participants or to have participants record their own answers (**Figure 22-1**).

An **interview** is the process of a researcher verbally asking a participant questions and recording that person's responses. A key advantage of using interviews to gather data is that trained interviewers record the responses, and they can ensure the accuracy and completeness of each questionnaire. However, interviews may require major time commitments from study personnel. One-on-one interviews may take a considerable amount of time per participant, and it may take months to schedule all of the needed interviews.

Interviews may be conducted in person or via telephone. For in-person interviews, transportation to the interview site may be a barrier to participation. For telephone-based interviews, hearing impairments and discomfort with using a telephone may be a challenge for some participants.

A **self-administered survey** uses a questionnaire form that participants complete by themselves, using either a paper-and-pencil version or an online version of the survey instrument. A major benefit of self-administered surveys is that they can allow for the cost-effective and time-efficient collection of data from a large number of participants. For example, a school-based survey could gather data from hundreds of students during one 20-minute period. Self-administered surveys may also be the best way to get honest answers to sensitive questions.

Self-administered questionnaires can be completed at a specific study site, such as a workplace or school or hospital, or they can be delivered by mail or the Internet and completed at a time that is convenient for the participant. Mailed surveys are often the least favorable option, because they incur direct costs related to photocopying, postage, and

FIGURE 22-1 Examples of Methods for Collecting Data

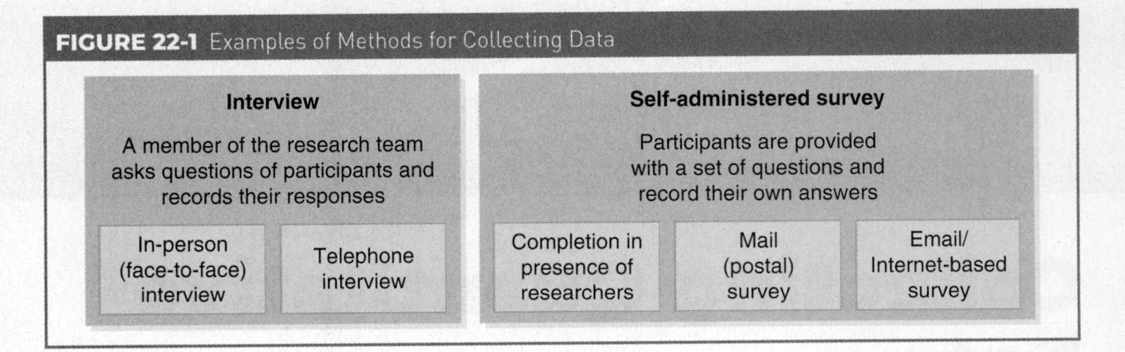
Interview
A member of the research team asks questions of participants and records their responses
In-person (face-to-face) interview
Telephone interview
Self-administered survey
Participants are provided with a set of questions and record their own answers
Completion in presence of researchers
Mail (postal) survey
Email/ Internet-based survey

data entry. The participation rate for mailed survey is often low. If only 1 completed questionnaire is returned for every 10 surveys that are mailed out, the costs per participant may be quite high. Additionally, mailed surveys may trickle in over an extended period of time and make it challenging for a researcher to know when to stop waiting for additional responses to arrive. Internet-based surveys may have relatively low costs if a free or low-cost survey-hosting website is used. However, Internet-based surveys are not an appropriate option to use when studying populations whose members are likely to have low literacy or to have limited Internet access or be unfamiliar with computers.

22.2 Recruiting Methods

Once a data collection method has been selected, the next step is to decide on an effective method for recruiting members of the sample population to be participants in the study. The best method for initiating contact with potential participants is often related to the intended data collection method.

If the plan is to interview people in person, the best recruiting method may be to have in-person conversations with potential recruits in a clinical setting, at work, at school, at a public venue, or at another appropriate location. Alternatively, if the contact details for sampled individuals are available from clinical care providers, employee databases, or other sources, an initial letter or email of invitation could be sent to sampled individuals. The researcher could then follow up by contacting those individuals again to schedule interviews.

If the plan is to interview participants by telephone, it may be possible to recruit some participants with cold calls. However, the participation rate will likely be higher if a letter of invitation is sent first. Sending a letter will also allow for the acquisition of signed informed consent forms prior to the interview, if they are required.

If the plan is to collect data via the Internet, then contacting potential participants via email or a website may be the most effective method.

22.3 Encouraging Participation

The goal of recruiting is to maximize the participation rate among members of the sample population so as to yield a study population that is reasonably representative of the source population. Ideally, the researcher should try to find a way to compare the characteristics of participants to the demographics of the source population as a whole. For example, in a school-based study, the proportion of participants by grade can be compared to the overall distribution of students by grade in the participating schools. A statistical test can show whether the study population skews old or young or is a close match to the source population.

Participation rates will likely be higher if recruits understand the importance and value of the research project. For example, suppose that the plan is to interview members of a selected organization by phone. The response rate is likely to be highest if interviewers start each phone conversation with potential participants by explaining why the participants are being contacted, how their contact details were acquired, and how completing an interview will assist the organization. The participation rate may be quite high, even for unscheduled telephone calls, because the importance and relevance of the study are addressed at the start of the call. Support for the study may be even higher if the study plans are shared ahead of time in an organizational newsletter or via an email to all members.

In contrast, several hundred calls made by **random-digit dialing**—calls to a computer-generated list of unscreened telephone numbers—may yield only a few people willing to participate in a study. Even then, many willing participants may turn out to be ineligible. Additionally, a growing problem with using random-digit dialing is that mobile phone numbers are often unlisted and are not necessarily indicative of the user's geographic location. These issues may further reduce the representativeness of study populations recruited by random-digit dialing. Nevertheless, using the first moments of a phone call to explain why a particular study will make a difference in the world or to a particular community may increase the willingness of randomly contacted individuals to participate.

Other ways to increase the participation rate are providing multiple invitations and opportunities to participate and making participation as easy as possible. Mailed survey packets should include a concise cover letter that explains the purpose and importance of the study and discloses any necessary information such as financial sponsorship and contact details for key members of the research team. The mailed packet should also include the survey instrument and a preaddressed stamped envelope so that the completed survey can easily be returned to the researcher. A few weeks after the initial mailing, a reminder postcard or a second copy of the questionnaire should be sent to those who have not yet responded. The follow-up mailing should reaffirm the study's importance and express gratitude to those who have already returned a completed survey as well as those who intend to do so. Similarly, multiple phone calls on different days of the week and at different times of the day may have to be made to reach potential participants by telephone. Multiple email invitations to complete an Internet-based survey may be required to get recruits to fill out an online questionnaire. Including a step-by-step guide for using the survey website may make participation more accessible to those who are willing to participate but uncomfortable with new technologies. Explaining how the privacy of participants will be protected may also be helpful for alleviating the concerns of potential participants.

Incentives such as small gifts or the opportunity to be entered into a drawing to win a prize may be an effective means of encouraging participation among those invited to be in a study. Any inducements, gifts, or compensation must be approved by an ethics review committee prior to being offered.

22.4 Data Recording Methods

Before starting data collection, the research team must decide how responses will be recorded and when they will be entered into a computer database. There are two basic options (**Figure 22-2**). One is to record the responses on paper and to enter or scan them into a computer later. The other is to have interviewers or participants enter responses directly into a database.

Paper questionnaires have several benefits. In some environments, they are required for the

FIGURE 22-2 Methods for Collecting and Recording Survey Data

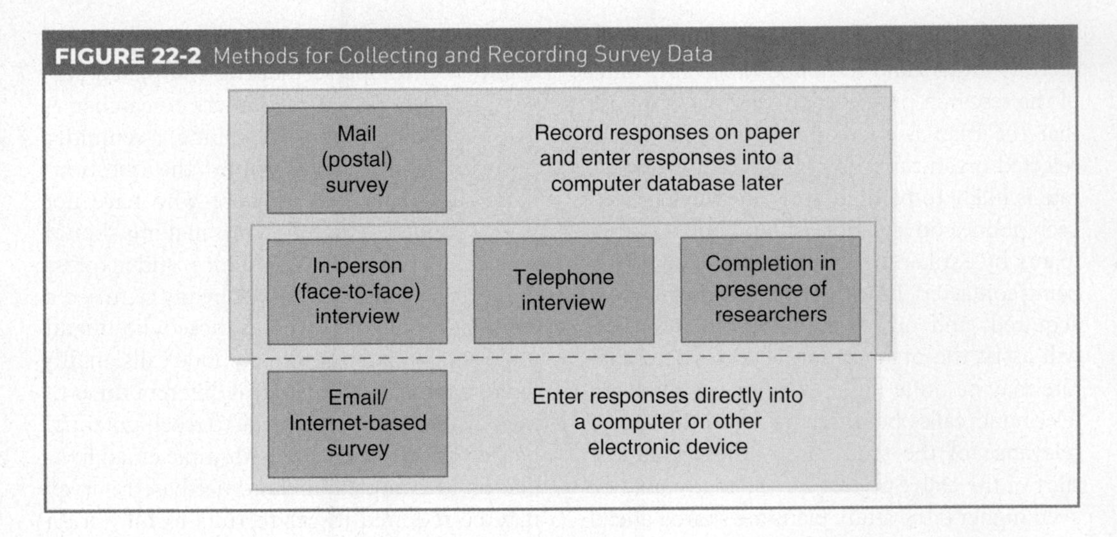

collection of data from a large number of participants at one time, as would be the situation when all students attending a school are asked to complete a questionnaire during the same 20-minute period. Paper instruments allow for the easy collection of signatures on informed consent statements, and some researchers value having paper records as a backup to electronic files. However, paper-based surveys also have a serious disadvantage: Unless somewhat expensive optical scan forms are used, all responses must be manually entered into a computer at a later time. Data entry is often a very time-consuming process, and that can become costly.

The major advantage of computer-assisted surveys is that they eliminate the need for later data entry. They may also simplify the questionnaire by automatically removing any questions not relevant to a particular study participant. For example, they may skip questions specific to females for participants who identify themselves as being males. The main limitation of computer-assisted surveys is that some populations are uncomfortable with computer technology. Discomfort with technology may be expressed in several ways. Older adults who have limited access to the Internet or do not routinely use computers may systematically choose not to participate in an Internet-based survey. There may also be cultural reasons not to use computers during in-person interviews. Some interviewees may be distracted by an interviewer entering responses into a computer as they give their responses, and they might not be similarly bothered by an interviewer with a clipboard jotting down their comments on a piece of paper.

For some data collection situations, having a limited number of computer terminals or portable electronic devices available for participant use may become a barrier to project success. For example, suppose that health fair attendees are being asked to complete an exit survey using handheld computers. Most people in the process of leaving a venue will not choose to wait 15 or more minutes for a tablet computer to become available to them unless there is a very desirable reward for participation. The participation rate may be very low, and the study population may not be representative of health fair attendees as a whole because only the most patient people will have recorded their answers. In this situation, having paper-based surveys, pens, and clipboards available may speed up the data collection process and increase the participation rate.

22.5 Training Interviewers

Interviewer bias is a form of information bias that occurs when interviewers systematically question cases and controls or exposed and unexposed members of a study population differently, such as probing individuals they believe to have the disease or exposure of interest for more information but not doing the same probing for participants they believe to be unexposed controls. To minimize the risk of interviewer bias, the interview process should be the same for all participants in a study. Uniformity is easiest to accomplish when all interviewers are provided with the tools they need to follow a standardized set of procedures.

Each interviewer should be given a comprehensive interviewer handbook that provides information about the purpose of the study, details about interview logistics, an annotated script for the in-person, phone, or online interview, and annotated copies of all study forms. The training and handbook should:

- Explain each step of the interview process.
- Specify exactly how to ask questions and record responses.
- Identify any prompts or follow-up questions that the interviewer must use, is allowed to use, or is not allowed to use.
- Emphasize any restrictions against asking for clarification about particular items.
- Provide checklists for handling problems that might arise during an interview, such as interruptions, disconnections, or uncomfortable or emotional participants.

All of this information should also be recorded in the study protocol.

All interviewers should undergo role-specific training and have an opportunity to practice their interview skills. In training sessions, the questionnaire is examined in detail so that all interviewers understand what each question is asking, how to pronounce all the words in each question, how to phrase the reading of each question, and how to present the possible answers for questions that are not open-ended. All paper response forms and/or computer-assisted data entry programs are closely examined so that every interviewer understands exactly how to record participant responses. Each interviewer then participates in several mock interviews from start to finish, including the informed consent process. Training sessions also emphasize the importance of strictly following the procedures spelled out in the interviewer handbook and make clear the absolute necessity of maintaining the confidentiality of all information that study participants share with interviewers. Interviewers may also need to complete institution-mandated research ethics training sessions prior to interacting with participants.

Clear guidelines and practice opportunities will help to create skilled, confident, and reliable interviewers (**Figure 22-3**). Well-trained interviewers will know how to make participants comfortable, how not to intentionally or unintentionally guide participants toward particular answers rather than letting participants provide candid responses, and how to complete all survey forms consistently and completely.

FIGURE 22-3 Characteristics of Well-Trained Interviewers

Characteristic	Actions That Demonstrate the Characteristic
Respectful	■ Communicates pleasantly and professionally with all study participants and members of the research team ■ Has practiced interviewing enough to be comfortable with both the script and the interview process ■ Asks supervisors for assistance when it is needed
Organized	■ Begins each scheduled interview session on time ■ Has all necessary materials on hand prior to the start of each interview session ■ Maintains meticulous records and completes all files and paperwork promptly
Considerate	■ Dresses and grooms appropriately for in-person interviews ■ Is alert to modifiable conditions that may make interviewees uncomfortable, such as loud background noises, cold or hot temperatures, or dim lighting ■ Allows adequate time for participants to respond to each question
Articulate	■ Speaks at an appropriate pace and volume ■ Enunciates clearly ■ Uses an appropriate tone of voice (and, for in-person interviews, appropriate facial expressions and gestures) ■ Rereads questions and/or the list of closed-ended responses when a participant does not understand the question or the acceptable responses
Consistent	■ Reads the script exactly as it is written ■ Probes for answers only when the script indicates that probing is approved ■ Does not provide explanations for any question unless an explanation is provided in the script or approved in the interviewer handbook
Impartial	■ Avoids verbal and nonverbal expressions of approval or disapproval ■ Does not express personal opinions ■ Avoids leading interviewees toward a particular answer (for example, by placing special emphasis on particular words in a question or by probing until receiving a particular desired response)
Honest	■ Does not fabricate or falsify reports ■ Records responses to open-ended questions verbatim, without rephrasing, paraphrasing, "correcting," or interpreting them
Careful	■ Completes all steps of the interview process in the correct order, as prescribed by the interviewer handbook ■ Documents informed consent prior to conducting an interview ■ Does not skip any component of the interview ■ Completes all response forms correctly

CHAPTER 23

Collecting Qualitative Data

Qualitative studies collect data using a variety of observational and participatory methods.

23.1 Overview

Quantitative data collection is often described using words like formal, impersonal, and detached. Qualitative data collection is more likely to be described using terms like informal, personal, and reflexive. In qualitative research, the researchers are allowed to express empathy with informants. Some qualitative data can be collected through open-ended questions on survey instruments, from existing documents and other artifacts, and by other means, but qualitative data collection in the health sciences usually involves in-depth interviews, focus groups, or other specialized data collection methods.

Because qualitative researchers are so closely engaged with participants, researchers need to reflect on how their backgrounds might bias their observations, and they need to be transparent about these potential biases when designing rigorous studies and reporting findings. The interpersonal aspects of qualitative research demand careful consideration of how best to recruit participants, acquire and document informed consent, and record observations of participants and interactions with them. These planned approaches should be described clearly in documents submitted to research ethics committees.

23.2 In-Depth Interviews

An **in-depth interview** is a qualitative research technique in which an interviewer spends 1 or 2 hours interviewing a key informant using open-ended questions. These conversations are usually in-person dialogues between one interviewer and one participant. Sometimes two researchers meet with one participant for safety or cultural reasons, or so that one researcher can ask questions while the other takes notes. Interviews are often audio- or video-recorded and then transcribed so that the exact words (and sometimes also the nonverbal expressions) can be coded and interpreted. Interviews are sometimes supplemented by other data collection methods, such as participant diaries or journals.

In a **semi-structured interview**, the interviewer starts with a list of open-ended questions that will be asked of each participant. However, these questions or lists of topics are merely starting points for eliciting responses from participants. **Probing** is an interviewing technique that prompts an interviewee to provide a more complete or specific response. During semi-structured interviews, a researcher can probe for more details about any response in order to gain fuller understanding of participants' experiences and perspectives.

Interviewers can also record their observations of the body language and other nonverbal communication conveyed by participants. At the same time, interviewers should be aware of how their own word choices, tone of voice, and body language may be interpreted or misinterpreted by interviewees. Interviewers should not ask leading questions and not move on to a new set of questions before full responses have been gathered from the current line of inquiry. If the interviewer is not certain what a participant meant by a particular statement, he or she can respectfully request clarification.

Audio recordings should be transcribed as soon as possible after an interview and reviewed for accuracy. All parts of the interview, including fillers (like "um" and "you know"), laughs, and other sounds should be included in the typed transcript.

23.3 Focus Group Discussions

A **focus group** is a qualitative data gathering technique in which approximately 8 to 10 people spend 1 or 2 hours participating in a moderated discussion. A focus group is especially valuable when interaction among participants stimulates richer responses and clarifies opinions. Focus groups often use **homogeneous sampling** to recruit participants with similar backgrounds, experiences, or perspectives. Focus groups can then be used to understand the norms of a group as well as to identify the diversity of perspectives that exist within a population. Focus groups do not work well when peer pressure inhibits disclosure of participants' true feelings and opinions.

Most focus groups are hosted by two researchers. One serves as the moderator who sets the agenda, facilitates the discussion, and keeps the conversation on track. The other serves as a note-taker while providing other support such as assisting with welcoming participants, collecting consent forms, operating recording devices and backup recorders, and time-keeping. Focus groups should be audio- or video-recorded so that complete transcripts can be created. The note-taker can track the key messages and themes that emerge during the discussion while also documenting observations like emotions and gestures that may not show up in the transcript.

A focus group session typically begins with the moderator establishing rapport with the group and setting the ground rules for the session. Participants need to understand that their involvement in the study is voluntary (even if they have been offered an incentive to participate) and they can leave at any time. The importance of privacy and respect must be emphasized. Participants are usually asked not to disclose to outsiders what others say during the group discussion. For sensitive topics, it may be helpful to assign pseudonyms to group members so that their true names are not disclosed to other participants.

Once everyone has agreed to the plan for the session, the facilitator will begin posing questions to the group, carefully keeping the conversation focused on the core discussion items and moving forward at an appropriate pace. Participants should be encouraged to interact with one another and to identify shared perspectives while not succumbing to groupthink. Follow-up questions posed by the moderator in a neutral manner can help clarify individual and shared perspectives. Times of silence are acceptable when participants need a few quiet moments to process their thoughts. The moderator must ensure that everyone has an opportunity to speak and that no one dominates the conversation or routinely interrupts other participants' comments. Being a good moderator requires practice. Mock sessions can be helpful training for the moderator role, and mentoring from an experienced facilitator is valuable for improving performance.

As soon as the focus group has concluded, the moderator and note-taker should hold a debriefing session to discuss what worked well and what can be improved in the next round. The transcript of the session should be typed up soon after the session and reviewed for accuracy by both facilitators.

23.4 Observational Methods

In qualitative research, **observational methods** are ones that involve systematic observations of human actions and interactions. **Naturalistic observation** occurs when the researcher unobtrusively observes study subjects in a natural setting, typically without the knowledge of the subjects. Data collection may involve a combination of listening and watching, with field notes recording both verbal and nonverbal cues. Naturalistic observation is contrasted with **controlled observation** in which study participants are observed in a laboratory setting and know that they are being observed. **Participant–observation** is a method of qualitative field observation in which a trained investigator seeks to understand a community by engaging with its members and immersing in its practices. Participant–observation methods are especially common in ethnographic research.

Field notes are observation records, interview transcripts, and other documents compiled during the qualitative research process. Researchers conducting observational research diligently record details about the observation scene in their field notes, including the geographic location, the appearance and sounds at the site, and the various events that occur during the observation period. They pay close attention to recording individual behaviors and interactions between people. The observer might also take notes about personal experiences or feelings during the observational period that might be affecting the observations and interpretations being recorded in the field notes. The observation process may be unstructured, or it may be a structured process in which specific events of interest are documented using templates or coding sheets. After each hour of observation, it may take several hours to complete the full process of reflective documentation.

Observational methods may be overt or covert. **Overt observation** occurs when the participants are aware that they are being observed and the researcher is transparent about the goals and methods of the study. Informed consent removes many of the ethical challenges associated with observational research. One limitation of overt observation is that people who know they are being observed may change their behaviors. **Covert observation** occurs when a researcher does not inform study subjects that they are being investigated. The lack of informed consent and possible use of active deceit in covert studies can raise serious ethical concerns, especially when the studies are conducted in places where people have a reasonable expectation of privacy. Few concerns would be raised about a study that observes the behavior of large groups of pedestrians using crosswalks at a busy intersection, but there might be significant concerns about covert observational studies that track individuals' behaviors over time at schools or workplaces or in private membership organizations. Debriefing with the subjects of non-anonymous covert observations after the study period might not remove feelings of betrayal. Both overt and covert observational studies typically require approval from an institutional review board prior to initiation of data collection. The review process seeks to protect the safety of the researcher as well as the study subjects.

23.5 Other Qualitative Research Techniques

Additional techniques that move beyond traditional interviews and observations may be used to ensure the comprehensiveness of information collected during qualitative studies. New methods are constantly being developed and shared with others in the research community.

Some qualitative studies collect and analyze the stories of individuals and communities. **Oral history** is the audiovisual recording of historical information about individuals, families, groups, or events. Oral histories may be complemented by other types of data. **Narrative inquiry** examines autobiographies, personal letters, family stories, interview tapes, and other records to understand how people frame their identities and social relationships. **Triangulation** is the process of using multiple different types of data, methods, and theories to better understand a phenomenon.

Some studies use the visual and performing arts as part of their inquiries. **Photovoice** is a qualitative research technique in which participants take photographs that they feel represent their communities and then they share what aspects of their lived experiences they intended to capture in those images. Related methods include asking participants to draw pictures about their experiences or to engage in point-of-view filmmaking or participatory theater.

A **vignette** is a brief written or pictorial scenario that is designed to elicit a response from participants in a qualitative study. Vignettes typically describe a hypothetical situation based on a real-life event that might raise a moral dilemma. The short story provides an opportunity for participants to talk about how they would react or make decisions if they encountered a similar situation. Participatory methods for groups might include discussion of vignettes, or they might include other exercises related to ranking, scoring, or mapping.

In a **think-aloud protocol**, or **talk-aloud protocol**, participants are asked to describe their thoughts and actions while they complete a task. This technique can be especially helpful when designing a new product or evaluating the usability of a new tool. For example, developers testing a new mobile health (mHealth) application might ask individuals to explain their thought processes as they figure out how to use the app, adjust the settings, and so on. This feedback can provide the developers with insights about how to make the user interface more intuitive.

23.6 Community-Based Participatory Research

Community-based studies often work best when they use research methods such as those developed for **Community-Based Participatory Research (CBPR)**, in which partnerships link academicians with community representatives, who together identify research priorities for the community, design and implement appropriate data gathering and analysis activities, and then apply those findings to the development of new policies and programs in the community. Under a CBPR model, community participants are partners rather than study subjects, and all partners are involved in decision making throughout the entire research process. CBPR projects often use mixed methods approaches. Community representatives are actively involved in all aspects of the data collection process, including planning and implementation.

23.7 Consensus Methods

The goal of some studies is to identify areas of consensus and areas of contention among individuals who are experts on a particular topic and/or a particular community or organization. The results of the deliberations are then used, for example, to select research priorities, to identify best practices, and to agree on plans of action. Several techniques have been developed for shaping these conversations and the resulting conclusions. For example, the **Delphi method** is a structured decision-making and forecasting process in which experts complete questionnaires, a facilitator summarizes and shares the responses, and panelists reconsider their perspectives after reflecting on the opinions expressed by others. The goal is for each iteration to move the panel of experts closer to agreement.

CHAPTER 24

Additional Assessments

A variety of tools can supplement the self-reported data gathered from individual-level surveys and interviews.

24.1 Supplementing Self-Reported Data

Self-reported data are a critical component of many health research projects, but questionnaires and interviews have some significant limitations. Respondents may not tell the truth. They may provide the answers they perceive to be the correct ones or the ones they think the interviewer wants to hear, rather than recording the responses that best represent their true thoughts or behaviors. Participants may not know or remember some of their health-related measures, such as their current weight or blood pressure or cholesterol level. They may skip those questions on a survey form, or they may report inaccurate guesses or numbers that are many years out-of-date. Data files generated from self-reports may have a lot of missing values, and the validity of many of the numbers provided may be of questionable validity.

Laboratory tests and other objective measures of the human body and the environment can supplement and validate self-reported data. These types of scientific data are usually collected in person by a member of the research team so that the measurements can be collected according to precise protocols. This chapter presents some of these additional types of data along with examples of the ethical considerations associated with using them.

24.2 Anthropometric Measures

Anthropometry is the measurement of the human body. Anthropometric measurements are often used in studies of nutritional status. Some of the most common body measurements are:

- Height (stature)
- Weight
- Waist circumference
- Hip circumference
- Mid-upper-arm circumference (MUAC)
- Skinfold measurements that estimate the body fat percentage

Standard methods should be used to take all anthropometric measurements. Any tools used for the measurements should be carefully calibrated to ensure accuracy and reliability. The individuals taking the measurements should be trained to use all equipment properly and to record results to the appropriate level of precision. Researchers should also ensure privacy for participants while the measurements are being taken, which typically requires conducting assessments in a designated examination room or behind a screen or room divider. It is often best for two members of the research team to be present when measurements are being taken. When a child is being measured, it may be necessary for a parent or guardian to be present with the child. Research ethics committees may be able to offer guidance on legal and safety requirements related to physical examinations.

24.3 Vital Signs

Vital signs are physiological measurements that provide clinical data about an individual's essential body functions. Most basic vital signs can be quantified accurately after minimal instruction. These include:

- Body temperature
- Blood pressure
- Heart rate (pulse)
- Respiratory rate (breathing frequency)

A thermometer measures body temperature. A manual or electronic sphygmomanometer (a blood pressure cuff) measures systolic and diastolic blood pressure. Measuring resting pulse and respiratory rates does not require any instruments other than timekeeping devices.

All assessors should be trained to use the same techniques. For example, guidelines for measuring blood pressure should specify the appropriate sitting position or other posture for the person being assessed as well as the way the back and arm should be supported, perhaps by resting the forearm on a surface. The measurement protocol should spell out the instructions that should be given to the participant about removing clothing from the arm, sitting back in the chair, putting feet on the floor without crossing the legs, and not talking or moving during the measurement of the blood pressure. The amount of resting time before taking the measurement after placing the cuff on the arm should be standardized. The research handbook should also state whether blood pressure should be measured in a particular arm or both arms and whether the blood pressure should be measured one time or several times. Two or three readings taken 1 minute apart and averaged may be more accurate than a single reading. Deviation from any of these procedures may cause a recorded blood pressure to be higher or lower than it would be if the protocol had been followed.

Standardization increases the precision and validity of the measurements. When multiple individuals are collecting data, tests of inter-rater reliability can be used to confirm that all assessors generate similar or identical results when they measure the same person.

24.4 Clinical Examination

A well-trained clinician can make accurate and reliable assessments of many health states that machines are unable to assess well. For example, a clinician can examine:

- Heart sounds
- Breath sounds and other respiratory functions
- Bowel sounds and the condition of the abdomen
- The range of motion and the condition of the joints
- The condition of the skin, hair, and nails
- The health of the eyes, ears, nose, and mouth
- Mental status

- The ability to conduct activities of daily living
- Other signs of health or disease

When a clinical examination is part of the data collection process, an assessment form should carefully describe each component of the examination, including the exact procedures to be used and the specific diagnostic criteria for each item on the assessment form, as well as the order in which these elements should be examined. Care should be taken to ensure the comfort, privacy, and safety of each person being assessed.

24.5 Tests of Physiological Function

Tests of physiological function can provide helpful data about health status. For example, spirometry measures lung function, electrocardiography measures heart function, electroencephalography measures brain function, and audiometry measures hearing acuity.

The costs associated with these tests must be considered when designing primary data collection protocols. Although some medically necessary tests may be covered by patients' insurance plans, tests conducted primarily for the benefit of researchers must be paid for by the research team. Because of cost considerations, secondary analyses of existing medical records may be the best option for researchers whose study questions require the use of expensive equipment.

When tests are conducted as part of a primary research protocol for research purposes rather than clinical purposes, the research team must decide ahead of time, in consultation with specialists in medical ethics, whether the results of the studies will be shared with patients and/or their healthcare providers. This decision must be disclosed to participants during the informed consent process prior to any measurements being taken.

24.6 Laboratory Analysis of Biological Specimens

Tests of blood, urine, stool, saliva, and other biological specimens may be helpful for identifying the presence of a disease or markers for a disease, the characteristics associated with having a disease, and the risk factors for a disease. Qualitative laboratory tests seek to confirm the presence or absence of disease, while quantitative laboratory tests seek to measure amounts (such as the count of blood cells or the titers for antibody tests).

Some immunologic, genetic, and other studies require the collection of body fluids or tissue biopsies from participants, either as part of routine clinical practice or specifically for the purposes of the research project. Before new specimens are collected, a research ethics committee must verify that the potential risks of sample collection will be minimized. Some studies may be able to make use of existing specimen banks. These samples may be fully anonymous, or they may be linked to other information about the donors. The use of new or existing samples requires ethics committee review and approval. Participants may have a right to the results of the laboratory tests conducted on their own biological specimens, and the protocol should discuss how notification will occur.

24.7 Medical Imaging

Medical imaging techniques are sometimes used to visualize parts of the human body. Examples are radiography (which uses x-rays), computed tomography scans, positron emission tomography, magnetic resonance imaging, and ultrasound. The resulting images may be useful to researchers for purposes of diagnosis and/or for the assessment of responses to therapies. Some findings can be read by radiologists, but advanced computational methods may be required for analysis and interpretation of research data.

24.8 Tests of Physical Fitness

Kinesiology is the study of the mechanics, physiology, and psychology of body movement, function, and performance. Many different tests can be used to measure physical fitness levels. For example:

- Cardiorespiratory fitness can be assessed using a 1-mile walking test, a 1.5-mile run test, a cycle ergometer test, and other tests of aerobic fitness.
- Muscle strength and endurance can be measured with tests like timed sets of curl-ups (sit-ups or crunches), push-ups, pull-ups, flexed arm hangs, bench presses, leg presses, and grip tests (using a handgrip dynamometer).
- Flexibility can be measured using a sit-and-reach test (often measured with a flexometer) and other activities that stretch the lower back, hamstrings, or other muscle groups.

Additional tests of fitness may assess agility, balance, coordination, gait, speed, power, and reaction time.

Researchers must make the safety of participants their top priority. Appropriate precautions must be taken to minimize hazards. Participants walking or running on a treadmill must be given clear instructions about how to step on and off the belt, they must wear appropriate footwear, they must use any automatic-stop safety clips and other devices recommended by the manufacturer, and they must be monitored throughout the test. The treadmill must be situated away from walls or other objects that could cause harm to someone falling off the treadmill. The use of a harness might be required for participants with poor balance. Study participants walking or running on an outdoor track must be alerted to any bumps, dips, or other hazards on the track and must not be allowed to be tested in conditions of extreme heat, humidity, or precipitation. Participants who have an existing injury or other impairment or condition that might make movements dangerous should not be allowed to participate without medical authorization, legal approval, and close supervision.

24.9 Environmental Assessment

The natural and built environments can have short- and long-term impacts on human health. Consider just a few of the many environmental factors that may affect the safety of a home:

- Is the entrance to the home accessible, or are there stairs or other barriers to access for people with mobility limitations? Is there adequate outdoor and indoor lighting?
- Are any stairs in the home loose or uneven? Do all stairs have handrails? Are all stairways free of clutter? Is any carpeting firmly affixed to the floor so that it will not slip?
- Does the bathtub or shower have a non-slip surface to prevent falls? Is the water heater set to prevent scalding and burns? Is the bathroom free of water damage, moisture, and mold?
- Do residents have reliable access to clean and safe drinking water?
- Is the kitchen free of pests and rubbish?
- Has the home been tested for toxic substances such as lead paint and asbestos? Is the home ventilated to prevent the buildup of radon gas? Are household chemicals, such as cleaning supplies, safely stored?
- Is the home equipped with working smoke alarms and carbon monoxide detectors?

- Does the home have adequate temperature control to prevent extreme heat and extreme cold?
- Are there sidewalks that facilitate safe walking near the home? Is the home located near a park, a playground, or another place where residents can safely engage in physical activity and recreation?

Similar lists of questions could be developed for schools, healthcare facilities, workplaces, and other locations.

Some of these questions can be answered by trained observers. These assessors may describe findings qualitatively, assigning ratings like "high" or "low" to observed conditions based on predefined lists of rating criteria. Other assessments require laboratory testing for environmental contaminants, such as tests of paint chips for lead and tests of basement radon levels. For some types of hazards, the exposure dose, frequency, and duration must be ascertained. Risk assessments may be conducted at one point in time or at several time points. Researchers must have the permission of owners and/or residents before entering a building, walking on private property, or conducting environmental assessments of a structure or lot.

24.10 Geographic Information Systems

A **geographic information system (GIS)** is a computer-based platform for mapping the locations of events, identifying spatial clusters, and testing complex spatial associations. The global positioning system (GPS) uses satellites to collect data about the latitude, longitude, and sometimes the altitude of locations. A GPS receiver can acquire the geographic coordinates for relevant locations, such as the homes of participants, nearby hospitals and other healthcare facilities, roads, schools, religious and social organizations, grocery stores, recreation facilities, water sources, and industrial sites. The coordinates for public locations can be collected by anyone, but permission from the owners or residents of private land may be needed before taking a GPS reading on their property. The GPS coordinates for the homes of participants are individually identifying information, so precautions must be taken to protect geographically linked personal data.

24.11 Monitoring and Evaluation

Monitoring and evaluation, often shortened to just **M&E**, is an important management tool that draws on a variety of qualitative and quantitative techniques. **Monitoring** is ongoing assessment to ensure that a project or program is staying on track toward achieving predefined targets. **Evaluation** is an assessment process that includes a variety of approaches for examining how well a project, program, or policy has achieved its associated goals, processes, and/or outcomes. The goal of M&E is to determine what is working well and what can and should be improved. M&E is typically done for specific performance purposes rather than as part of developing generalizable insights about the world. However, some M&E activities evolve from inquiry processes into research projects with broader value. Additionally, the methods of M&E can be useful for some types of research projects.

A **project** is a specific, time-limited set of activities. A **program** is an ongoing group of projects. **Program evaluation** is the systematic collection and analysis of data to answer questions about the effectiveness and efficiency of a program. There are many types of program evaluations that can be conducted, and they consider the various aspects of a project, including the inputs, processes, outputs, outcomes, and impacts. **Formative evaluation** describes the needs assessments and feasibility studies conducted as part of developing a new intervention or modifying an existing one.

Process evaluation (also called implementation evaluation) is the systematic analysis of an ongoing intervention in order to ensure that procedures are being implemented as planned. **Outcome evaluation** (sometimes called effectiveness evaluation) includes processes that examine whether an ongoing intervention is making good progress toward achieving stated objectives. **Impact evaluation** is the determination of whether an intervention achieved its objectives.

A **stakeholder** is a person who has an interest in the success or failure of a group and can influence or be affected by that group's decisions or actions. In the health sciences, a typical program evaluation begins with a meeting at which stakeholders describe the purposes of the program being examined, how it was intended to function, how it is actually functioning, and what they themselves hope to learn from the assessment. Based on these conversations, an evaluation approach is selected. Evidence is gathered from a variety of sources, possibly including a review of existing program documents, surveys of stakeholders, interviews with key informants, and observations at program sites. All the evidence is then reviewed and categorized, perhaps using a framework like realist synthesis or SWOT. **Realist synthesis** uses a systematic process to find and analyze evidence for the complex reasons why some programs succeed and others fail. **SWOT** identifies the strengths (internal organizational strengths), weaknesses (internal organizational limitations), opportunities (external strengths), and threats (external limitations, which might be political, economic, sociocultural, technological, environmental, or legal) of a program. Finally, practical suggestions are made based on the conclusions of the assessment.

A similar process can be used as a component of other forms of evaluative research, such as needs assessments, cost-effectiveness analyses, and **health services research**, which examines factors related to the types of health services and providers available to a population, the organization and financing of those health services, and the impact of governments and policies on population health.

CHAPTER 25

© DmitriyRazinkov/Shutterstock

Secondary Analyses

Some health research studies analyze existing clinical records, survey data, or population data rather than collecting new data.

25.1 Overview of Secondary Analysis

A **secondary analysis** is a study in which a researcher analyzes data collected by another entity. For secondary studies, the data collection stage of the five-stage research process is the step of acquiring an existing data set. The data file or files used for a secondary analysis may be publicly available individual-level or population-level data, privately held survey data, or electronic or paper health records. Whatever the data source, what makes a project a secondary analysis is that the researcher conducting the analysis has no involvement in collecting data from individuals. Secondary analysis is often an excellent option for researchers with strong statistical skills but limited time and/or data collection resources. A researcher conducting a secondary analysis contributes to scientific knowledge by analyzing and interpreting accumulated data that might otherwise remain unexplored.

25.2 Accessing Secondary Data

Understanding what existing data are available and what possible costs may be incurred when accessing them is an important early step to complete when planning a secondary analysis project. Sometimes a researcher can download spreadsheets from the Internet that are already cleaned and ready for analysis. At other times, data are available only as paper records or as electronic files from which the relevant information must be extracted and cleaned prior to analysis. Some data files are provided at no cost to the researcher. Sometimes researchers must pay to access data.

Some sponsoring organizations that have made their data sets available to the public at no cost to the user allow those files to be downloaded on demand from their websites. Sometimes there is a screening process. The researcher may be required to submit a formal proposal and have the research plan approved by an oversight committee before the requestor can be provided with a copy of the data by email or via a link to a password-protected download site. In rare cases, access to some data files may be limited to particular types of people, such as citizens or residents of the country in which the data were collected.

A researcher conducting secondary analysis needs to understand all the methods that were used for data collection and become familiar with all the variables in the data files. In addition to downloading the data files, the researcher should download and read all supporting documents, such as the project overview, protocol, or handbook; the questionnaire; the codebook; and any published articles that describe the origins of the data set and previous analyses of the variables in it.

25.3 Publicly Available Data Sets

A growing number of governmental agencies allow researchers access to anonymized individual-level data sets. An **anonymized data set** (also called a deidentified data set) is one that has been stripped of all potentially identifying information, such as names, street addresses, and personal identification numbers. **Deidentification** is the process of removing potentially identifying information from a data file so that the data can be shared with others without violating the privacy of the individuals whose data are included in the file. Governmental agencies often have expertise in collecting data but lack the resources to conduct a thorough statistical analysis of an entire data set before it becomes relatively obsolete. Sharing anonymized data with external researchers is therefore a cost-efficient way to extract as much information as possible from data sets, especially when the data were expensive to collect.

Available data sets are often listed on the websites of government health agencies. For example, the U.S. Centers for Disease Control and Prevention (CDC) website provides access to data from several nationwide cross-sectional studies, including the National Health and Nutrition Examination Survey (NHANES), the National Health Interview Survey (NHIS), the National Health Care Surveys, and the Behavioral Risk Factor Surveillance System (BRFSS). Other agencies within the U.S. Department of Health and Human Services also make data sets available to researchers, including the Administration for Community Living, the Agency for Healthcare Research and Quality, the Centers for Medicare and Medicaid Services, the Health Resources and Services Administration, the Indian Health Service, the National Institutes of Health, and the Substance Abuse and Mental Health Services Administration. The U.S. Environmental Protection Agency, the U.S. Department of Veterans Affairs, the U.S. Census Bureau (which sponsors the American Community Survey), USAID (which supports Demographic and Health Surveys in countries across the globe), and other agencies also participate in data sharing.

Additional data sets are available from other national, state, and provincial governments in addition to United Nations agencies like the World Health Organization. For example, Statistics Canada provides access to data sets such as the Canadian Community Health Survey via the Research Data Centres Program. Other data sets are available from compilation sites such as the Global Health Data Exchange (GHDx), which is hosted by the Institute for Health Metrics and Evaluation.

Some research teams supported by federal funding agencies and some private organizations are also required by their funders to make their data available to researchers upon request, and others voluntarily share their data. Investigators who make their data available to the public often do not expect to be coauthors on papers written by independent analysts. However, they may expect their contributions and/or the sources of funding and technical support to be acknowledged.

The supporting documents should state the expectations. If they do not, the researcher should ask a contact person for clarification. If the secondary analyst requires assistance from the individuals involved in designing the study and/or collecting and processing the data, those individuals may qualify for coauthorship even if the supporting documentation does not say that this is necessary. It is good practice to clarify coauthorship requirements and expectations prior to beginning secondary data analysis.

25.4 Private Data Sets

Individual researchers and small research teams may have data available that have not yet been analyzed. The researchers may have computerized data files that have not yet been fully explored, or paper records may have been set aside because they are not a current priority of the research team. Sometimes the original researcher or research team may have published the results of some portion of a data set but left unanalyzed some of the other potentially significant, interesting, and novel aspects of the study. In these situations, the original researchers may be open to a new researcher taking the lead on analyzing an underexplored portion of the data set and writing up the results for possible publication.

A request for access to a private data set is most likely to be granted when the new researcher has some existing connection to the original researcher. Students are most likely to have success asking their own professors for data sets to analyze. If students are interested in the work of a research group at another university or hospital, they may find it helpful to ask their professors to reach out to colleagues at the other institution. The ethics review committees of both institutions may need to approve the data sharing plan, especially if identifiable information might be included in the data file.

When privately held data are shared with a new investigator, the original researchers usually expect to be coauthors on any resulting publication. The roles and responsibilities of each party should be agreed on as early as possible in the research process, preferably before the data files are shared.

25.5 Challenges of Secondary Research

There are several major limitations associated with using already available data. The most important one is that the analyst is limited to exploring only the topics and specific questions included in the original survey or clinical records. There may also be quality control concerns. The analyst has to trust that the data were collected using valid and standardized methods and that the supporting documentation accurately describes the actual procedures used for data collection.

Another challenge arises when the analyst has questions about the data collection and management procedures that are not spelled out in the supporting documents. Finding someone who can answer those questions might be difficult. Some websites that provide access to research data do not list the name of a contact person, and some of the listed contacts may not have been integrally involved in the study design and data collection process.

A final issue is the risk of duplicating the analysis that someone else has done or is doing. A literature search may uncover related works that have been published or are in press, but it will not identify analyses in progress or papers under review by journals. The contact person for freely downloadable data sets may not know whether other researchers are conducting an analysis of the data or what topics other researchers are focusing on.

25.6 Clinical Records

Clinical records are a common source of data for case series. Individuals working in clinical settings often can apply to gain access to patient records for research purposes. Most clinical sites require researchers to submit an application to an oversight committee for review and approval prior to being authorized to access the data. The application must explain the goals of the study, the process that will identify eligible patient records, the specific details that will be extracted from each patient's file, the steps that will be taken to protect the confidentiality of the data set, and the analysis plan. Applicants must also provide evidence of having successfully completed both research ethics training and specific instruction about patient privacy laws and policies. For example, researchers working with patient records from the United States must be prepared to comply with the **Health Insurance Portability and Accountability Act (HIPAA)**, a set of regulations about patient protection that must be carefully followed.

Sometimes the relevant data can be extracted from an electronic database. The healthcare organizations may require the researcher to pay for the time it takes a database technician to extract the requested data. When electronic records are not available, a data extraction form can be created and used to compile the relevant details from each patient file. The extracted data can be entered directly into a computer database or recorded on paper for later data entry. Whenever possible, the data files should not contain any individually identifying information.

Some secondary analyses are conducted using deidentified records from registries. A **registry** is a centralized database containing information about people who have had a particular exposure or been diagnosed with a particular disease. For example, voluntary registries for many diseases with a relatively low prevalence have been established to facilitate the ability of researchers to access data about these conditions and to recruit patients for clinical trials.

A major limitation of using existing clinical records is that patient records are often incomplete. Researchers cannot make any assumptions about the missing information. For example, researchers cannot assume that the absence of information about a symptom means the patient did not experience the symptom. The patient might have had the symptom but failed to mention it to the clinician. Perhaps the clinician did not specifically ask whether the symptom was occurring. Maybe the patient did mention the symptom but the clinician did not record it, perhaps because the symptom did not seem especially relevant. Similarly, researchers cannot assume that the information in the medical records of one healthcare provider tells a complete story about those patients' health status. Consider medication use. Researchers cannot assume that a patient is not taking a particular medication just because that patient's records at one clinical site do not mention that the patient has been prescribed that drug. The patient might have been prescribed the medication by a clinician at some other site. Furthermore, even if the patient's records show that a prescription was written for a particular medication, that does not mean that the patient filled the prescription and took the drug. If the research question requires complete information about symptoms or medication usage or other details, a primary study design may be necessary.

25.7 Health Informatics, Big Data, and Data Mining

Health informatics is the application of advanced techniques from information science and computer science to the compilation and analysis of health data. **Bioinformatics** is the use of computer technologies to manage biological data. Bioinformatics often focuses on analysis of molecular-level data (or, less often, tissue-level data). Clinical informatics and public health informatics usually focus on patient- or population-level data. The tools of health informatics can be used to create novel data sets for research purposes.

Big data refers to data sets that are so large and complex that they must be analyzed using powerful hardware and special statistical software applications. These data sets may include data for many thousands or even millions of individuals. Clinical databases are one source of large data sets. An **electronic medical record (EMR)** is a digital version of a patient's medical history and other details recorded at one healthcare provider's office. An **electronic health record (EHR)** is a digital version of a patient's health data that is designed to be shared among different healthcare providers. Other big data sets might relate to consumer behavior, such as the data collected and stored by credit card companies and the developers of smartphone applications. Large data sets can also be generated by collating social media posts or content from other Internet-based sources.

Codes within large data files enable relevant data to be extracted. EHR and EMR systems often use SNOMED CT (Systematized Nomenclature of Medicine Clinical Terms) as a standard terminology. Billing records often use ICD codes (International Classification of Diseases codes) based on diagnoses or CPT codes (Current Procedural Terminology codes) based on procedures. Laboratory records often use LOINC codes (Logical Observation Identifiers Names and Codes). Medication records often use NDC codes (National Drug Code identifiers).

Data mining is the process of examining big data sets to identify patterns and develop new knowledge. Text mining and other forms of data mining can be used to extract particular phrases from large sets of records. Clinical informatics projects might use data mining techniques to explore hospital records. Public health informatics projects might use data mining and computational linguistics to explore social media events. Big data approaches have the power to reveal patterns and trends that are not apparent in smaller data sets analyzed with traditional statistical methods. Specialized training is usually required before researchers are prepared to implement data mining and other big data methods.

25.8 Ethics Committee Review

Use of hospital records for research purposes always requires review by one or more research ethics committees. If the data for a secondary analysis come from a private source, the analyst usually must obtain clearance from his or her own institution and perhaps also from the institution that houses the data. This permission must be secured prior to even looking at the data set. The application for permission to analyze existing data is often shorter than the application required for primary studies, and review usually can be expedited. It is better to err on the side of submitting an unnecessary proposal than to erroneously presume that a project is exempt from review without confirming the validity of this assumption.

Most publicly available data, especially those collected by government agencies or federally sponsored researchers, were collected under protocols approved by one or several research ethics committees and then the files were stripped of all personal identifiers prior to being shared. Additional approval by an ethics committee at the institution where the secondary analysis will be conducted is often not required when several conditions are met:

- The data were collected after approval by a trusted organization's research ethics committee.

- The data set contains no individually identifying information.
- The data to be analyzed are publicly available.

However, researchers are responsible for becoming familiar with the requirements of their own institutions and ensuring that their work is compliant with all institutional policies. When there is any doubt about whether review is required, the institutional review board should be consulted.

CHAPTER 26

Systematic Reviews and Meta-analyses

A systematic review is the careful compilation and summary of all publications relevant to a particular research topic, and a meta-analysis creates a summary statistic for the results of systematically identified articles.

26.1 Overview of Tertiary Analysis

Tertiary analysis involves the review and synthesis of existing knowledge about one well-defined topic. Most synthesis research in the health sciences takes the form of a systematic review. For some study questions, it is appropriate to create a summary statistic by pooling data from the included studies. However, meta-analysis is not required for most systematic reviews.

Figure 26-1 illustrates the systematic review process. The data collection process for tertiary research studies involves searching research databases using carefully selected keywords, screening potentially relevant abstracts and full-text articles for eligibility, and then extracting data from all of the eligible publications. This literature review process allows the researcher to gather all of the information needed to understand, summarize, and synthesize the current state of knowledge about the topic.

26.2 Search Strings

Once a well-defined study question has been selected, the next step in a systematic review or meta-analysis is composing appropriate

FIGURE 26-1 Systematic Review Process

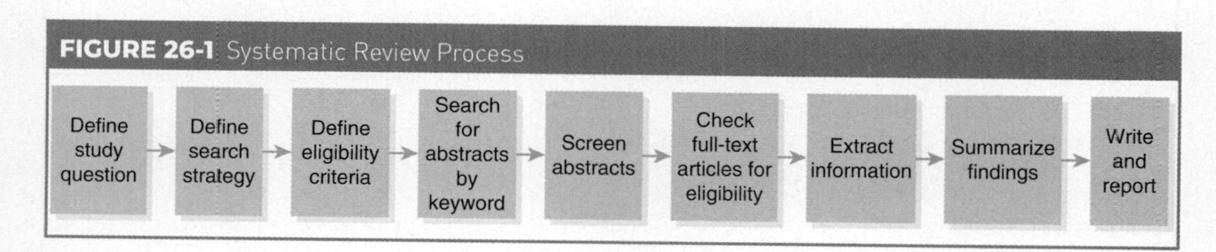

search strings. A helpful first task is an exploration of the MeSH dictionary (available through the PubMed website) in order to identify the definitions of key terms as well as synonyms and related terms. For example, a search for "health care costs" shows that synonyms for this term include "treatment cost" and "medical care costs." Subheaders (sometimes called child terms) for "health care costs" include "direct service costs," "drug costs," "employer health costs," and "hospital costs." "Health care costs" itself is a child term for several categories (sometimes called parent terms), including "health care economics and organizations," "health care quality, access, and evaluation," and "delivery of health care."

The next step is to begin building candidate search phrases using keywords or MeSH terms. **Boolean operators** are conjunctions such as AND, OR, and NOT that can define relationships between search terms. Most abstract databases allow parentheses, square brackets, or other notation to indicate the start and end of a search string or a component of a search string. The word OR expands the number of results that will be generated by a search (**Figure 26-2**). A search for [a OR b] will find any abstract that includes "a" or "b" or both. The word AND shrinks the number of results from a search. A search for [a AND b] will yield only abstracts that include both terms. More complex search strings can also be used. For example, a search for [a AND (b OR c)], which will find any abstract that includes both "a" and "b" or includes both "a" and "c" or includes all three search terms. The exact search string(s) used for a systematic review should be presented in the methods section of the resulting research report. (If a variety of lengthy search options, such as the names of all countries in the world region of interest, were applied to various databases that use

FIGURE 26-2 Examples of Boolean Operators in Search Strings

Search String	Approximate Number of "Hits" in PubMed
schistosomiasis	27,000
"schistosomiasis"[MeSH]	23,000
Schistosoma mansoni	14,000
cancer	4 million
bladder cancer	80,000
bladder cancer NOT schistosomiasis	75,000
bladder cancer OR schistosomiasis	105,000
bladder cancer AND schistosomiasis	700
cancer AND schistosomiasis	1600
bladder cancer AND *Schistosoma mansoni*	40
(bladder cancer OR colorectal cancer) AND schistosomiasis	850
bladder cancer AND colorectal cancer AND schistosomiasis	20

Note: The PubMed database is constantly adding new abstracts, so the numbers in this table will not exactly match the results of a new search.

slightly different syntax for searches, it may be appropriate to provide the search parameters in an appendix rather than in the main text.)

Understanding the language used by MEDLINE and other databases allows for the design of a database-appropriate search string. For example, in MeSH language a "child" is defined as a person who is 6 to 12 years old. Individuals who are 2 to 5 years old are classified as "preschool children," and those who are 13 to 18 years old are "adolescents." A keyword search of [child]—that is, a search for the word "child" in all of the titles and abstracts of articles indexed in PubMed—will yield hundreds of thousands more hits than a search for ["child"[Mesh]] that only searches for articles indexed with "child" as a MeSH keyword. The researcher must be attuned to these particularities when designing search procedures.

To check the appropriateness of search terms, the researcher can identify several articles known to be relevant to the study question and then confirm that the search string captures all of those articles. If the search misses one or more of those key references, then the search strategy needs to be modified. However, this process must not be used to exclude disliked articles, which would cause the inclusion bias that systematic reviews seek to minimize.

Once a validated system for identifying all of the potentially eligible articles is in place, the selected abstract databases are systematically searched for articles that might meet the inclusion criteria. If the topic is appropriately narrow, then keyword searches can often reduce the number of abstracts and/or articles that must be screened for eligibility to a reasonable number, often several hundred articles rather than many thousands of articles.

26.3 Search Limiters

Researchers must be cautious about artificially limiting the number of articles that will be identified during a literature search based on publication year, database, or language. Any limiters must be justified. For example, many systematic review protocols limit searches to include only recently published studies, but this must be done with great caution. In many situations, older papers will still be relevant to the study question and should not be excluded based solely on presumed obsolescence. A study seeking to characterize current prevalence rates or examining a recently implemented public health law might reasonably exclude studies that are more than 10 years old, but a study examining risk factors might not be justified in excluding older papers. When the year can be justified as a relevant inclusion criterion, the year of data *collection* should usually be used for determining eligibility rather than the year of *publication*. It would be inappropriate to include a study that collected data in 1995 and published the results in 2010 while excluding a study that collected data in 2008 because its findings were published in 2009.

Researchers must be able to answer the following types of questions:

- Why did the search include only articles published after 2010?
- Why was the search restricted to articles indexed in MEDLINE rather than using a more diverse set of databases (such as CINAHL, PsycINFO, and SciELO in addition to MEDLINE)? If no scientific justification is obvious, then the search should include multiple databases.
- Why was the search restricted to English-language papers rather than including a more comprehensive set of articles? This question is especially important for reviews covering the global population and not just English-speaking countries. A researcher's lack of fluency in other languages is not an acceptable justification for including only English-language articles. Online translation programs can assist with multilingual searches. For studies focused on countries

with rich literatures in non-English languages, a collaborator who is fluent in those languages can be recruited.

- Does the use of a broad search term like "adult" or "United States" help or hurt a search, given that many papers reporting on these populations do not include these descriptors as keywords or even mention them in their abstracts?

Researchers should be especially cautious about using the built-in filters available in some abstract databases. For example, PubMed allows researchers to use filters to restrict results to particular types of articles (such as clinical trials or reviews), particular species (such as human-only studies), and particular age groups (such as infants or adults aged 65+ years). These limiters work only if an article was indexed appropriately by the submitting journal. Because many articles about humans do not add "human" as a keyword and many studies do not include keywords for the ages of participants or the study design, the built-in limiters often exclude many studies that would otherwise be eligible for the review. It is usually better to use study-specific exclusion checklists to screen out abstracts that are ineligible rather than to artificially limit the number of screened abstracts using filters.

26.4 Supplemental Searches

Three additional strategies may be used to complement database searches: snowball sampling, searching the grey literature, and hand searching. The methods used for these expanded searches must be disclosed in the research report, and any documents found through these supplemental search methods must meet all of the inclusion criteria for the study.

Snowball sampling, or "snowballing," is a literature searching technique that involves looking up every article cited by eligible articles in order to identify additional sources that might be relevant even if they are not indexed in the selected databases.

The **grey literature** (or gray literature) consists of research reports that are available in a format that is not indexed in databases of journal article abstracts. The grey literature for a health science study might include government reports that have been scientifically vetted and published, dissertations and theses that have been successfully defended and are available in repositories, and early versions of research manuscripts that have been posted on preprint sites for review but have not yet been accepted for publication in a peer-reviewed journal.

Hand searching is conducted by scanning every article in the table of contents of selected volumes of relevant journals to see if any of those articles might be eligible for inclusion in a review.

26.5 Eligibility Criteria

The decision about an article's eligibility for inclusion in a systematic review or meta-analysis is based on predetermined lists of inclusion and exclusion criteria. These eligibility criteria should ensure that all of the included studies pertain to the main research question. If the study question includes specific exposures, diseases or outcomes, and/or populations, the eligibility criteria should ensure that all of the included studies match those "EDPs." The eligibility criteria can also impose some requirements about the study design and sample size, if those restrictions are scientifically justifiable. Studies of causality are often appropriately limited to reviews of experimental studies, but reviews that are not focused on causation generally do not need to exclude observational studies.

The protocol for a systematic review typically defines a list of inclusion criteria and a complementary list of exclusion criteria. Suppose that an analysis of the connections between tobacco use and lung cancer included the following as some of its inclusion criteria:

- The study used a case–control or cohort study design.
- The article reported an odds ratio or rate ratio for the relationship between cigar smoking and lung cancer.
- The study included at least 50 participants with lung cancer.

This study would then have the following as some of its exclusion criteria:

- The study used a cross-sectional design, an experimental design, or any design other than a case–control or cohort study.
- The article did not include information about cigar use.
- The article did not report a statistic for the association between cigar smoking and lung cancer.
- The study included fewer than 50 participants with lung cancer.

To be eligible for inclusion in the tertiary analysis, an article would need to meet all of the inclusion criteria and none of the exclusion criteria.

Researchers must be able to justify each of the inclusion and exclusion criteria they select for a systematic review. This means that they must be able to satisfactorily answer a variety of methodological questions. For example:

- Why were only randomized controlled trials included rather than also considering case–control, cohort, and other observational studies?
- Why were articles published before the year 2010 excluded?
- If a quality assessment was part of the inclusion criteria, what evaluation tools were used to assess the quality of the studies and the risk of bias in their results? Was this a fair mechanism to use to decide which studies merit inclusion in the review?

The screening process begins during the searches of the selected databases. The title and abstract for each article found during the search are reviewed so that an initial determination can be made about whether the article is likely to be eligible for inclusion in the systematic review. For example, a search for studies about hepatitis B might yield many articles about other types of viral hepatitis, such as hepatitis A and hepatitis C. The titles and abstracts of those studies will make it clear that they did not look at hepatitis B, and the irrelevant articles can be removed from the list of potentially eligible items. Similar determinations can often be made about the study population. If a systematic review is focused on Canada, abstracts that are obviously about research studies conducted in other countries can be excluded during the initial screening phase. When an abstract is not available or the abstract does not allow a determination about ineligibility to be made, the full text of the article must be read.

Most systematic reviews end up with approximately 10 to 25 included articles after screening, although some have many more than that. The full text of each of these articles must be read to confirm eligibility. Ideally, each article should be assessed by at least two independent reviewers. **Figure 26-3** summarizes this process. The count of articles at each step—identification, screening, checks of eligibility, and inclusion in the analysis—should be included in the research report. The PRISMA flow diagram and similar tools facilitate complete reporting of the searching and screening process.

26.6 Quality Assessment

Not all published research is generated from equally rigorous methods. Tertiary analyses typically conduct some type of quality assessment before including studies in their analysis. A fair and transparent process must be used to determine which studies merit inclusion in a systematic review and which appear to lack

FIGURE 26-3 Systematic Search Strategy and Counts to Report

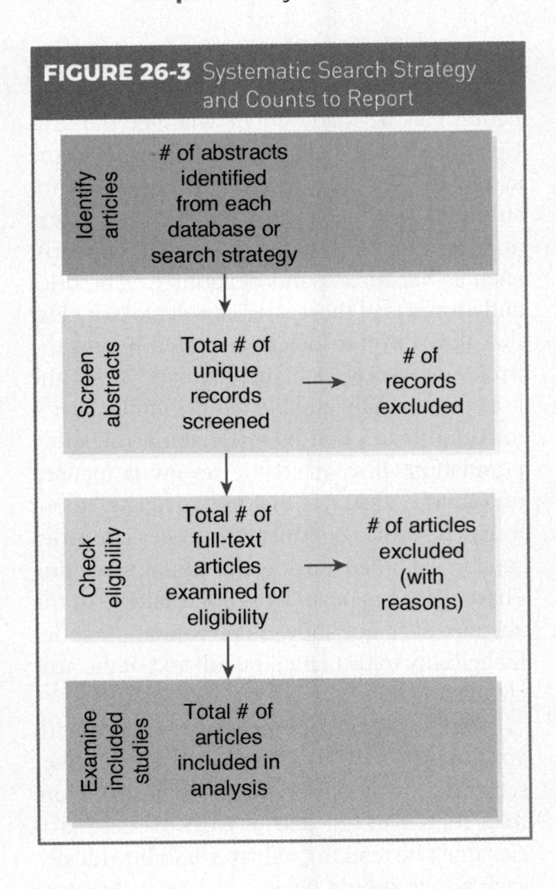

internal validity or external validity and must therefore be excluded. For example, the critical appraisal process might consider whether a study's methods are likely to have measured exposures and outcomes precisely, whether the study results are likely to represent a true effect or are likely to be due to some type of bias, and whether there are any unresolvable concerns about vagueness or inconsistencies pertaining to the methods and results.

A second type of quality assessment applies to considerations of the body of knowledge as a whole. Suppose that a systematic review process is being used to examine whether a particular intervention is effective at improving patient outcomes. If many high-quality randomized controlled trials have shown a meaningful improvement in outcomes for the intervention group, that is strong evidence for the value of the intervention. If the only publications about the intervention are a few small case series, that is weak evidence. The conclusion might be that the available data are insufficient, and it is not possible to make a recommendation for or against the intervention. A variety of quality assessment scales, checklists, and other approaches can help guide this process.

26.7 Data Extraction

Once all eligible articles are identified, the content of these articles is summarized in a data extraction table that lists descriptive characteristics such as:

- The study location
- The years of data collection
- The study design
- The study population and sample size
- The definitions used for key exposures and outcomes
- The key findings of interest, including both quantitative (numeric) results and qualitative conclusions
- An evaluation of the quality of the study

A data extraction table allows for easy compilation and comparison of observations relevant to the study question. A condensed version of the table is usually included in the research report.

26.8 Systematic Review Results

When interpreting the results of a systematic review, studies that find no statistically significant results for an item of interest are just as valuable as those that find a significant association. The researcher should record and report both statistically significant findings ($p < .05$) and statistically insignificant findings ($p \geq .05$) that are related to the main study question. A report may state, for example, "Five of 40 published studies of the association between exposure A and disease B found an increased rate of disease B among those exposed to A; the remaining 35 studies found no association." That is a more accurate depiction of the literature than a report that merely says, "Five studies found an increased risk of disease B in those exposed to A." The latter statement incorrectly implies a consensus that exposure A is significantly associated with disease B. One of the primary contributions of systematic reviews to the health science literature is the ability to identify both areas of consensus and areas of disagreement and uncertainty that need to be further examined.

Systematic review reports also need to address the possible influence of publication bias on the findings. **Publication bias** occurs when articles with statistically significant results are more likely to be published than those with null results. If 10 studies look at the association between the same exposure and disease, the 1 study that finds the exposure to be risky is much more likely to be published than the 9 null result studies. Proving that publication bias has occurred may not be possible, but consensus in a systematic review should be conservatively interpreted when only a limited number of studies have been published on a topic or the eligible studies include a mix of statistically significant and insignificant results.

26.9 Pooled Analysis

A meta-analysis creates a summary statistic by pooling the results of studies identified during a systematic review. Only comparable statistics from similar studies can be pooled. For example, a summary estimate of efficacy can be estimated from several high-quality randomized controlled trials with the same active intervention, the same type of control, and similar population groups. The results from studies using different study designs, different interventions, or dissimilar population groups should not be pooled.

Before pooling data from independent studies, the researcher must show that the results of the studies are comparable. **Homogeneity** is a synonym for similarity. **Heterogeneity** means dissimilarity. Homogeneous studies can be combined into a summary statistic, but caution should be used if the studies are heterogeneous. The amount of variability in a measure across studies is often examined using the Q statistic and the I^2 statistic. **Cochran's Q statistic** examines heterogeneity among the studies included in a meta-analysis by calculating the weighted sum of the squared differences between individual and pooled effects across studies. A large value for Q indicates that there is more variation across the studies than among participants within the included studies. The **I^2 statistic** adjusts the Q statistic based on the number of studies being pooled. I^2 is reported as a percentage from 0% to 100%. When I^2 is 0%, all of the variability can be explained by chance. When I^2 is greater than 75%, there is a high level of heterogeneity between studies. When a large number of studies or a very small number of studies are included in a meta-analysis, other statistical tests may be more appropriate.

If the examinations of variability across studies suggest that it is appropriate to generate a summary statistic, the next step is to select a model that will be used for creating a pooled estimate of the effect size. **Effect size** is the magnitude of the difference in the value of a statistic in independent populations. Many types of statistics can quantify effect sizes, including odds ratios, rate ratios, correlation coefficients, and difference in means measures. The effect size is important for determining whether a statistically significant difference is also a meaningful one. For example, an odds ratio of OR = 1.05 might be statistically significant in a study with a large number of participants, but that value is so close to OR = 1.0 that it might not represent a meaningful difference in risk. By contrast, an odds ratio of OR = 3.0 is likely to be statistically significant and represent a meaningful difference in risk. OR = 3.0 is a greater effect size than OR = 1.05, because it is farther from the null value (OR = 1.0). Statistical tests can assist with the evaluation of the impact of effect sizes. For example, a statistic called **Cohen's *d*** evaluates whether the result of a *t* test comparing means is significant enough to be meaningful in applied practice.

There are two main choices of models to use for meta-analysis. A **fixed effects model** can be used to create a pooled estimate when there is little variability among the included studies. A **random effects model** can be used to create a pooled estimate when tests of heterogeneity show that there is considerable variability among the included studies. The point estimate for the summary measure will be similar for both model types. However, a random effects model will result in a wider 95% confidence interval for the summary statistic because the random effects model will adjust for the variability among the included studies.

A specialized computer software program can use the selected model to estimate the value of the pooled statistic (such as a pooled Mantel-Haenszel adjusted odds ratio) and its confidence interval. **Weighting** is a statistical method that adjusts for sampling methods, demographic differences between a study population and a source population, different sample sizes among the studies included in the meta-analysis, or other circumstances. In meta-analysis, the contribution of each study to the pooled estimate is usually weighted based on its sample size, but other approaches to weighting can be used. The methods sections of reports about meta-analyses should explain how the authors determined that a meta-analysis was justifiable, how they investigated heterogeneity among the studies, and why they selected a particular weighting method for the analysis. Step-by-step guides to meta-analysis techniques are available from Cochrane and other research groups.

26.10 Forest Plots and Funnel Plots

The statistics from contributing studies and the pooled statistic generated from a meta-analysis are often displayed using a **forest plot**, a graphical display of the effect sizes of the included studies and the pooled statistic (**Figure 26-4**). A forest plot usually has:

- A horizontal axis showing effect size.
- A vertical line showing the effect size that indicates no effect (such as an odds ratio of 1).
- A row for each included study that uses a square or other marker to indicate the

FIGURE 26-4 Example of a Forest Plot

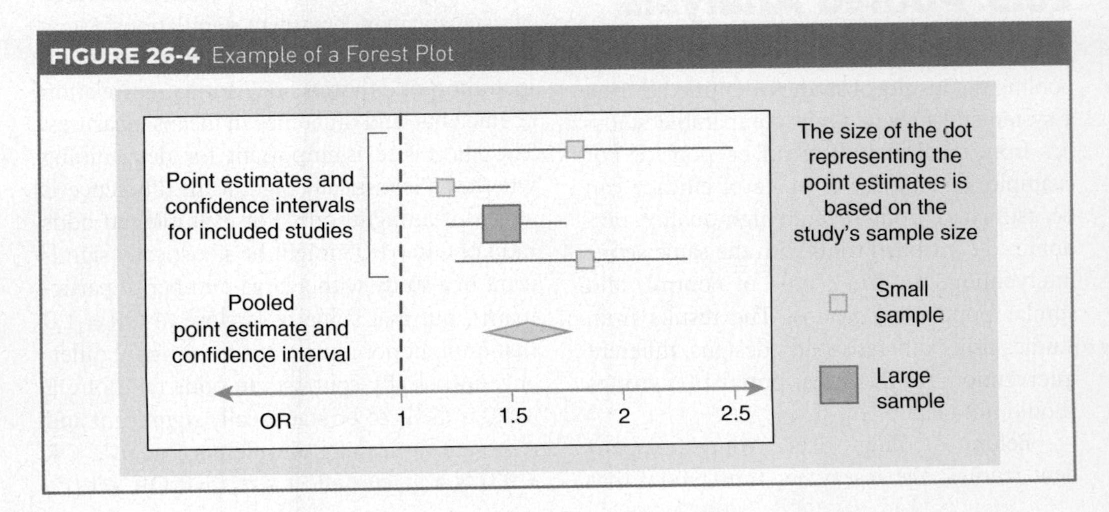

point estimate for the effect size and uses a horizontal line to show the 95% confidence interval. Markers for the point estimate are often presented in different sizes that illustrate how each study was weighted in the meta-analysis. Small markers usually indicate studies with small sample sizes, and large markers usually indicate studies with large sample sizes.

- A representation of the summary measure and its confidence interval, often shown using a diamond shape.

There are two main threats to the validity of a meta-analysis: poor quality of included studies and publication bias. The selection criteria used during the systematic review process can eliminate studies of questionable validity. The possibility of the preferential publication of studies reporting a statistically significant or favorable outcome can be examined using a funnel plot. A **funnel plot** is a graphical display of the results of the studies included in a meta-analysis that reveals the likelihood that publication bias has kept relevant studies with null results out of the formal literature. A point for each study included in the meta-analysis is plotted on a graph that shows the effect size on the x-axis and the number of participants on the y-axis (**Figure 26-5**). If no publication bias has occurred, the points for the included studies will form a triangle. If publication bias has reduced the number of publications with statistically insignificant results, part of the triangle will be missing. In that situation, the pooled estimate is likely to have overestimated the true effect size.

FIGURE 26-5 Example of a Funnel Plot

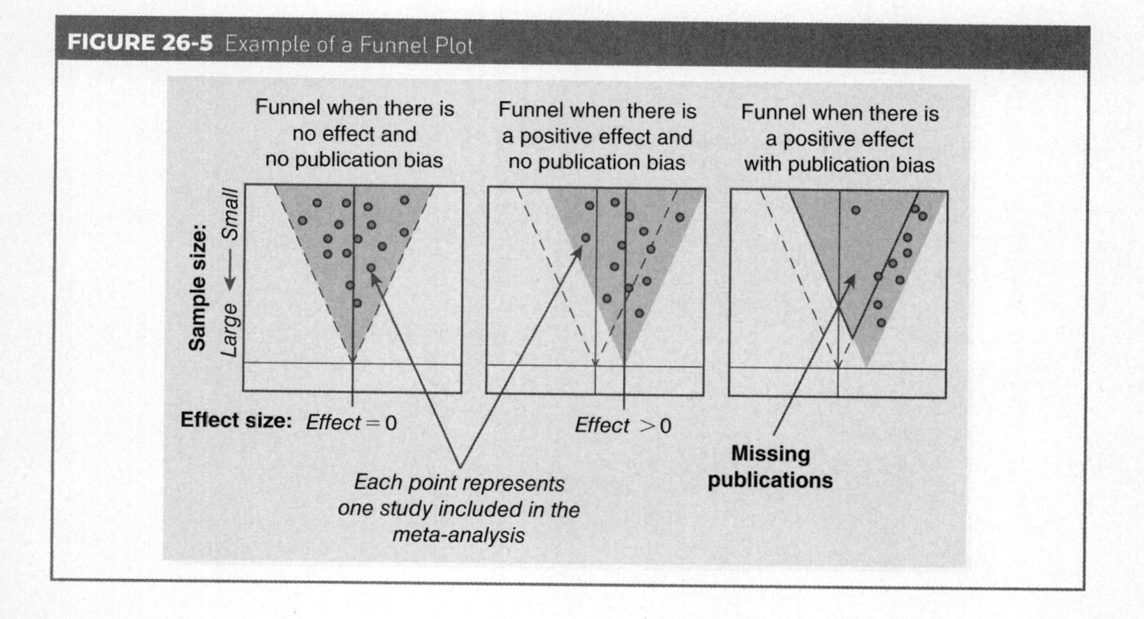

© DmitriyRazinkov/Shutterstock

CHAPTER 27

Writing Grant Proposals

A proposal written to request funding or approval for a new research project must demonstrate that a new research question is important and that the research plan will yield an answer to that question.

27.1 Preparing to Write a Proposal

A **proposal** is a written request for approval of or funding for a research project. A formal research proposal is commonly written for two purposes. One is to seek approval for a project from a supervisor or a review panel, as occurs when a student submits a proposal to a thesis or dissertation committee for review, feedback, and eventual approval. The other is to apply for grant funding.

Much of the planning and design work for a project must be completed before a proposal can be written. The study goal and specific aims must be finalized, and the researcher must be able to justify the value of the proposed study. Critical decisions about the study design and methodologies must have already been made, and there must be evidence to support the feasibility of the project. The author of the proposal must also be prepared to explain how the proposed study aligns with the expectations of the host institution or the mission of the funding entity.

27.2 Identifying Grant Opportunities

Although not all research projects require financial support, projects sometimes need outside sources of money, or, at a minimum, they would be significantly enhanced by monetary support. The main sources of funding for research include universities and colleges, governmental agencies, private foundations and nonprofit organizations, and businesses. An **internal grant** describes research funds provided by the researcher's school or employer. An **external grant** is a grant funded by an organization outside the researcher's institution.

A diversity of resources may be useful when searching for funding opportunities:

- Supervisors and mentors may be able to offer advice about sources of internal and external funding that are appropriate for the particular project under development.
- The grants management offices of colleges, universities, healthcare systems, and other organizations may offer consultations

to researchers affiliated with those institutions.

- The websites of funding agencies provide details about the types of research they fund and the eligibility criteria for applicants.
- The newsletters and websites of some professional organizations include lists of new and ongoing funding opportunities relevant to people working within that discipline.
- Some subscription databases compile information about grant opportunities, and these may be accessible through library websites or other institutional offices.

There are several factors researchers should consider when selecting which grant opportunities to apply for. These questions include:

- What research areas and types of research methodologies are supported by the granting organization? Some funders support only projects focused on a particular disease or a very specific population, while others are much more general in scope. Some fund only technology-intensive and laboratory-based studies, while others support only library-based research.
- How much money is available? Some student-focused awards may allow budgets of only a few hundred dollars, while some government agencies offer millions of dollars to established researchers.
- When is the submission deadline? Some funders offer rolling submission deadlines, while others have only one funding cycle per year or one cycle every other year. If a project must be completed within a particular time frame, those scheduling demands must align with the funder's timeline.
- How long after submission of a proposal will an award decision be made? When will funds be available to grantees? Some granting organizations make decisions about funding as proposals are submitted, while others may require nearly a year to make a decision about whether to fund a project.
- How competitive is the award? Some grants (usually ones open only to students in a particular program) are available to nearly everyone who applies for them, while other agencies may fund less than 1% of submitted proposals.

27.3 Requests for Proposals

A **request for proposals (RFP)**, alternatively called a **request for applications (RFA)**, is a notice distributed by a funding organization that is seeking applications from researchers who want to conduct research on topics of interest to the funder. The RFP specifies the research areas of particular interest to the funder and describes the types of projects the selection committee will consider supporting. The instructions will also specify whether individuals should submit their own grant proposals or whether submissions must be made by an institution's grants management office.

A funding group may ask all applicants to submit full proposals, or they may use a multistage application process. A **letter of intent** (LOI) presents a preliminary research plan and states the intention to submit a full proposal. Some funding organizations require researchers to submit an LOI several weeks or months before the deadline for the full proposal. Receiving this information from likely applicants enables the funding agency to prepare a process for reviewing all of the planned submissions. A **preproposal** is a brief research plan required by a funding organization that wants to confirm alignment between the funder's vision and the proposed research plan before inviting a full proposal to be written and submitted.

Sometimes a funding organization does not circulate RFPs, and instead invites potential applicants to contact the funder and ask about current organizational priorities. A **letter of inquiry** is a letter a researcher sends to a potential funding organization to ask about whether a particular research idea might be of interest to the funder. If there is alignment between the project and the goals of the funder, the researcher might be invited to submit a full proposal.

When a researcher submits a proposal in response to an RFP, that is considered to be an unsolicited proposal because the funding group is open to receiving applications from a diversity of eligible candidates. A **solicited proposal** is a request for funding submitted by a researcher after a funder has contacted the researcher to invite that researcher to submit a proposal. When a funder solicits a proposal, the organization might offer a contract rather than a grant. A **contract** is research funding that requires the researcher to deliver an agreed-upon product to the funder. A contract usually requires that a particular product, such as a commissioned report, be submitted to the funding agency by the end of the contract period. The term **deliverable** describes a tangible or intangible object produced to fulfill the terms of a contract-funded research project. The organization might stipulate that the final payment on the contract will not be disbursed until after a satisfactory deliverable is submitted by the researcher.

27.4 Research Proposal Components

Research proposals typically include a standard set of components (**Figure 27-1**):

- An abstract or a short summary of the proposal
- A background (sometimes called a literature review) that explains what is known and what is not known about the proposed study area and justifies the importance of the proposed project
- A statement defining the overall research goal and the specific aims
- A research plan or **project narrative** that describes the methods that will be used to answer the research question and explains why those methods are appropriate
- A plan for the dissemination of the study's findings
- A timeline
- A budget with a justification for each item
- Details about the researchers

Additional components may be required, such as an abstract written for a nontechnical audience, a description of the facilities and other resources available to the researcher, a statement about the broader impacts of the research, and letters of support from collaborators.

Guidelines from funding agencies and review committees usually specify how a proposal should be organized, what content should be included in each section of the proposal, and how long each section should be. For example, the instructions may allow only the lead investigator to submit a two-page CV (or résumé), or they may require each collaborator to submit a **biosketch** (a brief summary of the individual's professional and educational accomplishments) that follows a template from the funding agency. The guidelines may restrict the number of references that are allowed to be cited, and they may dictate how those citations should be formatted. It is important for applicants to carefully follow all the instructions of the organization to which the application materials will be submitted. Neglecting to be compliant with all of the rules about formatting, layout, word length, components, and other details can be grounds for cursory rejection of the proposal. Poorly written, disorganized, and typo-filled applications are also easy to dismiss.

FIGURE 27-1 Typical Proposal Content

Background ■ Brief summary of what is already known about the topic that includes (1) a literature review citing the previous work of other researchers and (2) a summary of the researcher's own previous work on the topic and any preliminary results, if applicable ■ Purpose of the new project ■ Significance and importance of the new project ■ Definition of key terms
Goal(s) and specific measurable or testable aims, objectives, or hypotheses
Methods and procedures ■ Study design ■ Source population (for new data collection) or data source (for analysis of existing data) ■ Sampling methodology and expected sample size ■ Recruiting procedures (for new data collection) ■ Definition and measurement of key variables ■ Data collection procedures ■ Laboratory procedures (if applicable)
Analysis plan ■ Data management plan ■ Data analysis plan
Dissemination plan
References
Timeline
Budget and budget justification
Researcher information (such as a biosketch, CV, or résumé)
Optional appendices ■ Letters of support ■ Questionnaire or other survey instruments. ■ Research ethics review application and supporting documents

27.5 Writing a Research Narrative

An effective research proposal clearly answers three questions:

- What is the problem that the project will examine? (What are you going to study?)
- Why is the problem important? (Why do you want to do the study?)
- How will the proposed project help solve the problem? (How will you do the study?)

The answers to all three of these question areas should be clear in the abstract and the narrative. Every component of the submission should answer at least one of these three questions. Each section of the proposal should contribute to substantiating the importance of the health issue that will be studied, expressing how the proposed work will contribute to advancing knowledge and improving health, and/or justifying the validity of the planned methods.

For funding proposals, the applicant must also convince the funding organization that the researcher will successfully answer the study question if given money to implement the project.

Every component of the proposal—the title, the summary, the narrative, the budget, the biographical sketch, and all other items—should express why the researcher should be given money for the project. The background needs to demonstrate that the health concern that will be studied is a serious problem that is worth supporting financially. The methods section needs to convince the reader that the proposed approach to answering the study question is a valid and efficient one that is worth bankrolling. The budget needs to show that the researchers will make good use of the money given to them. The biographical information about the researchers needs to prove that the research team has the experience necessary to see the project through to successful completion so that funds given to them will not be wasted.

Funding applications also need to clearly connect the research idea with the goals of the sponsor. However, it is not a wise idea to use language declaring that the application is a perfect fit or an ideal proposal for the funder. Instead, provide the information the reviewers will need to make that determination for themselves.

When writing a research proposal, the researcher should usually assume that readers of the proposal will not know much about the particular research area. It is the writer's responsibility to provide the background information necessary for the proposal to be understood by diverse technical and nontechnical audiences.

27.6 Funding Criteria

If a funding organization provides information about the criteria that will be used to evaluate proposals, applicants should address those criteria specifically and overtly in the application packet. If possible, key points related to areas of evaluation can be bolded in the text to draw attention to them. Figures, tables, or diagrams can be used to illustrate critical methodologies.

Research proposals submitted to the U.S. National Institutes of Health (NIH) are scored as a function of significance, investigators, innovation, approach, and environment (**Figure 27-2**). If a funding organization does not make its criteria available to applicants, these five domains provide a starting point for strengthening the competitiveness of a research proposal. In addition to these five areas, the application should show how the proposed

FIGURE 27-2 NIH Review Criteria

Significance	Does the application show that the proposed project will answer an important research question that can lead to improvements in individual or population health status?
Investigator(s)	Does the application provide evidence that the research team has the training, experience, and skills required to successfully complete the proposed project?
Innovation	Does the application explain what is novel and exciting about the proposed research question and methodologies?
Approach	Does the application demonstrate that the proposed methods for the study are rigorous, ethical, and feasible and will answer the research question? Are the methods ones that can be implemented with the proposed budget?
Environment	Does the application provide evidence that if the proposal is funded the research team will have access to the resources required to successfully complete the proposed project (such as data sets, library services, laboratory equipment, statistical software, and so on)?

project aligns with the goals of the sponsoring agency and its typical funding level.

27.7 Budgeting

Granting agencies prioritize funding for research projects that will answer well-defined and significant study questions using a budget appropriate for the work that will be done. The budget should cover all the essential costs of the research project without being excessive in total amount or in any category. Each line in the budget may need to be accompanied by an explanation of why the item is necessary and a description of how the budget for that item was determined. A student or trainee applying for a small grant may be limited to requesting funding for only basic expenses, such as photocopying study materials or obtaining a license for data collection or analysis software. Other studies may become quite expensive if they require travel to a distant field site, laboratory testing or other clinical assessments, lengthy durations of data collection, and the hiring of interviewers and data entry personnel.

Direct costs are the specific monetary expenses associated with a particular research project. A large grant proposal may request support for a variety of direct costs, such as:

- Salaries (or partial salaries) and benefits for key personnel
- Stipends for consultants and support staff, such as interviewers and laboratory technicians
- Funds for expenses related to data collection, such as providing beverages and snacks to study participants, offering gift cards to volunteers, and reimbursing interviewers and interviewees for mileage and parking or covering the costs of public transportation to and from the study site
- Funds for the purchase of equipment and supplies (such as computers, smartphones, software programs, storage devices or services, and laboratory equipment and consumables)
- Funds for communication and office expenses directly related to the project (such as photocopying, postage, and Internet access)
- Support for dissemination activities like presenting at conferences (which requires payment of registration fees, transportation, hotels, and meals) and paying for publications to be made open access

Allowable costs are expenses that are approved for a funded grant or contract as opposed to items that are not acceptable according to the terms of the grant or contract. For example, an organization might mandate that equipment be purchased from particular vendors, or it might allow meals but not alcohol to be purchased when traveling for grant-related work.

Overhead describes the institutional costs of maintaining research infrastructure, operating research facilities, purchasing library resources, and administering research functions such as ethics reviews and compliance reporting. **Indirect costs** are the general research-related expenses that institutions incur but cannot attribute to specific research projects. These expenses are sometimes called facilities and administrative costs, often abbreviated as **F&A costs**. In addition to providing money to pay for the direct costs of research, some funding agencies allow the host institution of the researchers to request that a portion of the grant budget be allocated to cover indirect costs. Foundations may allow an indirect rate of 10% to 20%, although applications for small grants typically do not allow any overhead to be included in the budget. Federal grants may allow a substantial F&A rate to be applied to the grant, sometimes exceeding 50% of the direct costs.

When preparing a budget, it is important to carefully read the funding agency's guidelines for what direct costs are allowable and what indirect rate (if any) can be requested. If there is a cap on the funding amount, determine whether the cap applies to just direct costs or if it applies to the total direct plus indirect costs. A funder that allows a 25% F&A rate may indicate that the overhead should be part of the total budget. For a funder with a maximum award of $100,000, the budget would allocate $75,000 for direct costs and $25,000 for indirect costs. Another funder that allows a 25% F&A rate might allow a maximum of $100,000 in direct costs and then add $25,000 to cover indirect costs, making the total value of the award $125,000.

When developing a budget, money and materials are not the only resources to consider. For many studies, the most important resources are the individuals who are available to contribute their time, expertise, and connections to the project. Nonmonetary resources may include:

- Access to potential study participants
- Access to data sets
- Use of existing laboratory space, office space, and meeting rooms
- Availability of existing equipment, such as computers and scanners

These existing resources can be highlighted in a grant proposal as part of the description of the research environment available to the researcher.

It is not unusual for the funding cap for a student research award to be lower than the actual amount required for a project. In this situation, the researcher should show in the grant application which expenses will be covered by the new grant, if funded, and which will be supported by other sources. The researcher should be prepared to turn down offers of partial financial support if they are inadequate for the project. For example, if the direct costs of a project will be $1400 and funding is secured for only $250, the researcher may decide not to undertake the project rather than needing to invest personal money in the project along with time and energy.

27.8 Financial Accounting

Grant awardees accept responsibility for the financial management of their projects and must carefully update and maintain accounting records. Regular internal appraisals of accounting paperwork, equipment logs, and other documentation must be conducted. The primary investigator should always be prepared for a possible external **audit**, a systematic check of financial records and other actions and decisions that is conducted to confirm accuracy and compliance with standards of practice. **Reconciliation** is the process of resolving any discrepancies between the researcher's financial records and the reports produced by the institution hosting the researcher's grant or contract accounts. Accounts should be checked at least monthly and reconciled to ensure that the balance in the account matches the balance in the financial records. Working closely with a budget officer at the host institution can prevent costly mistakes.

All primary investigators and others with responsibilities for grant operations should follow best practices for money management, including:

- Adhering to all of the policies, regulations, and laws of the funding organization, the host institution, and the government
- Maintaining impeccable records of all project-related activities (including time sheets, if salaries or stipends are part of the budget)
- Confirming that every expenditure is allowable before making a purchase
- Keeping receipts for all purchases
- Maintaining a log of all equipment and conducting regular inventory checks
- Consulting with the granting agency and host institution authorities before reallocating any portion of the approved budget to another area

Financial reports may be required to be submitted to the host institution and the funder on a monthly, quarterly, or yearly basis. In addition to financial reports, technical or performance reports may be required on a quarterly, semiannual, or annual basis. These reports typically provide details about the research activities completed under the grant since the previous report, including summaries of major findings and details about any project-related presentations or publications.

27.9 Grant Management

Funders that have decided to support a research project will send an award letter or other notification to the grantee that specifies the amount of money being offered, the opening and closing dates for the grant, and the obligations of the grantee. The initial paperwork establishes the acceptance of the terms and conditions of the grant. Grant-related documentation is then required at assigned times throughout the months or years the grant is active. Interim reports provide updates on finances and scientific progress. Final paperwork for the closeout of the grant typically includes a final accounting report and a final project outcome report. Some funders also request updates on all grant-related dissemination activities that occur in the years after the grant ends. The content of the various types of reports is dictated by the sponsor, and these updates are in addition to any paperwork required by the host institution, such as annual reviews of human subjects research protocols.

If a portion of the budget has not been spent as the end date nears, some (but not all) funding agencies will allow an extension of the timeline for spending the budgeted money. A **no-cost extension** postpones the closing date for the grant but provides no additional funding. It merely allows more time to spend already-allocated funds. Other funders have a "use it or lose it" model and will take back any money that has not been appropriately disbursed by the end of the grant period. Some funders offer the opportunity for grantees to apply for additional funding. A **grant renewal** or **grant continuation** is an extension of a grant that provides additional funding to continue a research project and expand it in new directions.

Closeout is the process that determines that all applicable administrative actions and all required work for an award have been completed by the grantee. Closeout paperwork must usually be submitted shortly after the closing date for the grant. All grant-related paperwork must be retained by the primary awardee for at least several years after closeout, in case the funder, the host institution, or a governmental agency requires an audit to be conducted.

27.10 Unfunded Research

Although receiving a research grant is an accomplishment worth celebrating, it is important to remember that getting a grant (or signing a contact) is not the same as actually implementing the research plan. A grant is merely the opportunity to conduct a particular research project with funding support. Not all research projects require funding. Many projects can be successfully completed with minimal or no costs to the researcher beyond the researcher's time. For example, a secondary analysis of existing data or a review of the published literature may require only access to a computer, a statistical software program, and a decent collection of electronic journals. Some primary studies that collect new data incur only minor expenses, such as the cost of printing flyers that explain how volunteers can access an online survey form. Although funding may open up opportunities to conduct more elaborate research studies, researchers without grant funding still have many options for doing meaningful research.

© DmitriyRazinkov/Shutterstock

STEP 4

Analyzing Data

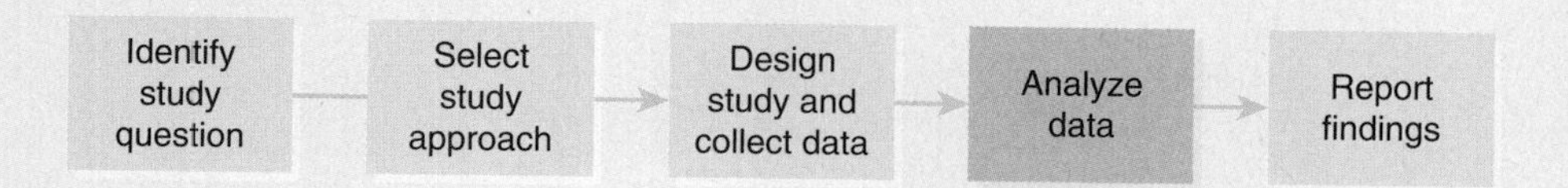

The fourth step in the research process is compiling and analyzing the data that were collected during step 3. Most research projects require only the use of descriptive and perhaps some comparative statistics, but others benefit from the use of advanced analytic methods.

- Data management
- Descriptive statistics
- Comparative statistics
- Regression analysis
- Qualitative analysis
- Additional analysis tools

CHAPTER 28

Data Management

Data entry, data cleaning, and recoding are important preparatory steps for data analysis.

© DmitryRazinkov/Shutterstock

28.1 Data Management

Data management is the entire process of record keeping before, during, and after a research study. Data management refers to extracting data from patient charts for a case series, logging the responses to a cross-sectional or case–control survey, recording all the results of clinical assessments conducted during a longitudinal cohort or experimental study, or tracking articles considered for eligibility in a systematic review. Data managers must take care to protect the confidentiality of protected data and to ensure the integrity of data sets. Once data are entered into a database or spreadsheet, the files typically need to be cleaned before beginning analysis.

28.2 Codebooks

A **codebook** is a guide written for a particular study that describes each variable and specifies how the collected data will be entered into a computer file. It is useful to create a codebook prior to beginning data entry (**Figure 28-1**). For quantitative surveys, numeric or alphabetical codes can be assigned to the options for the closed-ended response options provided on the questionnaire form. For open-ended questions and qualitative studies, a codebook provides clear instructions for how to organize and classify free-response comments.

In addition to providing specific instructions about how each piece of data should be entered into the computer file, the codebook should specify:

- The name of each variable (which usually employs only a limited number of capital letters or a combination of capital letters and numbers and avoids starting with a symbol such as an underscore)
- The variable type
- The wording of the question that was asked
- The options listed on the survey instrument as possible answers to the question
- The specific instructions for how answers should be entered into the computer database
- How to handle missing responses

The codebook is also the place to describe how anticipated data problems will be handled. For example, what should be done if

FIGURE 28-1 Example of Codebook Entries

Question Number	Variable Name	Question	Variable Type	Variable Length	Codes
1	INTDATE	{Date of interview}	Date	8	Enter as DD-MM-YYYY
2	AGE	What is your age in years?	Numeric	3	■ Enter number ■ *Missing = 999*
3	SEX	What is your sex?	Text	1	■ Male = M ■ Female = F ■ Other/Prefer not to answer = 9 ■ *Missing = {leave blank}*
4	WORK	Which of the following categories best describes your current employment status?	Text	10	■ Working full time = FULLTIME ■ Working part time = PARTTIME ■ Unemployed but want to work = UNEMP ■ Retired = RETIRED ■ Student = STUDENT ■ Homemaker = HOME ■ Other = OTHER ***➔ If OTHER go to 4b; otherwise skip to 5***
4b	WORK_ OTHER	Other occupational description	Text	50	{Enter text as reported by respondent; only enter for those for whom WORK = OTHER.}
5	STUDENT	Are you currently enrolled in school?	Text	1	■ Yes = Y ■ No = N ■ *Don't know/Missing/Refused = D*
6	FISH	In the typical month during the past year, how often did you eat fish?	Numeric	1	■ Never = 0 ■ Less than once per month = 1 ■ About 1 to 3 times each month = 2 ■ About once per week = 3 ■ Several times per week = 4 ■ Every day or almost every day = 5 ■ *Don't know = 8* ■ *Refused/Missing = 9*
7	STD_EVER	Has a doctor ever told you that you had a sexually transmitted disease?	Numeric	1	■ Yes = 1 ■ No = 0 ■ *Don't know = 7* ■ *Refused to answer = 8* ■ *Missing = 9*

a respondent marks two answers in a multiple-choice list when the instructions explained that only one item should be selected? What if the handwriting on a form is illegible or the person doing the data entry is not absolutely certain about what the words say or which box was checked? If unanticipated quandaries arise, the codebook should be amended to state how the situation was addressed, so that there is a record of the decision and problems with subsequent entries can be resolved in ways that are consistent with that original decision.

The codebook will also specify for each variable whether missing answers should be left blank in the computer file, indicated with a numeric code (such as entering a 9 if the expected entry code is 0 or 1 for a dichotomous variable), or marked with the word "MISSING." The statistical analysis process may need to account for the ways that missing data were handled. For example, if missing responses for age are entered as 999, all of the "999" entries will need to be removed prior to analysis or else the mean age will end up artificially high and the corresponding standard deviation will be very large.

28.3 Data Entry

Two types of software programs are used for entry of quantitative data: spreadsheets (like Microsoft Excel or Google Sheets) and databases (like Microsoft Access). A **spreadsheet** is a file that stores data in the cells of a row-by-column table. A **database** is a data management system that stores data in tables in which each row represents one record, and related records in different tables can be linked. Both database and spreadsheet files can be uploaded into statistical software programs for analysis.

If data are entered into a spreadsheet, variable names should be entered in the first row, with one variable per column. Each individual's data should be in a new row, with the first line of data in the second row of the spreadsheet. The advantage of this data entry approach is that it does not require creating a data entry form, defining fields and variable names, and testing the data entry system. The disadvantage is that it is easy to input inconsistent codes or even to accidentally enter new data over an existing row of data. Spreadsheets often require significant cleaning prior to analysis.

Databases have built-in features that help ensure the consistency of entries and the completeness of the file. A database can limit the values that are acceptable for each variable, force fields to be completed before a record is saved, and automatically leave blank any fields that are irrelevant for a particular record. Databases also tend to have better data security features than spreadsheets, and some systems facilitate data sharing among users.

Double-entry is a method for ensuring the accuracy of a data file by having two individuals enter the same data into separate computer files (or the same person enter the data into two different files), comparing the two files for agreement, and resolving any discrepancies. File comparison software programs can facilitate the creation of a clean final data file. The entries in the two data entry files are linked by an ID number or another unique variable and then the comparison program identifies all discordant entries. The researcher can look through each discrepancy and select the best response for the final clean data file after consulting the original survey forms. For example, suppose that one of the two database files indicates that a participant was 32 years old, and the other says that the participant was 42 years old. The original form completed by the participant may show that the true age is 42, and 42 can be selected as the correct entry for the cleaned file.

When a paper-based survey form is used and the responses are then typed into a computer file, it may be valuable to do double-entry of at least some of the completed forms (often a minimum of 10% of them) to confirm the accuracy of data entry. If the agreement between the two files is not extremely high, then double-entry of all records is probably required to ensure the accuracy of the final data file. This type of validity check is not necessary when respondents complete online survey forms or record answers on optical answer sheets (like the bubble sheets used for Scantron tests) that are scanned into a computer.

28.4 Data Cleaning

Data cleaning is the process of correcting any typographical or other errors in data files. **Figure 28-2** shows how errors such as extra spaces, typos, and the use of lowercase instead of capital letters can be corrected so that all of the responses follow the rules spelled out in the codebook. Most of these types of typographical errors do not occur when computerized data entry forms force responses to be selected from a limited list of precoded options. However, when paper-based data collection methods are used, these types of errors can occur frequently. Fixing incorrect entries sometimes requires rechecking original survey forms. For example, while an "m" or "N" for SEX might reasonably be assumed to be a mistyped "M" (for male), it is not appropriate to assume without checking that an "R" for STUDENT was intended to be an "N" (for no). Missing values in a computer database may also require rechecking the original survey forms to confirm that no details written on the survey forms were overlooked by the data entry person. When there is any doubt about the validity of an entry after paper-based data have been typed into a computer file, the respondent's original paperwork should be consulted.

The data cleaning step is also an appropriate time to remove extremely unreasonable responses. For example, suppose a participant's age in years is listed as 192. This number can reasonably be assumed to be a typo. If a paper-based survey form is available, it should be consulted for the true age. If the survey form lists the age as 192—or if the data collection process was computer-assisted and there is no paper trail—this value must be excluded from

FIGURE 28-2 Example of Data Cleaning

Before Cleaning			After Cleaning	
Variable	**Response**	**Frequency**	**Response**	**Frequency**
SEX	F	498	F	498
	M	493	M	497
	m	3		
	N	1		
STUDENT	N	899	N	903
	N	2	Y	89
	R	1	[Missing]	2
	Y	87		
	y	1		
	[Missing]	4		

analysis because it is clearly an impossible age. However, a study of adults could possibly include an individual with an age of 102 years. A value of 102 would not be reasonable to delete or ignore, but it would be worth checking the original survey form (if available) for agreement with the entry in the database.

Data cleaning should also ensure that duplicate entries are removed from the computer file and that the electronic records are complete, with all data from all participants entered into the computer file that will be analyzed.

28.5 Data Recoding

A **derived variable** is a new variable created during data analysis from existing variables in the data file. **Recoding** is the process of generating values for a new variable based on one or more existing columns of data in a file. For example, values for new categorical variables can be assigned based on the values of existing numeric or categorical variables. The variable AGE could be used to create a new variable ADULT that is coded as 0 (no) for any participant younger than 18 years and 1 (yes) for any participant who is 18 years or older (**Figure 28-3**). New numeric variables can be calculated using mathematical operators. For example, height and weight variables can be used to calculate a derived variable for the body mass index (**Figure 28-4**). Other types of mathematical operators can be used to generate the number of days between two dates and to conduct other types of calculations. Derived variables can be created before or during data analysis. Creating these variables prior to analysis is often the easiest approach.

A few simple, routine practices will help protect a data file during the cleaning and coding process. Always save a backup of the data file before creating new variables in a duplicate copy of the file. A saved file backed up elsewhere allows the researcher to start anew if a file is damaged during recoding. Never recode into the same variable, because that process replaces the original values with the new recoded values. Instead, always code into a different (new) variable. Having both the original variable and the new variable in the file enables the researcher to compare the original and derived values and confirm that the coding was done correctly.

28.6 Statistical Software Programs

There are many statistical software programs available to assist with coding and analyzing quantitative data, including R, SAS, SPSS, and Stata as well as Epi Info, MATLAB, Python,

FIGURE 28-3 Example of Recoding

Original Variable	Derived Variable	Example of Coding
AGE	ADULT	
6	0	
29	1	
43	1	IF AGE < 18, THEN ADULT = 0 IF AGE > 17, THEN ADULT = 1 IF AGE = [MISSING], THEN ADULT = [MISSING]
14	0	
91	1	
50	1	

FIGURE 28-4 Example of Calculating

Original Variable	Original Variable	Derived Variable	Example of Coding
HT_IN	WT_LB	BMI	
66	155	25.0	BMI = (WT_LB / (HT_IN * HT_IN)) * 703
73	253	33.4	
59	112	22.6	
63	159	28.2	
70	180	25.8	
61	98	18.5	

and others. Most statistical software programs are able to run all of the common statistical functions, including simple descriptive and comparative statistics as well as linear and logistic regression. The results generated from different programs are identical or nearly identical, because they use the same underlying equations.

The decision about which software program to use is often based simply on the preference of the analyst. Some programs are more user-friendly than others and have a point-and-click user interface rather than requiring programming. Some generate excellent data visualizations, while others create less impressive graphical outputs. Some are open-source tools (like R), while others are expensive if they cannot be accessed as part of an institutional license. If particular types of advanced statistics are required, such as some types of complex samples functions and regression modeling techniques, programs that include the necessary functions are desirable.

28.7 Data Security

Data security is the process of protecting computer files with passwords and other mechanisms for restricting unauthorized access and use. Researchers are legally and ethically required to maintain the confidentiality of any potentially identifiable personal information participants disclose with the expectation that the data will be used only for agreed-upon purposes. It is especially important to be careful with **protected health information (PHI)**, any information about an individual's health history or health status that by law must be kept confidential. One way to maintain confidentiality is to safely store paper records, including signed informed consent statements, in a locked and secure room. Another is to destroy individually identifying information once the records are no longer needed (such as after the data have been entered into a computer file and the files have been thoroughly cleaned) and a research ethics committee has approved the secure disposal of consent statements and other documents.

Data protection also requires the creation of secure computerized data files. In general, no individually identifying information (such as a name or national identity card number) should be included in an electronic file containing other information about participants (such as responses to survey questions or the results of laboratory tests). If there is a need to link records to individuals—something which is rarely necessary unless the study is following participants forward in time across multiple assessment periods—then two separate files linked by a unique study identification number should be maintained. One should contain participant names and contact details. The other should include all other study data. The file containing identifying information should be securely stored separately from the file that contains the other participant data. Access to all files containing sensitive information should be password-protected, and access to them should be limited to essential research personnel. A consultation with an information technology expert prior to data collection can help ensure the security of participant information.

CHAPTER 29

Descriptive Statistics

Descriptive statistics such as means, medians, proportions, and standard deviations are used to characterize the distributions of quantitative variables.

29.1 Analytic Plan by Study Approach

Biostatistics is the science of analyzing data and interpreting the results so that they can be applied to solving problems related to biology, health, environmental science, or related fields. Various types of statistics can be used to tell a complete and compelling story about the quantitative data collected during a health research study. For most research reports, especially those written by researchers with limited training in advanced statistical methods, the goal of analysis should be to use the simplest statistics possible to make the results of the study clear to the researcher and the intended audience.

The typical analytic plan for each of the major study design approaches is shown in **Figure 29-1**. Each plan starts with a description of the study population. **Univariate analysis** describes one variable in a data set using

FIGURE 29-1 Analytic Plan

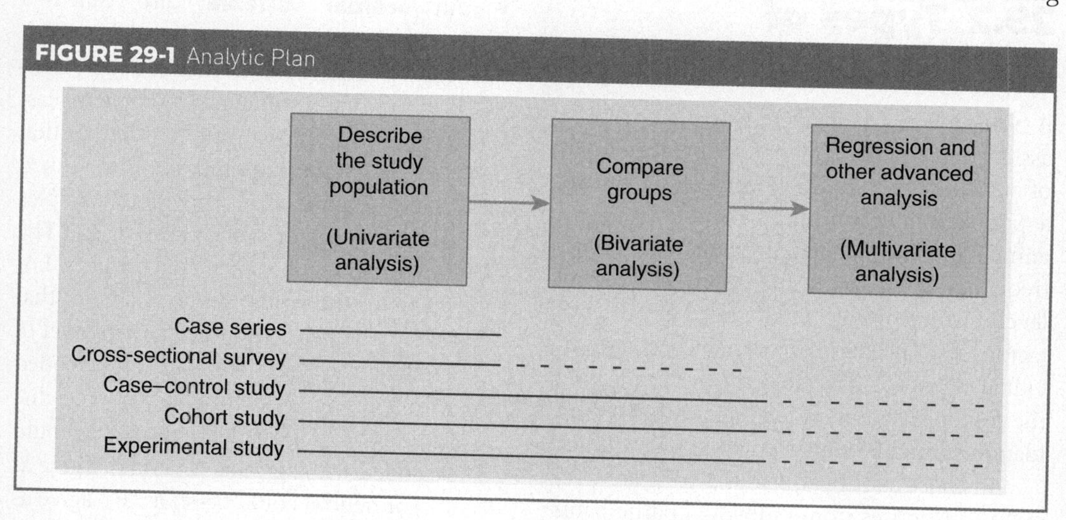

simple statistics like counts (frequencies), proportions, and averages. For studies with no comparison group, such as case series and cross-sectional studies, univariate analyses describing each of the key variables may provide a sufficient set of study results.

Bivariable analysis uses rate ratios, odds ratios, and other comparative statistical tests to examine the associations between two variables. For studies comparing two or more populations, the analysis must include both descriptions of the key variables and statistics that compare the groups. For example, a cohort study must compare the rates of disease incidence among exposed and unexposed groups, and a case–control study must compare the odds of exposure among the groups with and without disease.

Multivariable analysis encompasses statistical tests such as multiple regression models that examine the relationships among three or more variables. Most research studies do not require the use of complex statistics. Advanced statistics should be used only when they are appropriate for the study question and the analyst knows how to use and interpret them correctly.

29.2 Types of Variables

A **variable** is a characteristic that can be assigned to more than one value. Examples of variables that could be examined during a population health study include age, sex, annual income, languages spoken at home, frequency of alcohol ingestion, cholesterol level, history of chickenpox, and use of contact lenses. The value of a variable for an individual does not have to vary over time, but the response among individuals within a population should be something that might differ.

In most statistical analysis software programs, responses from individual participants are displayed in the rows of a data table and each column represents one variable. If one column presents the data for sex, one value for sex—such as an F or 0 for females or an M or 1 for males—will be listed in each row of that column. Another column may represent age in years, and one value for age—usually a whole number—will be listed in each row.

There are several ways to classify variables (**Figure 29-2**).

- A **ratio variable** is a numeric variable that can be plotted on a scale on which a value of zero indicates the total absence of the characteristic. For example, if height is measured in feet, a measurement of 0 feet tall means there was no height. As a result, the ratio of heights is meaningful. A person who is 6 feet tall is twice as tall as a person who is 3 feet tall, yielding a ratio of 2 to 1.
- An **interval variable** is a numeric variable for which a value of zero does not indicate the total absence of the characteristic. An outside temperature of 0°C does not mean there is no heat. If the weather turns colder, the temperature may fall to –10°C or lower. A day with a high temperature of 40°F is not twice as hot as a day with a maximum temperature of 20°F.
- An **ordinal variable**, also called a **ranked variable**, is a variable with responses that span from first to last, from best to worst, from most favorable to least favorable, or from always to never, or that are expressed using other types of ranked scales. (Figure 21-4 provides examples of other types of ranked responses.) The rank order can be assigned a number. For example, the responses to a survey that asks participants to indicate their level of agreement with a statement can be coded with agree as "3," neutral as "2," and disagree as "1." Alternatively, responses could be coded with agree as "1" and disagree as "3," or neutral could be set as "0," agree as "1," and disagree as "–1." No matter what

FIGURE 29-2 Types of Variables

Variable Type	Definition	Examples
Ratio	Numbers on a scale for which zero indicates the complete absence of the characteristic	Blood pressure, height, weight (The ratio of 20 kg to 10 kg is meaningful because the weight doubles when it increases from 10 kg to 20 kg.)
Interval	Numbers on a scale for which zero does not indicate the complete absence of the characteristic	Temperature (°F or °C) (The heat does not double if the temperature increases from 20° to 40°, because 0° does not represent the absence of all heat.)
Ordinal/ranked	An ordered series that assigns a rank to responses (from first to last in the series) but for which the numbers assigned to the values are not meaningful	Highest educational level completed, scales for never (1) to always (5), scales for strongly disagree (1) to strongly agree (5)
Nominal/categorical	Categories with no inherent rank or order	Employment sector, blood type
Binomial	Categorical variables for which only two responses are possible	yes/no, male/female, case/control

the scale is, the order of the responses is indicated by their numeric values.

- A **nominal variable**, also called a **categorical variable**, has values that represent no inherent rank or order. For example, there is no obvious way to numerically rank the favorite recreational sports activities of participants or their blood types. A **dichotomous variable** is a subtype of categorical variable with only two possible answers. A **binomial variable** is a dichotomous variable that has been coded as having values of only "0" and "1," such as coding yes as 1 and no as 0 or coding adults as 1 and children as 0.

Ratio and interval variables can be further classified as either continuous variables or discrete variables. A **continuous variable** is a numeric variable that can take on any value within a range. For example, although height is often rounded to the nearest inch when it is measured, a person's height could actually be 59½ inches or 68¾ inches or 77.1529 inches. A **discrete variable** is a numeric variable that is not continuous. Discrete variables often are generated by counting items, so there are gaps between the acceptable values. For example, a family can own 2 egg-laying chickens or 17 chickens, but cannot own 2½ chickens or 5¼ chickens.

29.3 Measures of Central Tendency

Descriptive statistics are often used to describe the average value of a variable in a population. For numeric variables, **central tendency** refers to various types of average values. There are several ways to report the average (**Figure 29-3**).

- A sample **mean** for a ratio or interval variable is calculated by adding up all the values for a particular variable and dividing that sum by the total number of individuals with a value for the variable.

FIGURE 29-3 Example of a Mean, Median, and Mode

Values Reported by Participants	Measure of Central Tendency	Value	Calculation
25 40 30 50 30 62	Mean	39.5	(25 + 30 + 30 + 40 + 50 + 62) ÷ 6 = 237 ÷ 6 = 39.5
	Median	35	The two middle values from 25-30-30-40-50-62 are 30 and 40; 35 is halfway between 30 and 40.
	Mode	30	Two participants provided a response of 30; no other response was listed more than once.

- The **median** for a numeric variable is identified by putting all the values for a particular variable in order from least to greatest and finding the middle number. Half of the responses in a data set will be greater than the median and half will be lesser than the median.
- The **mode** is the most frequently occurring value for a particular variable in a data set.

For ratio and interval variables, the central tendency can be described using means, medians, and modes. For ordinal variables, a median or mode can be reported. A mode can be reported for categorical variables.

29.4 Range and Quartiles

Variability describes the extent to which the values for a particular variable deviate from the average value of that variable in the data set. Means and medians provide information about the central value of a data set, but they do not provide information about how much variability is present. For example, the participants in a study of adults with a mean age of 50 years may all be 50 years old, or they could range from 18 to 104 years old. That information about the study population is very important when interpreting the results. Measures of **spread**, also called **dispersion**, describe the variability and distribution of values for a numeric variable.

For a particular variable, the response with the lowest (least) numeric value is the **minimum** and the response with the highest (greatest) value is the **maximum**. The **range** for a variable is the difference between the minimum and the maximum values in the data set. For example, if the youngest participant in a study is 18 years old and the oldest is 104 years old, the range is 104 – 18 = 86 years.

The median marks the value that divides the responses into two halves with equal numbers of observations after sorting the values from least to greatest. **Quartiles** mark the three values that divide a data set into four equal parts. The **interquartile range** (IQR) captures the middle 50% of values for a numeric variable. Other divisions can be made following the same pattern. For example, tertiles divide a data set into 3 equal parts, quintiles divide a data set into 5 equal parts, and deciles divide a data set into 10 equal parts.

29.5 Displaying Distributions

The best way to display the responses to a variable depends on the type of variable being visualized. A **histogram** is a graphical representation of the distribution of ratio or interval data in which the *x*-axis shows the values of responses and the *y*-axis displays the count of

FIGURE 29-4 Sample Histogram

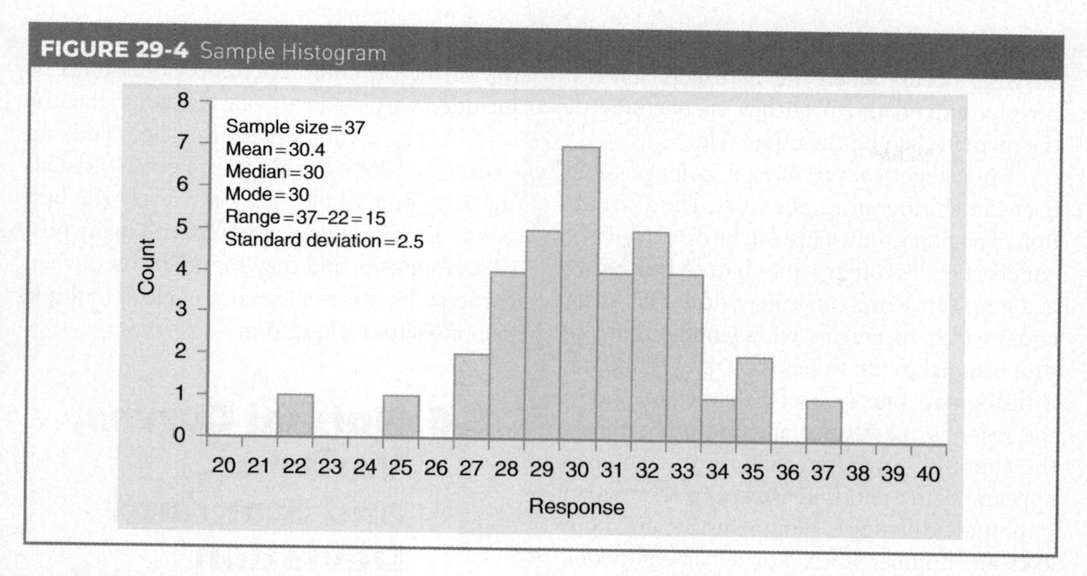

the number of times each response appears in the data set (**Figure 29-4**). For a graph to be considered a histogram, each bar must be the same width. There should be no gaps between the bars in the middle of the distribution, except for gaps that indicate values of the variable with a count of 0 responses.

A **boxplot** (also called a box-and-whisker plot) is a graphical depiction of a numeric variable that displays the median, the interquartile range, and any outliers. Boxplots can display the distribution of both ratio/interval and ordinal/ranked variables (**Figure 29-5**). Boxplots can be especially helpful for displaying

FIGURE 29-5 Sample Boxplot

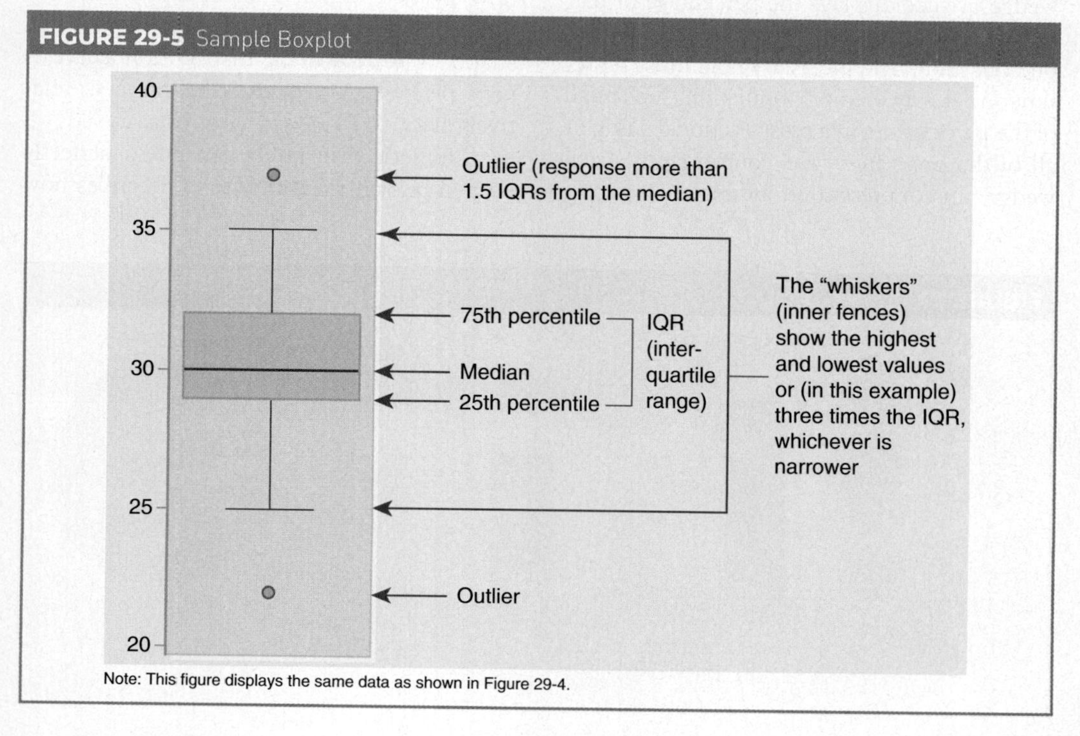

Note: This figure displays the same data as shown in Figure 29-4.

responses when the distribution is skewed. Skewing occurs when the "whiskers" on the boxplot extend much farther on one side of the median than on the other side.

For categorical variables, it is not possible to create a histogram or a boxplot. The distribution of responses must instead be displayed in a bar chart or, less often, a pie chart. A **bar chart** is a graph that presents categorical data using equal-width rectangles with lengths that are proportional to the values they represent. Like a histogram, one axis of a bar chart shows the response categories and the other shows the count of how often each set of responses appears in the data set. However, there is an important difference. For a histogram, both axes are number lines. For a bar chart, one axis is a number line and the other displays categories. The categories represented by the bars may appear in any order (**Figure 29-6**). The bars in bar charts can be displayed vertically or horizontally, and there are usually spaces between the bars.

A **pie chart** is a circle in which each wedge or slice displays the percentage of participants who provided a particular answer to one question. The pie as a whole must represent one clearly defined population, such as all of the participants in a cross-sectional study or all of the cases in a case–control study. Each wedge must represent an independent subset of the individuals included in the denominator for the pie chart. Each individual must be included within exactly one wedge, so that the sum of the percentages for the slices adds up to exactly 100%. Pie charts are used to visualize parts of a whole. They are rarely the best option for displaying variables with many possible responses, and they cannot be used when participants were allowed to select multiple responses to one question.

29.6 Normal Curves, Variance, and Standard Deviation

If the data for a numeric variable have a **normal distribution** (or **Gaussian distribution**), a histogram of the data will show a bell-shaped curve with one peak in the middle. Normal and nearly normal curves do not all look identical (**Figure 29-7**). **Kurtosis** describes how peaked or flat a bell-shaped distribution is. A mesokurtic curve has a typical bell shape. A **leptokurtic** distribution curve is very peaked. A **platykurtic** curve is relatively flat.

Numeric data rarely generate a perfectly shaped bell curve. **Skewness** describes how

FIGURE 29-6 Sample Bar Chart and Pie Chart

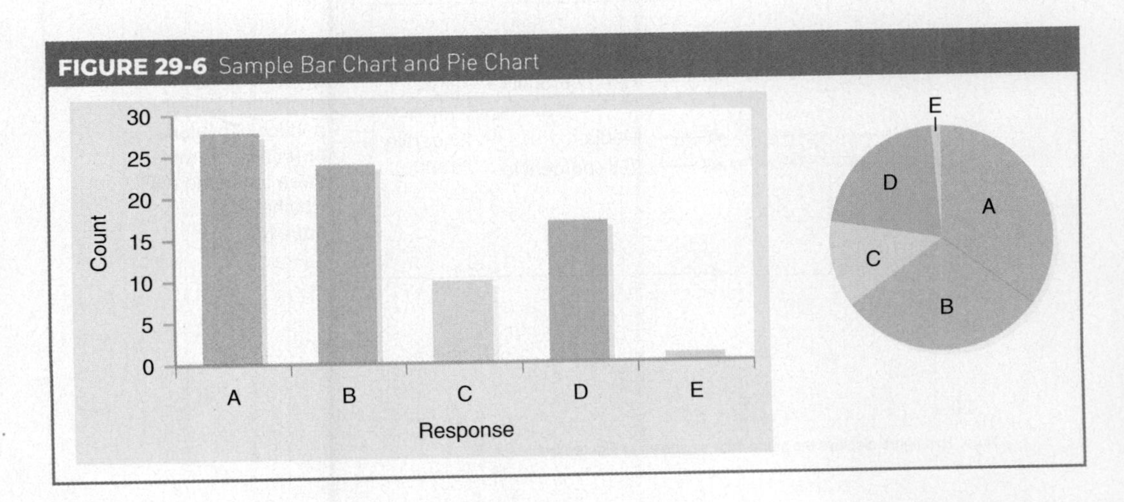

FIGURE 29-7 Kurtosis and Skew

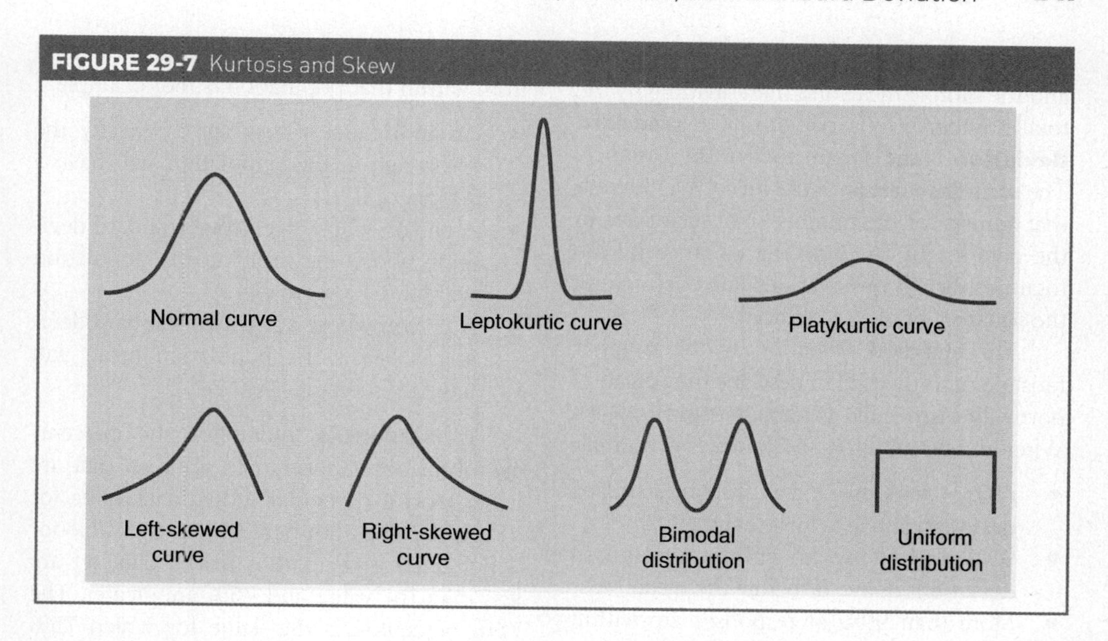

asymmetrical a nearly normal distribution is. If the responses extend farther to the left than to the right, the curve is left-skewed. If the "tail" extends farther to the right than to the left, the curve is right-skewed.

Additionally, not all numeric data generate a **unimodal** distribution with just one peak. A histogram might show a **bimodal** distribution with two peaks, or it might show a **uniform distribution** that appears rectangular because approximately equal numbers of responses were provided for each allowable value of the numeric variable.

For variables with a normal or approximately normal distribution, there are several ways to quantify the narrowness or wideness of the distribution (**Figure 29-8**). The **variance** is calculated by adding together the squares

FIGURE 29-8 Example of Variance, Standard Deviation, and Standard Error

Values Reported by Participants	Measure of Spread	Value	Equation	Calculation
25 40 30 50 30 62	Variance	202	The sum of the squares of the differences between each point and the mean divided by the sample size minus 1 (in a sample drawn from a larger population)	$[(39.5 - 25)^2 + (39.5 - 40)^2 + (39.5 - 30)^2 + (39.5 - 50)^2 + (39.5 - 30)^2 + (39.5 - 62)^2]/(6 - 1) = [14.5^2 + (-0.5)^2 + 9.5^2 + (-10.5)^2 + 9.5^2 + (-22.5)^2]/5 = [210.25 - 0.25 + 90.25 + 110.25 + 90.25 + 506.25]/5 = 1007.5/5 = 201.5$
	Standard deviation	14.2	The square root of the variance	$\sqrt{201.5} = 14.1951$
	Standard error of the mean	5.8	The square root of the variance divided by the sample size	$\sqrt{\frac{201.5}{6}} = 5.7951$

of the differences between each observation and the sample mean and then dividing by the total number of observations. The **standard deviation** is the square root of the variance. The **standard error** of the mean is a measure that adjusts for the number of observations in the data set by dividing the variance by the total number of observations and then taking the square root of that number.

The standard deviation is the number most commonly used to describe the spread of normally distributed variables (**Figure 29-9**). When the distribution of responses is normal:

- 68% of responses fall within one standard deviation above or below the mean
- 95% of responses are within two standard deviations above or below the mean
- More than 99% of responses are within three standard deviations above or below the mean

A small standard deviation indicates that most responses were fairly close to the mean. A large standard deviation indicates that the distribution of responses was wide.

A ***z* score** is a number that indicates how many standard deviations away from the sample mean the response for an individual from within that population is. For example:

- An individual whose age is exactly the mean age in the population will have a z score of 0.
- A person whose age is one standard deviation above the mean in the population will have a z score of 1.
- A person whose age is two standard deviations below the population mean will have a z score of –2.

A **percentile** quantifies the percentage of all observations in a data set that are lesser than a particular individual's value for a variable. The 25th percentile is the value for which 25% of responses in the data set are less than the value and 75% are greater. The 75th percentile is the value for which 75% of responses in the data set are less than the value and 25% are greater. The interquartile range extends from the 25th percentile to the 75th percentile. A z score of $z = 0$ corresponds with the 50th percentile; $z = -2$ is at the 2nd percentile, $z = -1$ is at the 16th percentile, $z = 1$ is at the 84th percentile, and $z = 2$ is at the 98th percentile.

FIGURE 29-9 Standard Deviations for Normally Distributed Numeric Variables

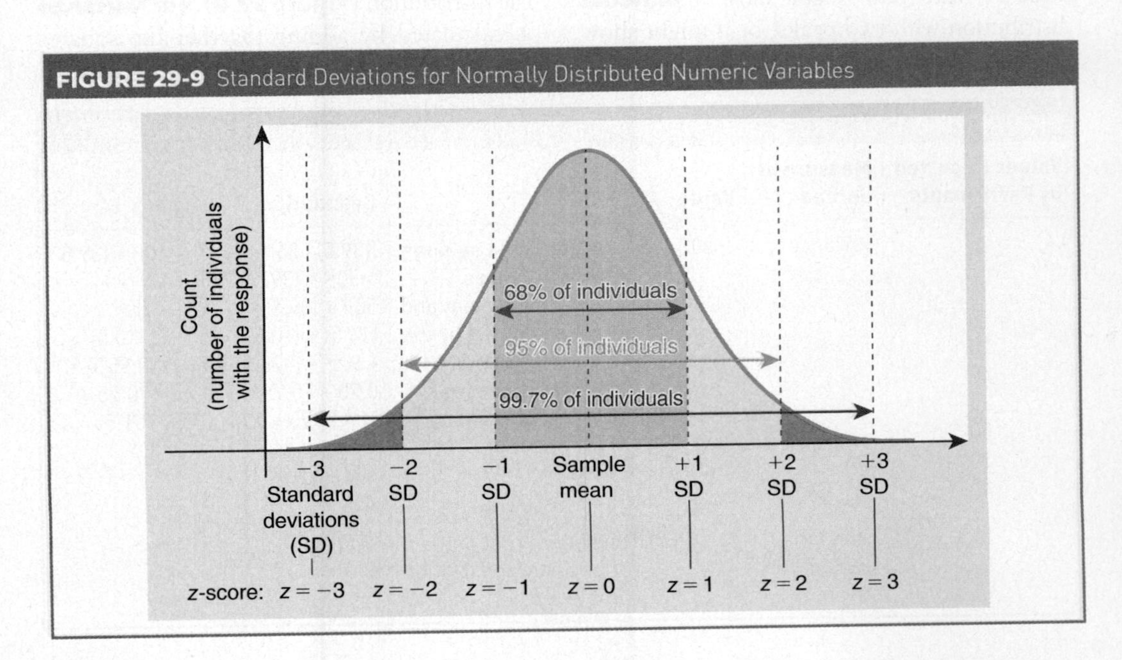

29.7 Reporting Descriptive Statistics

Descriptive statistics are statistics that describe the basic characteristics of quantitative data, such as means and proportions. The goal of descriptive statistics is to accurately describe the responses to a variable (**Figure 29-10**).

- For ratio and interval variables with normal or nearly normal distributions, both the mean and the standard deviation are typically reported.
- For ordinal variables and for ratio and interval variables with non-normal distributions, the median and interquartile range are often reported.
- For categorical variables, the proportions of participants who provided particular responses are usually used to describe the population.

29.8 Confidence Intervals

Confidence intervals (CIs) provide information about the expected value of a measure in a source population based on the value of that measure in a study population (**Figure 29-11**). For example, if the mean age in a study population of 100 people randomly sampled from all workers at a large company is 30 years, the researcher should not assume that the mean age of all employees is exactly 30 years. The 95% CI states how close to 30 years the mean age in the source population (the company) is expected to be. If the 95% CI for the mean age in the study population extends from 26 to 34, a researcher can be 95% confident that the mean age of all employees is between 26 and 34 years.

The width of the interval is related to the sample size of the study. A larger sample size will yield a narrower CI. If every member of the source population is included in the study population, then a CI is not needed because the exact value for the source population is known.

A 95% CI is usually reported for statistical estimates, and that 95% CI corresponds to a significance level of $\alpha = 0.05$ for a statistical test. This means that 5% of the time a 95% CI is expected to miss capturing the true value of a measure in the source population. Using a 99% CI ($\alpha = 0.01$) makes the CI wider, which means that it is more likely that the value in the source population will be captured within the CI. However, a 99% CI also makes it more difficult to classify a result as statistically significant, because fewer results will be classified as extreme. Alternatively, a 90% CI ($\alpha = 0.10$) could be used. A 90% CI is narrower than a 95% CI, which makes it easier for a result to be deemed statistically significant because more results will be classified as extreme. However, a

FIGURE 29-10 Common Descriptive Statistics by Variable Type

Variable Type	Common Measure of Central Tendency	Common Measure of Spread	Common Graphical Display
Ratio	Mean	Standard deviation	Histogram
Interval	Mean	Standard deviation	Histogram
Ordinal/ranked	Median	Interquartile range	Boxplot
Nominal/categorical	Mode	—	Bar chart, pie chart
Binomial	Mode	—	—

FIGURE 29-11 Interpreting Confidence Intervals

Statistic	Result with 95% CI	Interpretation
Mean age of all participants (years)	30 (26, 34)	Based on the mean age in the study population (30 years), we are 95% confident that the mean age in the source population is between 26 and 34 years.
Proportion of all participants with a disease (%)	9.0 (7.3, 10.7)	Based on the proportion of individuals in the study population who had the disease (9.0%), we are 95% confident that the prevalence of disease in the source population is between 7.3% and 10.7%.

90% CI will be less likely than a 95% CI to capture the true value in the source population.

29.9 Statistical Honesty

Researchers are obligated to describe their data accurately and to correctly report the results of statistical tests. To do otherwise is a form of research misconduct. Three of the most serious forms of research misconduct are fabrication, falsification, and plagiarism.

- **Fabrication** is the creation of fake data, such as creating fictitious rows of data in a spreadsheet for people who never completed a questionnaire or never participated in an experiment.
- **Falsification** is the misrepresentation of results, such as modifying extreme data values to improve the results of statistical tests, manipulating photographs or other images collected during laboratory work, or intentionally misreporting a study's methods to make the study look more rigorous than it was.
- Plagiarism is the use of other people's ideas, words, or images without permission and proper attribution.

Statistical honesty requires avoiding fabrication, falsification, and plagiarism. It also requires conducting statistical analyses according to established standards that ensure the rigor and validity of results. Statistical analysis should aim to discover the truth about a data set, not to creatively manipulate analysis toward a preferred result.

It is not acceptable to run a dozen different types of statistical tests on a data set, hoping that one of them will happen to yield a statistically significant result to feature in a report. Instead, the researcher must select the correct test for the question being asked and the variables being examined.

It is not appropriate to recode ratio variables into categorical variables by preferentially selecting the cutoff values that yield statistically significant results for tests of the derived variable. In general, it is better to use quartiles or other preselected divisions when recoding into new categories.

An **outlier** is a value in a numeric data set that is distant from other observations and outside the expected range of values. It is not permissible to ignore outliers when there is not a valid and widely accepted reason for doing so. For example, a recorded birth weight of 80 pounds may be reasonably assumed to be an error in the data file, and it can be removed from analysis if the true value cannot be ascertained. (Before removing the entry from the data set, the researcher should try to track down the true value. Sometimes the values in computer files do not match those listed in clinical records or provided by study participants who submit written answers to questions.) However, it is not reasonable to remove an 80-pound adult from the data file, because an adult could weigh 80 pounds. That value must be included in the analysis.

29.10 Statistical Consultants

Ideally, a researcher who will be collecting quantitative data should consult with a statistician during the study design process to ensure that the sampling methods and sample size are appropriate, the questionnaire will yield usable data, and the analysis plan is a reasonable one. Checking with a statistical expert for the first time later in the research process increases the risk of unfixable flaws in the study data. If elaborate analytic techniques will be required in order to answer the study question, an expert in that technique should be invited to serve as a collaborator and as a coauthor on the resulting research report. This invitation should be made as early as possible in the project and in consultation with other coauthors. (See Chapter 6 for information about working with coauthors.)

A research mentor can provide information about the institutional resources (if any) that are available to researchers. Some universities offer a statistical consulting service that provides a limited number of consulting hours for free but requires payment for more involved consulting services. Coauthorship status is based on contributions to the project, not payment status, so both expectations for pay and for coauthorship should be discussed with consultants.

For students, the supervising professor can clarify what types of outside assistance are acceptable and unacceptable for honors projects, theses, and dissertations. Many institutions, departments, and degree programs have rules requiring students to run all their own statistical tests. Students may be allowed to consult with experts about their analytic plans, but they cannot outsource their work to others. It is important for students and employees to understand the policies that apply to research conducted within their institutions. Unauthorized assistance may result in the same types of harsh penalties (such as expulsion from school) that are applied to other violations of academic integrity.

© DmitriyRazinkov/Shutterstock

CHAPTER 30

Comparative Statistics

Comparative statistics such as rate ratios, odds ratios, t tests, and chi-square tests are used to compare groups of participants by exposure status, disease status, and other characteristics.

30.1 Comparative Analysis by Study Approach

Comparative statistics are tests that compare the characteristics of two or more independent populations or compare the before-and-after characteristics of a study population being followed forward in time. For example, the analysis of a case–control study requires first using comparative tests to show that the cases (people with the disease) and controls (people without the disease) in the study are similar in terms of age distribution and other demographic characteristics and then using additional comparative tests to determine whether the exposure histories of cases and controls are different. Similar types of comparative tests can compare the disease outcomes of exposed and unexposed participants in a cohort or experimental study. **Figure 30-1** summarizes the uses of comparative statistical tests for several common study approaches.

FIGURE 30-1 Analytic Plan for Comparing Groups

Study Approach	First Step	Key Analysis
Case–control study	Show that cases and controls are similar except for disease status	Use odds ratios to see whether cases and controls have different exposure histories
Cohort study	Show that the exposed and unexposed are similar except for exposure status	Use rate ratios to see whether the exposed and unexposed have different rates of incident disease
Experimental study	Show that the individuals assigned to the intervention and control groups are similar except for exposure status	Use tests of efficacy and other types of statistics to see if the intervention and control groups have different outcomes

A **parameter** is a measurable numeric characteristic of a population. A **statistic** is a measured characteristic of a sample population. **Inferential statistics** use statistics from a random sample of members of a population to make evidence-based assumptions about the values of parameters in the population as a whole. For example, the sample mean age ($\bar{x}$) is a statistic that is intended to provide insight about the population mean age (μ) in the population from which the sample was drawn. Similarly, the sample standard deviation (s) for age is intended to provide insight about the population standard deviation (σ) for age. Comparative statistics are inferential because the researcher does not study an entire population but instead makes educated guesses (inferences) about parameters in the full population based on a sample of members of the population.

30.2 Hypotheses for Statistical Tests

Comparative statistical tests are designed to test for difference rather than for sameness. Accordingly, the questions driving the selection of hypotheses for statistical tests are usually phrased in terms of differences: Are the means different? Are the proportions different? Are the distributions different? Each question about statistical difference has two possible answers: The values are different, or the values are not different.

A **null hypothesis** (H_0) is a statement describing the expected result of a statistical test if there is no difference between the two or more values being compared. Null means nothing or zero. The term **null result** describes a statistical test that shows no statistically significant differences between populations or over time. An **alternative hypothesis** (H_a) is a statement describing the expected result if there truly is a difference between the two or more values being compared (**Figure 30-2**). For example, for a test comparing the mean ages of cases and controls in a case–control study, the hypotheses could be:

- H_0: There is *no* significant difference in the mean ages of the two populations.
- H_a: There *is* a significant difference in the mean ages of the two populations.

A test to compare the distribution of responses to a categorical question in two groups would have as hypotheses:

- H_0: There is *no* significant difference in the distribution of responses in the two populations.
- H_a: There *is* a significant difference in the distribution of responses in the two populations.

30.3 Rejecting the Null Hypothesis

Because statistical tests do not ask questions about sameness, the answers provided by statistical tests do not allow a researcher to say conclusively whether two values are the same. Instead, a researcher must make a decision about whether the results of a statistical test indicate that values are different or not different. The language used to describe this decision is that the researcher will either "reject the null hypothesis" or "fail to reject the null hypothesis."

- Rejecting the null hypothesis means concluding that the values are different by rejecting the claim that the values are not different.
- Failing to reject the null hypothesis means concluding that there is no evidence that the values are different. Functionally, this is like saying that the values are close enough to be considered similar, but failing to reject the null hypothesis should never be taken as evidence that the values are the same.

Chance describes a random event that occurs by happenstance rather than design. The decision to reject or fail to reject the null

FIGURE 30-2 Examples of Hypotheses for Statistical Tests

Goal	Statistical Question	Null Hypothesis (H_0)	Alternative Hypothesis (H_a)
Test whether the average ages of cases and controls in a case–control study are similar	Are the mean ages of cases and controls *different*?	The means are *not* different.	The means *are* different.
Test whether the mean age of participants drawn from a population with a mean age of 40 years is close enough to 40 years that the study population can be considered representative of the source population	Is the mean age in the study population *different* from 40 years?	The mean is *not* different from 40.	The mean *is* different from 40.
Test whether the proportion of responses to a categorical question about the frequency of hand washing was similar for male and female participants in a cohort study	Are the distributions of responses from males and females *different*?	The distributions are *not* different.	The distributions *are* different.
Test whether participants, on average, had different scores on a pretest administered prior to an intervention and a posttest administered after the intervention	Are the "before" scores of participants *different* from the "after" scores?	The scores are *not* different.	The scores *are* different.

hypothesis is based on the likelihood that the result of a test was due to chance. One way to understand chance events is to consider the variability in sample populations. When a sample population is drawn from a source population, the mean age in the sample population is usually not exactly the mean age of the source population. The range of expected values for the mean age of sample populations drawn from a source population can be estimated using statistics (**Figure 30-3**). Some sample populations will have mean ages that are very close to the mean in the source population; other sample populations will have mean ages that are quite far from the mean in the source population. No set cutoff defines what will be considered extremely far from the mean age in the source population, but the standard is to say that the 5% of sample means farthest from the true mean are extreme. Thus, by chance, 5% of the samples drawn from a source population will be expected to have an extreme mean.

Similarly, if two sample populations are drawn from the same source population, the mean ages for those two sample populations will not be identical even though their members are drawn from the same pool of individuals. Comparative statistical tests accommodate this expected difference when testing whether two groups within a study population are different. For example, a test that compares the mean ages of cases and controls in a case–control study accounts for the fact that there will be some difference between the mean ages of cases and controls even when the cases and controls are sampled from source populations with identical mean ages. The test determines whether the mean ages are so far apart that if the cases and controls were drawn from source populations with the same mean age the difference between the mean ages of the cases and the controls would fall among the most extreme differences expected to occur by chance. When the statistical test shows that

FIGURE 30-3 Example of the Distribution of Mean Ages for Sample Populations Drawn from a Larger Source Population

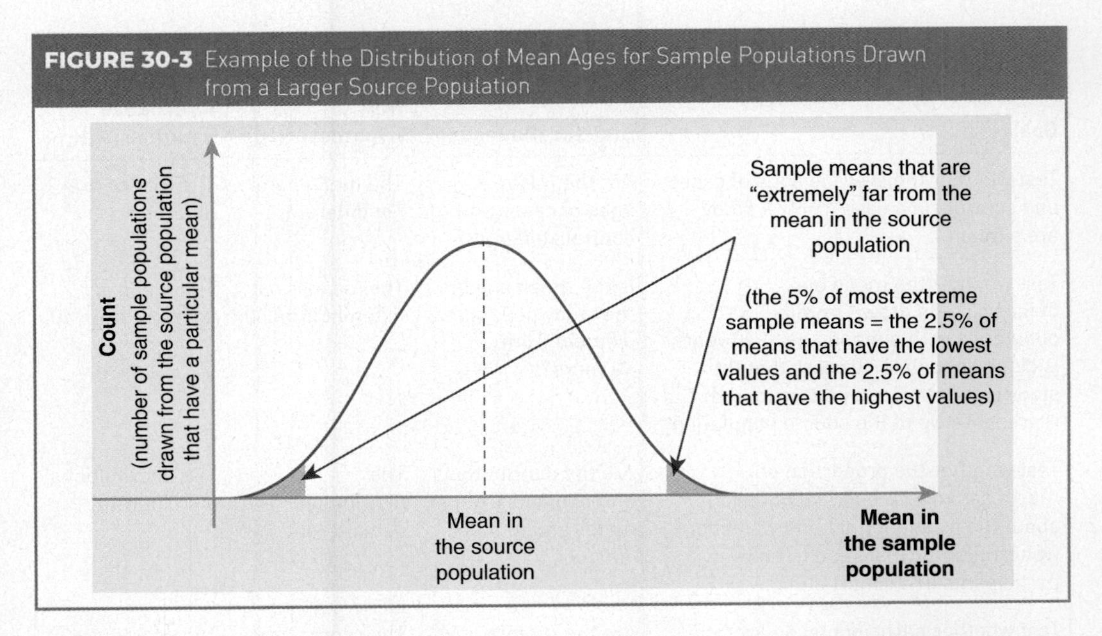

the mean ages of cases and controls are fairly close, the researcher will fail to reject the null hypothesis and will conclude that the means are not different. When the difference between the mean ages of cases and controls is extreme, the statistical test will show that it is highly unlikely that the group means are not different. The researcher will therefore reject the null hypothesis and conclude that the mean ages of the cases and the controls are different. The difference between the mean ages of cases and controls in the study population will be taken as evidence that the mean age of individuals in the source population for cases and the mean age of individuals in the source population for controls are different. This conclusion assumes that the difference between the source populations is reflected in the sample of cases and controls that happened to be drawn from their respective source populations.

30.4 Interpreting *p* Values

A ***p* value**, short for probability value, describes the likelihood that a test statistic as extreme as or more extreme than the one observed would occur by chance if the null hypothesis were true. The *p* values generated during statistical analysis help researchers decide whether the results observed in a study are likely to reflect real differences between groups of interest. A very small *p* value means the observed test result is highly unlikely to have occurred by chance.

The interpretation is similar for all statistical tests: the *p* value for the study determines whether the null hypothesis will be rejected. The **significance level**, represented by the Greek letter alpha (α), is the *p* value at which the null hypothesis is rejected. A test result is classified as having **statistical significance** if the *p* value is less than α. The standard is to use a significance level of $\alpha = 0.05$, or 5%. Any statistical test with a result that is in the 5% of most extreme responses expected by chance when the null hypothesis is true will result in the rejection of the null hypothesis (**Figure 30-4**). Some studies choose to use a more rigorous significance level, such as $\alpha = 0.01$.

In comparative statistics, a type 1 error occurs when a test indicates a significant difference between two or more populations even though the null hypothesis is true. This type of

FIGURE 30-4 Interpreting *p* Values

Null Hypothesis (H_0)	Conclusion when $p < .05$* = Reject H_0	Conclusion when $p \geq .05$* = Fail to Reject H_0
The means are not different.	The means are different.	The means are not different.
The proportions are not different.	The proportions are different.	The proportions are not different.
The distributions are not different.	The distributions are different.	The distributions are not different.

*Assuming $\alpha = 0.05$.

error will happen 5% of the time if the analyst has selected a 5% significance level. Choosing a significance level of 1% ($\alpha = 0.01$) will reduce the probability of a type 1 error but will make it more difficult to reject the null hypothesis. Choosing a significance level of 10% ($\alpha = 0.10$) will increase the probability of a type 1 error, making it more likely that a test will yield a statistically significant result. When $\alpha = 0.10$ is used as the threshold, 1 in 10 tests will conclude that there is a difference between study groups even when there really is no difference in the source population.

Although most statistical tests use an alternative hypothesis that simply expresses difference (such as "the means are different"), some tests allow for an alternative hypothesis that states the direction of the difference (like "males have a higher mean age than females") (**Figure 30-5**). If a direction is specified in the alternative hypothesis, then all of the extreme values (all of the shaded area shown in Figure 30-3) will be on one side of the distribution (either all on the left of the distribution or all on the right). A **one-sided *p* value** is the probability value that is used for a statistical test when a direction is specified in the alternative hypothesis. A **two-sided *p* value** is the probability value that is used for a statistical test when a direction is not specified in the alternative hypothesis. Most comparative statistical tests use a two-sided *p* value to make the decision about rejecting or failing to reject the null hypothesis.

30.5 Measures of Association

Some of the most common types of comparative statistics used in the health sciences are measures of association such as the correlation

FIGURE 30-5 Examples of One-Sided and Two-Sided Alternate Hypotheses

Null Hypothesis	Two-Sided Alternative Hypothesis	Example of a One-Sided Alternative Hypothesis
The means are not different.	The means are different.	The mean among cases is higher than the mean among controls.
The proportions are not different.	The proportions are different.	The proportion in the intervention group is lower than the proportion in the control group.
The scores are not different.	The scores are different.	The after scores were, on average, higher than the before scores.

used for aggregate studies, the odds ratio (OR) used for case–control studies, and the rate ratio (RR) used for cohort studies. ORs and RRs compare responses to two variables that have each been divided into two groups. Prior to using a computer to calculate an OR, RR, or other type of 2×2 analysis, variables that are not already divided into two categories must be recoded into binomial variables (often coded numerically as yes = 1 and no = 0). In some situations, the cutoff points for the categories are obvious, such as those that divide an ordinal variable into categories for disagreement (strongly disagree or disagree) and agreement (agree or strongly agree). Sometimes the population can be divided into groups of relatively equal sizes using the median or other sample-based cutoff points. Alternatively, biologically or socially meaningful cutoff points can be defined, such as using the 18th birthday to divide a study population into children and adults in countries where legal adulthood begins at 18 years of age. The selected cutoff value may influence whether the exposure and outcome have a statistically significant association. Accordingly, the decision about how to define categories should be based on appropriate scientific or social justifications.

The results of 2×2 analyses are often presented using tables like the one shown in **Figure 30-6**. The reference group for an OR or RR should be specified, such as stating that males are being compared to females (the reference group) and those with a waist

FIGURE 30-6 Example of Odds Ratios for a Case–Control Study

Exposure		Percentage of Cases (Acute Myocardial Infarction [AMI]) (*n* = 150)	Percentage of Controls (no AMI) (*n* = 250)	Odds Ratio (95% Confidence Interval)	Interpretation
Sex	Female	39.3%	42.4%	Reference group	Cases and controls in the study did *not* have significantly different proportions of males.
	Male	60.7%	57.6%	1.14 (0.75, 1.72)	
Waist circumference > 35 inches	No	37.3%	51.2%	Reference group	Cases had greater odds than controls of having a waist circumference greater than 35 inches.
	Yes	62.7%	48.8%	1.76 (1.16, 2.67)*	
Tobacco use	Never smoked	68.0%	73.6%	Reference group	Cases and controls in the study population did *not* have significantly different smoking histories.
	Former smoker	11.3%	10.8%	1.14 (0.58, 2.18)	
	Current smoker	20.70%	15.60%	1.43 (0.84, 2.44)	

*Statistically significant at $\alpha = 0.05$ level.

circumference greater than 35 inches are being compared to those with smaller girths (the reference group). In the example shown in the figure, the variable for tobacco use had three possible responses instead of just two, so two 2×2 tables were created and an OR was calculated for each of the two tables. One compares former smokers to never smokers, and the other compares current smokers to never smokers.

The 95% confidence interval (CI) provides information about the statistical significance of the tests. For example, the 95% CI for the OR comparing the sex distribution of cases and controls contains OR = 1. This means that it is not clear from the test whether cases are more likely or less likely than controls to be male. The conclusion is, therefore, that there is no statistically significant difference in the proportion of cases and controls by sex.

30.6 Interpreting Confidence Intervals

CIs provide more information about a statistic than can be conveyed by a p value. For example, the CIs for ORs and RRs indicate whether the populations being compared are different or not different, and they also provide information about how well the test statistic from the study population captures the true value of that measure in the source population (**Figure 30-7**). A 95% CI corresponds to a significance level of $\alpha = 0.05$. For ORs and RRs, a 95% CI that does not overlap 1 (that is, one for which the full range is below 1 or the full range is above 1) is equivalent to having a p value of $p < .05$. If the CI overlaps 1, that is the equivalent of $p > .05$.

Suppose that the point estimates for the incidence RRs calculated from two cohort studies are both RR = 3.0. One study has 200 participants and a 95% CI of 1.6 to 5.8. This is usually written as RR = 3.0 (1.6, 5.8). The other study has 1000 participants and its RR = 3.0 (2.2, 4.0). In the first study, the researcher can be 95% confident that the true value of the RR in the source population is a number between RR = 1.6 and RR = 5.8. There is a 5% likelihood that the actual RR is less than RR = 1.6 or greater than RR = 5.8. In the second study, the one with the larger number of participants, the 95% CI is narrower, and there is more certainty about the true value. Because the 95% CIs for both of these studies do not overlap 1, the p values for both are $p < .05$. For both studies, the conclusion is that the incidence rates in the two populations being compared are different. However, there

FIGURE 30-7 Interpreting Confidence Intervals

Statistic	Result with 95% Confidence Interval	Interpretation
Odds ratio (OR)	1.7 (0.6, 5.3)	Based on the OR in the study population (OR = 1.7), we are 95% confident that the OR in the source population is somewhere between 0.6 and 5.3. Because this overlaps with OR = 1, we conclude that there is no association between the exposure and disease status.
Rate ratio (RR)	1.6 (1.1, 2.4)	Based on the RR in the study population (RR = 1.6), we are 95% confident that the RR in the source population is between 1.1 and 2.4. Because this confidence interval does not overlap with RR = 1, we conclude that the exposure is associated with an increased risk of disease.

is more certainty about the level of difference in the larger study.

Other types of CIs can also be calculated. A 90% CI corresponds to a significance level of $\alpha = 0.10$. A 99% CI corresponds to $\alpha = 0.01$. A 90% CI for an OR is less likely to overlap OR = 1 than a 99% CI. The 90% CI is also less likely than the 99% CI to capture the true OR in the source population. The 90% CI is more likely than the 99% CI to be deemed "statistically significant" and to lead to the conclusion that there is an association between the exposure and the outcome being examined (**Figure 30-8**).

30.7 Selecting an Appropriate Test

For statistical comparisons more complex than 2×2 analysis, analysts must select tests that are appropriate for the goals of the analyses and the types of variables being analyzed. The steps for identifying and using a statistical test are summarized in **Figure 30-9**.

First, the variables to be compared are selected and the goal of the test is clearly stated. The goal could be:

- To compare the mean ages of males and females. The key variables for this test are age (a ratio variable) and sex (binomial).
- To see whether the proportion of cases and controls with various blood types is similar. The key variables for this test are blood type (a nominal variable) and disease status (binomial).
- To compare the responses of older and younger adults to a question about how often the participant eats dark chocolate as per a 5-point frequency scale ranging from "never" to "every day." The key variables are age group (binomial) and the frequency of chocolate consumption (an ordinal variable).
- To determine whether participants with higher levels of high-density lipoprotein (HDL) cholesterol tend to have lower resting heart rates. Both of these variables are continuous variables (ratio variables).

Once the variables are selected, a test that is appropriate for the types of variables being examined must be selected. An **assumption** is a premise that is presumed to be true. Some statistical tests are based on assumptions that the variables being examined have particular distributions or other characteristics. For example, some tests are appropriate to use only when the data being examined have a normal distribution, some tests require two numeric variables to have a linear association, and many comparative tests require the groups

FIGURE 30-8 90%, 95%, and 99% Confidence Intervals for the Same Odds Ratio

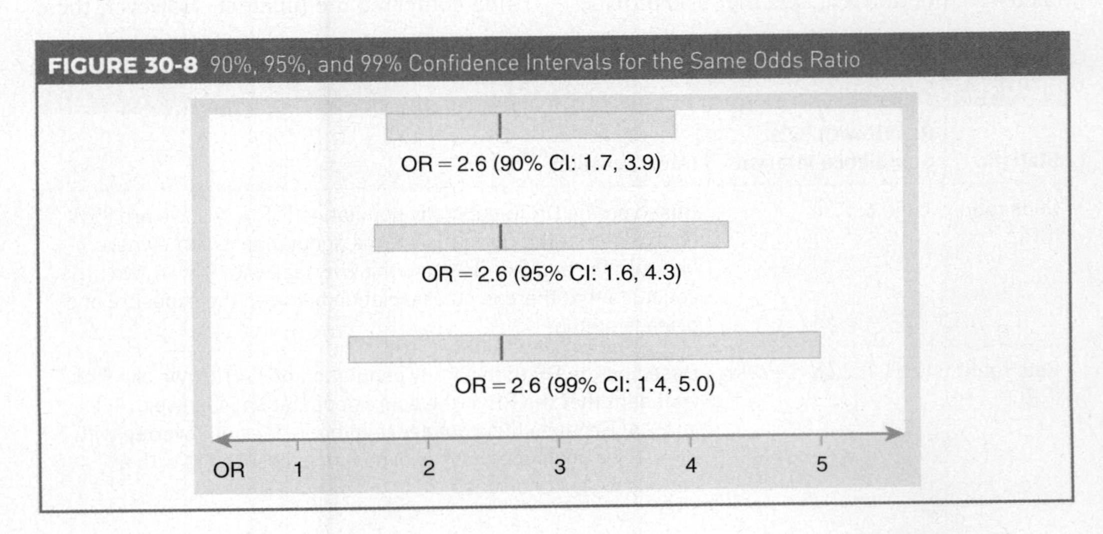

FIGURE 30-9 Plan for Hypothesis Testing

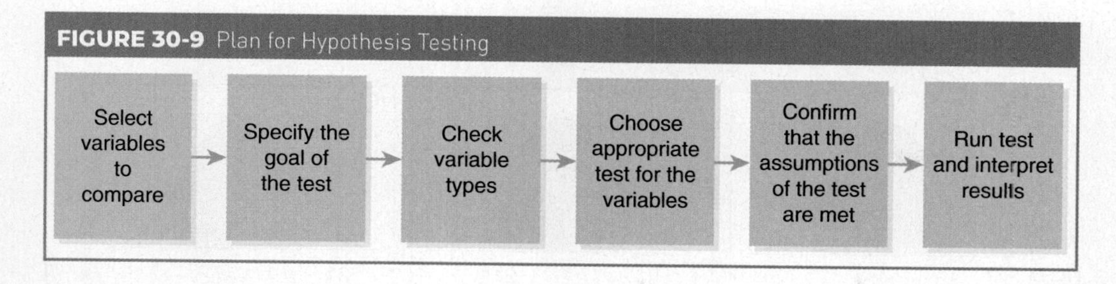
Select variables to compare
Specify the goal of the test
Check variable types
Choose appropriate test for the variables
Confirm that the assumptions of the test are met
Run test and interpret results

being compared to be independent, with no members in common. A researcher must confirm that the variables meet the assumptions of a statistical test prior to running the test and interpreting the output.

30.8 Parametric and Nonparametric Tests

Statistical tests are often classified as being either parametric or nonparametric. The basic difference between these two types of tests is that parametric tests make more assumptions about the variables being examined than nonparametric tests.

- A **parametric test** assumes the variables being examined have particular distributions, often requiring the variables to have normal or approximately normal distributions. Parametric tests may also require that the variance of the variables of interest—the spread of observations around the mean—be equal or at least similar in the population groups being compared.
- A **nonparametric test** does not make assumptions about the distributions of responses.

Parametric tests are typically used for ratio and interval variables with relatively normal (bell-shaped) distributions of responses. Parametric tests tend to be more statistically powerful than nonparametric tests, so the preference is to use a parametric test whenever the variable being examined fits reasonably well with the assumptions the test makes about sample size, distribution, and the equality of variances.

Nonparametric tests are often used for ranked variables, such as the responses to surveys that ask participants to indicate preferences using scales from 1 (strongly disagree) to 5 (strongly agree), and for categorical variables, including variables with just two groups (such as cases and controls, males and females, or children and adults). They are also used when the distribution of a ratio or interval variable is non-normal.

30.9 Comparing a Population Statistic to a Set Value

The goal of some statistical tests is to compare the value of a statistic in a study population to some set value. A **one-sample *t* test** is a statistical test that compares the mean value of a ratio/interval variable to a selected value. A **binomial test** compares the proportion expressed by a binomial variable to a selected value. A **chi-square goodness-of-fit test** compares the proportion of responses to a nominal variable to a selected value.

Suppose that participants in an experimental study are students at a university where the mean age of undergraduate students is 21 years. The research team intended for the study volunteers to be reasonably

FIGURE 30-10 Comparing the Sample Mean to Some Other Value (One-Sample *t* Test)

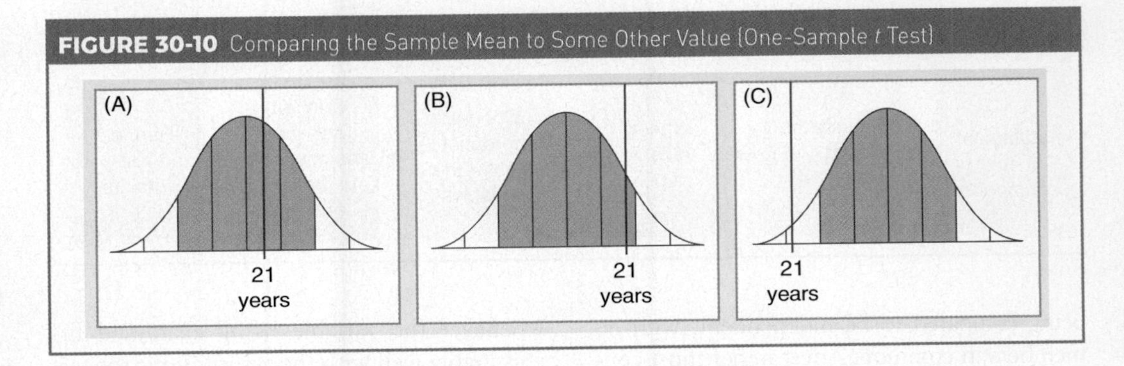

representative of the undergraduate student population as a whole. One way to examine whether the volunteers are representative is to determine whether the mean age of the study participants is close to 21 years. If the distribution of ages in the study population looks like the distribution in Box A of **Figure 30-10**, then 21 years is captured within two standard deviations of the study's mean age. The conclusion would be that the sample mean is not so far from 21 that the means would be considered different. In other words, the sample shown in Box A fails to reject the null hypothesis that the means are not different. The *p* value for this one-sample *t* test is p > .05. The conclusion is that the means in the study population and the university student population as a whole are not significantly different. Box B also captures 21 years within the 95% CI, even though the mean age of study participants is farther from 21 than it was in Box A. The *p* value for this test is also $p > .05$. In Box C, however, the study participants were several years older than the average undergraduate student at the university, and 21 years does not fall within the 95% CI. The *p* value for this test is $p < .05$. In this situation, the null hypothesis is rejected, and the conclusion is that the study population mean is different from 21 years. This test result shows that the study population may not be adequately representative of the university's undergraduate student population.

30.10 Comparing Independent Populations

Independent populations are populations in which no individual is a member of more than one of the groups being compared. For example, if the populations being compared are divided by age, the population of adults ages 18 to 49 years will not overlap with the population of adults ages 50 to 79 years. Each individual participant in the study population can be assigned to no more than one of these groups, so the populations are independent.

Many statistical tests compare independent populations. The appropriate test to use depends on the type of variable being examined (**Figure 30-11**). An **independent-samples *t* test** (also called a **two-sample *t* test**) compares the mean values of a ratio/interval variable in two independent populations, such as comparing the mean ages of cases and controls participating in a case–control study. A **Mann-Whitney *U* test** (also called a **Wilcoxon rank sum test**) compares the median values of an ordinal/rank variable in two independent populations. A **Fisher's exact test** compares the values of a binomial variable in two independent populations, such as examining whether the proportions of males in the exposed and unexposed groups of a cohort study are similar. A **chi-square test**, often written as χ^2 test, of independence compares

FIGURE 30-11 Common Tests for Comparing Two or More Groups

	Type of Variable Being Examined			
	Ratio/Interval (Parametric Tests)	**Ordinal/Rank (Nonparametric Tests)**	**Binomial**	**Nominal Categories**
Statistic being evaluated	Mean	Median	Proportion	Proportions
Test for whether the statistic in one population is different from a hypothetical value	One-sample *t* test	One-sample median test	Binomial test	Chi-square goodness-of-fit test
Test for whether the statistic differs in two populations	Independent-samples (two-sample) *t* test	Mann-Whitney *U* test (Wilcoxon rank sum test)	Fisher's exact test	Chi-square test of independence
Test for whether the statistic differs in three or more populations	One-way ANOVA (*F* test)	Kruskal-Wallis *H* test	Chi-square test	Chi-square test of independence

the values of a nominal variable in two or more independent populations, such as determining whether the distributions of participants by race or ethnicity are similar for the intervention and control groups of an experimental study. The **Yates correction** improves the validity of a chi-square test statistic when the sample size for the test is small.

When running statistical tests, it is often beneficial to create a table of basic information about the variables of interest for each of the comparison groups and the results of statistical tests comparing those populations. **Figure 30-12** shows sample output for tests of whether responses differed for the male and female participants of a cohort study. In this example, the males have a significantly greater average age than the females because the p value for the independent-samples t test is $p < .05$. However, the proportion of males and females who smoke is not significantly different because the p value for Fisher's exact test was $p > .05$. Some results tables add columns for additional details about the particular statistical tests that were used. A **test statistic** is a value calculated from study data for a hypothesis test, such as the t stat used for t tests and the F stat used for one-way ANOVA. The **degrees of freedom (df)** for a test is the number of values in the calculation of a test statistic that are free to vary.

The table shown in Figure 30-12 includes more information than is usually included in published manuscripts. However, it allows the researcher to double-check that the correct tests were used and the correct interpretations were made. It also facilitates the writing of the statistical methods portion of the research report. A more succinct comparison table is usually prepared for the final report. A sample results table for publication is shown in **Figure 30-13**. When a simplified table is presented, details about the statistical tests used for the analysis are expected to be provided within the text of the report.

FIGURE 30-12 Examples of Tests for Comparing Subpopulations Within a Study Population

Variable	Report	Males (n = 200)	Females (n = 200)	Variable Type	Test of Comparison	p Value for Test	Interpretation
Age	Mean (Standard Deviation)	43.7 (7.8)	40.1 (8.1)	Ratio (normal)	Independent-samples t test	<.01	The means are different.
Current smokers	%	12%	10%	Binomial (yes/no)	Fisher's exact test	.52	The proportions are *not* different.
Home district	n (%)			Nominal	Chi-square test	.86	The proportions are *not* different.
North		90 (45%)	87 (44%)				
Central		50 (25%)	48 (24%)				
South		60 (30%)	65 (33%)				

FIGURE 30-13 Simplified Version of Figure 30-12

Characteristic		Males (n = 200)	Females (n = 200)	p Value
Age	Mean (Standard Deviation)	43.7 (7.8)	40.1 (8.1)	< .01*
Current smokers	%	12%	10%	.52
Home district	n (%)			.86
North		90 (45%)	87 (44%)	
Central		50 (25%)	48 (24%)	
South		60 (30%)	65 (33%)	

*Statistically significant at α = 0.05 level.

30.11 Multivariable Comparisons of Means

Several tests compare means in independent populations, including t tests and more complex types of analysis (**Figure 30-14**). **ANOVA** is an acronym for analysis of variance, and it compares the mean values of a continuous variable across independent populations. **One-way ANOVA** compares the mean values of one interval/ratio predictor variable across independent groups of people, such as comparing the mean age of patients at three different family practices who do not share any patients. One-way ANOVA typically uses an ***F* test** to determine whether the mean values of an interval/ratio variable are different or not different across three or more independent populations. A **Kruskal-Wallis *H* test** compares the median values of an ordinal/rank variable in three or more independent populations.

The Latin term *post hoc* means "after the event." A **post hoc** test examines paired

FIGURE 30-14 Examples of Tests for Comparing Means in Two or More Groups

Name	Independent Variable(s)	Dependent Variable(s)
One-way ANOVA (analysis of variance)	1 nominal variable	1 ratio/interval variable
Two-way ANOVA (analysis of variance) = factorial ANOVA	2 nominal variables	1 ratio/interval variable
ANCOVA (analysis of covariance)	1+ nominal variable and 1+ ratio/ interval and/or nominal covariate	1 ratio/interval variable
One-way MANOVA (multivariate analysis of variance)	1 nominal variable	2+ ratio/interval variables
Two-way MANOVA (multivariate analysis of variance)	2 nominal variables	2+ ratio/interval variables
MANCOVA (multivariate analysis of covariance)	1+ nominal variable and 1+ ratio/ interval and/or nominal covariate	2+ ratio/interval variables

comparisons after an omnibus (overall) test comparing three or more populations shows differences among the populations. For example, **Tukey's test** is a post hoc test that examines all of the possible pairwise comparisons across the three or more populations included in an ANOVA. Suppose that an *F* test shows that the mean ages of three groups (A, B, and C) are generally different. Tukey's test would examine the differences in the mean ages in each pair of groups: A and B, A and C, and B and C. If four groups had been compared (A, B, C, and D), Tukey's test would examine six different pairs (A and B, A and C, A and D, B and C, B and D, and C and D).

Two-way ANOVA, also called **factorial ANOVA**, compares the mean values of an interval/ratio variable across groups that are defined by two different variables. For example, two-way ANOVA could compare mean ages by sex and smoking status (never smoker, past smoker, current smoker). This analysis would involve six comparison groups: female never smokers, male never smokers, female past smokers, male past smokers, female current smokers, and male current smokers.

Both of these types of ANOVA require several assumptions to be met before the tests are used. The populations being compared must be independent, with each study participant assigned to only one of the groups being compared. The dependent variable—the variable for which the mean values are being compared—must be approximately normally distributed. There can be no significant outliers in the variables included in the ANOVA analysis, and **Levene's test** must demonstrate the homogeneity of the variances across the different groups. Suppose that mean ages of patients at a family practice and a pediatric practice are being compared. The family practice is likely to have a wide spread of ages, while the pediatric practice will have a narrow distribution of ages among its patients. In this example, Levene's test would show that the variances were different ($p < .05$), so ANOVA would not be an appropriate test of comparison to use.

Several extensions of ANOVA allow for more complex analyses of the differences of means in independent populations. Variance is a measure of how much a single variable varies. **Covariance** is a measure of the joint variability between two random variables. **ANCOVA**, analysis of covariance, compares the means of a ratio/interval variable in two or more independent groups while controlling for one or more additional ratio/interval or nominal variables. **MANOVA**, multivariate analysis of variance, compares differences in group means across multiple dependent variables. **MANCOVA**, multivariate analysis of covariance, compares differences in group means across multiple dependent variables while controlling for one or more additional ratio/interval or nominal variables.

30.12 Correlation Analysis

A variety of measures of correlation can be used to examine the relationship between two variables. The correct test for the types of variables being compared must be used. The Pearson correlation coefficient (r), also called Pearson's product-moment correlation, examines the association between two ratio/interval variables. Spearman's rho (ρ_s) and **Kendall's tau** (τ) are measures of the degree to which changes in the value of one ordinal/rank variable predict changes in the value of another ordinal/rank variable. The **phi coefficient** (φ) is a statistical measure of the degree to which changes in the value of one binomial variable predict changes in the value of another binomial variable. **Cramér's *V*** quantifies the degree to which changes in the value of one categorical variable predict changes in the value of another categorical variable. Some tests of correlation allow for comparison of different types of variables. Eta (η) and eta squared (η^2) are measures of the correlation between one ratio/interval variable and one nominal variable. The point-biserial r_{pb} correlation calculates the correlation between one ratio/interval variable and one binomial variable. The Glass rank biserial r_{rb} measures the correlation between one ordinal/rank variable and one binominal variable. Epsilon squared (ε^2) quantifies the correlation between one ordinal/rank variable and one nominal variable.

30.13 Comparing Paired Data

Paired data are variables linked together for analysis because they were gathered from individuals who were matched on specific characteristics (such as genetic siblings) or they were gathered from one individual at two or more points in time (such as at baseline and after an intervention). Special comparative tests are used when the goal is to compare before-and-after results in the same individuals or to examine other types of relationships among paired data (**Figure 30-15**).

A **matched-pairs *t* test** compares the values of an interval/ratio variable in members of one population measured twice or among individually matched pairs from two different groups. For example, a matched-pairs t test might be used to examine whether, on average, a participant in a cohort study gained weight between the baseline exam and the 1-year follow-up exam. A **Wilcoxon signed-rank test** compares the values of an ordinal/rank variable in one population measured twice or among individually matched pairs from two different groups. **McNemar's test** compares the values of a binomial or nominal variable in one population measured twice or among individually matched pairs from two different groups. For example, suppose that a researcher wants to examine whether a safe driving course improves the pass rates for a driving licensure exam. McNemar's test uses the number of participants who switched from failing a pretest to passing a posttest, the number who switched from passing a pretest to failing a posttest, and the number who had no change in status to generate a test statistic that indicates the likelihood that the course had a significant impact on exam pass rates. **Repeated-measures ANOVA** compares the values of an interval/ratio variable across several time points or in several individually matched populations. A **Friedman test** compares the

FIGURE 30-15 Common Tests for Comparing Matched Populations

	Type of Variable Being Examined			
	Ratio/Interval (Parametric Tests)	**Ordinal/Rank (Nonparametric Tests)**	**Binomial**	**Nominal Categories**
Test for whether value of the variable is different in one population measured twice (such as "before" and "after" in the same population) or in two paired groups	Matched-pairs (paired, dependent) *t* test	Wilcoxon (matched-pairs) signed-rank test or sign test for matched pairs	McNemar's test	McNemar's test
Test for whether the value of the variable is different in three or more matched groups	Repeated-measures ANOVA	Friedman test	Cochran's *Q* test	Cochran's *Q* test

values of an ordinal/ratio variable across several time points or in several individually matched populations. **Cochran's Q test** compares the values of frequencies or proportions in three or more matched sets of binomial or nominal data.

Figure 30-16 shows sample output for paired tests. In this example, the typical participant in a 3-month exercise program lost weight during the study period because the p value for the matched-pairs t test was $p < .05$. However, the participants did not increase their ability to run 1 mile in less than 10 minutes, because the p value for McNemar's test was $p > .05$, which indicates that there was no difference in this variable from the start to the end of the study period.

FIGURE 30-16 Examples of Tests for Comparing Pretest and Posttest Results

Variable	Report	Pretest	Posttest	Difference	Variable Type	Test of Comparison	*p* Value for Test	Interpretation
Sample size	*n*	40	40	0	Count	—	—	—
Weight (pounds)	Mean (Standard Deviation)	178 (19)	172 (18)	–6 (6)	Ratio (normal)	Matched-pairs *t* test	< .01	Individuals, on average, had pre- and posttest weights that were significantly different.
Able to run 1 mile in less than 10 minutes	*n* (%)	12 (30%)	16 (40%)	6 no-to-yes, 2 yes-to-no, 32 no change	Binomial	McNemar's test	.30	The pre- and posttest ability for an individual participant to run 1 mile in less than 10 minutes was, on average, *not* different.

CHAPTER 31

Regression Analysis

Linear and logistic regression models are among the most commonly used advanced statistical techniques.

31.1 Regression Modeling

A **regression model** is a statistical model that seeks to understand the relationship between one or more independent variables and one dependent variable. An **independent variable** (or **predictor variable**) is a variable in a statistical model that predicts the value of some outcome variable. A **dependent variable** (or **outcome variable**) is a variable in a statistical model that represents the output or outcome for which the variation is being studied. When multiple independent variables are included in the model, the effect of one predictor variable on the outcome can be examined while controlling for other predictor variables by keeping those other values constant.

The two most commonly used types of regression are linear regression and logistic regression. A **linear regression** model is used when the outcome variable is a ratio or interval variable. A **logistic regression** model is a probability-based regression model used when the outcome variable is binomial. The steps for model fitting are similar for both types of models and are summarized in **Figure 31-1**.

Once the general type of model has been selected, most statistical software programs require the analyst to select a variety of specifications for the model. For example, analysts must select the particular estimation technique for the model, such as ordinary least squares, generalized least squares, or maximum likelihood estimation.

The analyst must also choose the method the computer will use to select variables for inclusion in the model. **Parsimony** is the principle that when two models are equally good, the one that is simpler or more economical should be used. In regression modeling, parsimony means that additional components should not be added to a model if the additions do not significantly improve the fit of the model. A **simultaneous multiple regression** model includes all predictor variables selected by the analyst in the model, even if some of the independent variables are not making a significant contribution to model fit. A **stepwise multiple regression** model

FIGURE 31-1 Steps in Fitting a Regression Model

Step 1	Select one outcome (dependent) variable.
Step 2	Identify the appropriate type of regression model for the outcome variable, such as linear regression or logistic regression.
Step 3	Select one or more predictor (independent) variables.
Step 4	Check to make sure that any assumptions required for the model are met, such as the variable types or the distributions of the outcome and predictor variables.
Step 5	Choose the methods that the computer will use to determine which set of predictor variables is the best at explaining the relationship between the independent variables and the outcome variable.
Step 6	Examine the model for potential problems. For example, examine residuals for possible autocorrelation, check for possible interaction between predictor variables (such as the multicollinearity that might occur when two predictor variables are highly correlated), and look for other potential problems that might need to be addressed.
Step 7	Interpret the results of the regression model, and consider whether they fit with the theoretical framework for the analysis (for example, confirming that all necessary covariates are included and all illogical ones are excluded).

systematically adds or removes predictor variables to find the most parsimonious model that provides a good fit. A forward stepwise method instructs the computer to add the best predictor variables to the model one at a time until adding an additional variable does not significantly improve the overall fit of the model. A backward stepwise method deletes predictor variables from the model until deleting a variable significantly reduces the overall fit of the model.

Most statistical software programs do not automatically confirm that all of the assumptions of a model are met, and most do not provide assistance with correctly interpreting the results. Once the results are generated by the software program, the analyst must carefully check to be sure the model is a valid one. One step in examining the validity of a regression model involves examining a model's residuals and the results of goodness-of-fit tests. A **residual** is the difference between the observed value in a data set and the value predicted by a regression model. A **goodness-of-fit** test examines how well real data match the values predicted by a model. Analysts without extensive training in applied statistics benefit from consulting a statistics reference guide or a statistician about these and other advanced analytic techniques.

31.2 Simple Linear Regression

A **simple linear regression** model examines whether there is a linear relationship between one ratio or interval predictor variable and one ratio or interval outcome variable. The relationship between the predictor and outcome variables in a simple linear regression can be visually displayed using a scatterplot. Each point from a data set is plotted on the graph, and a best-fit line is drawn through those points.

Ordinary least squares (OLS) is a linear regression modeling approach that finds the line that minimizes the average vertical distance from each point in a data set to the fitted line. Suppose that a regression model is being fit for the relationship between body mass index (BMI) on the *x*-axis and cholesterol levels on the *y*-axis, and the fitted line describing that relationship indicates that a person with a BMI of 24 is predicted to have a total cholesterol level of 180. If one of the study participants being analyzed had a BMI of 24 but a cholesterol level of 195, that individual's data point will be 15 units of cholesterol distant from the line. That distance of 15 units—the vertical distance from the data point to the best-fit line—is the residual for the data point. OLS calculates the residual for each data point in the full data set, squares each distance, and then adds up all of the squared residuals. The best-fit regression line is the one that minimizes the sum of the squared residuals (often shortened to SSR).

Figure 31-2 provides an example of how to interpret the results of a simple linear regression. The coefficient for the predictor variable (often designated as β or beta in the output of statistical software programs) is the slope of the line. The constant in the regression model is the y-intercept for the line. These values can be used to write an equation for the best-fit line, and that equation can be used to predict the expected value of the outcome variable for different values of the predictor variable. The r^2 for the model, which is the square of the correlation coefficient, provides information about how well the regression model predicts the variation in the values of the outcome variable. The value of r^2 ranges from 0 to 1, with values closer to 1 indicating a better model fit.

31.3 Simple Logistic Regression

Logistic regression models predict the probability of a particular dichotomous outcome occurring. Logistic regression is commonly used in case–control studies, for which the outcome variable is usually case status, with case = 1 and control = 0. For outcome variables that are other types of yes/no variables, it is typical to let yes = 1 and no = 0. Predictor variables for a logistic regression model can be categorical (if the categories are coded with numbers) or continuous.

FIGURE 31-2 Example of a Simple Linear Regression Model

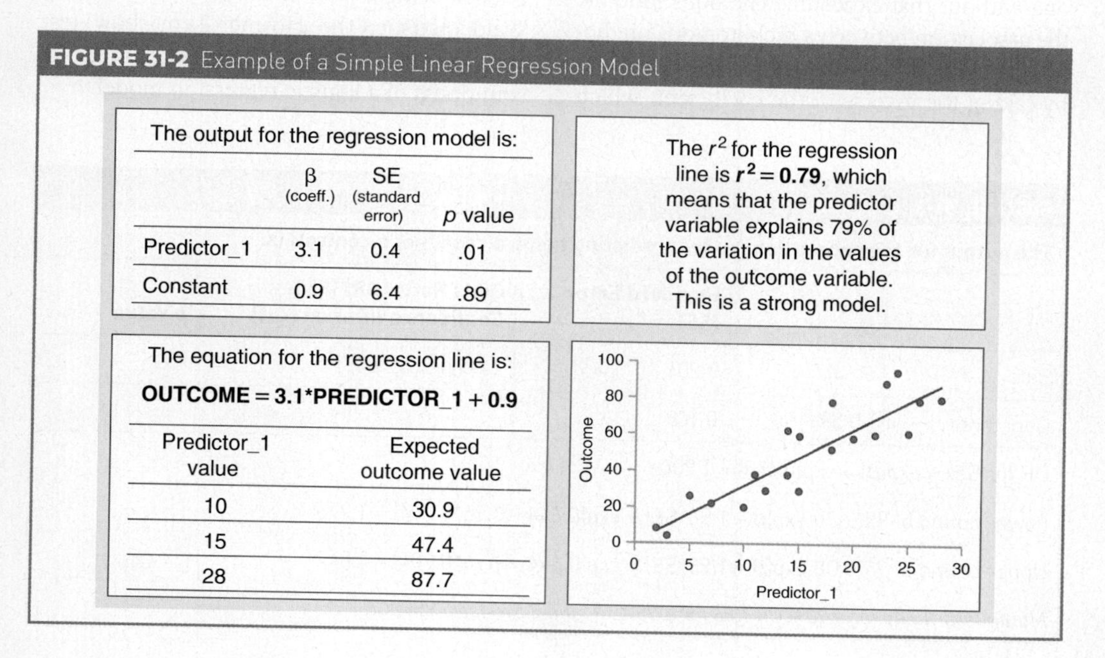
The output for the regression model is:

	β (coeff.)	SE (standard error)	*p* value
Predictor_1	3.1	0.4	.01
Constant	0.9	6.4	.89

The r^2 for the regression line is $r^2 = 0.79$, which means that the predictor variable explains 79% of the variation in the values of the outcome variable. This is a strong model.

The equation for the regression line is:

OUTCOME = 3.1*PREDICTOR_1 + 0.9

Predictor_1 value	Expected outcome value
10	30.9
15	47.4
28	87.7

Logistic regression models are based on probability. When the outcome being modeled is case status, the probability of being a case is p, the probability of not being a case is $1 - p$, and the odds of being a case are $\frac{p}{1-p}$. Logistic regression models are sometimes called **logit regression** models because they use a logit link function of $\ln\left(\frac{p}{1-p}\right)$, which is also written as just logit(p), as their outcome. "Logit" is an abbreviated version of "logistic unit." The "ln" in the equation represents the natural logarithm. (By contrast, a probit regression model also has a binary outcome but uses the inverse of a logit link function as its outcome.)

The **maximum likelihood estimate (MLE)** is the value of a coefficient in a logistic regression model that gives the model the greatest probability of matching the observed data. MLE is one of the most common approaches used for fitting logistic models. The coefficient (the β value) for a dichotomous predictor variable in a logit model represents the difference in "log-odds" of the outcome for a person with that predictor characteristic compared to someone without that exposure. The odds ratio for the association between a dichotomous predictor variable and case status can be calculated by taking the inverse of the coefficient, which is written as e^{β} or $\exp(\beta)$. The odds ratio for each predictor variable represents the change in the odds of the outcome—typically, the odds of being a case or being classified as a "yes"—for a 1-unit change in the predictor variable. The confidence interval for the odds ratio can be calculated using the value of the coefficient and its standard error.

Figure 31-3 provides an example of the output for a simple logistic regression model. The value of the coefficient for sex, which was coded as female = 0 and male = 1, is $\beta = 0.644$. The point estimate for the odds ratio for sex and case status is the exponential of beta: $e^{\beta} = \exp(\beta) = \exp(0.644) = 1.90$. The 95% confidence interval can be calculated using beta, the standard error, and a multiplier of 1.96, which is used because 95% of the area under a normal curve falls within 1.96 standard deviations of the mean. Because the confidence interval of (1.28, 2.84) does not overlap with OR = 1, sex is considered to have a statistically significant association with being a case. In this example, being male rather than female (that is, a 1-unit increase in the value of the sex variable) is associated with 1.90 times greater odds of being a case. Likelihood ratio tests, the Wald statistic, the Hosmer–Lemeshow test, and other goodness-of-fit tests can confirm the soundness of a logistic regression model.

FIGURE 31-3 Example of a Simple Logistic Regression Model

The output for the regression model predicting being a case (not a control) is:

	β	Standard Error (SE)	Odds Ratio (OR) (95% Confidence Interval [CI])	*p* Value
Sex	0.644	0.204	1.90 (1.28, 2.84)	.002
Constant	−0.584	0.109		

OR for Sex = exp(β) = exp(0.644) = 1.904

Lower bound of 95% CI: exp(β – 1.96*SE) = exp(0.644 – 1.96*0.204) = 1.276

Upper bound of 95% CI: exp(β + 1.96*SE) = exp(0.644 + 1.96*0.204) = 2.842

Multipliers: 1.645 for a 90% CI, 1.96 for a 95% CI, and 2.576 for a 99% CI

31.4 Dummy Variables

The predictor variables in regression models can take a variety of forms, but they must have numeric responses. Nominal categorical variables have responses that cannot be ordered, so they cannot be assigned a rank. However, a set of dummy variables can be created to convert categorical responses to a series of dichotomous variables that can all be included in the same regression model. **Dummy variables** are derived variables created by recoding one variable with *n* categorical responses into a set of *n – 1* dichotomous (0/1) variables. Dummy variables can also be used to convert ratio/interval variables into a set of derived categories, so that a series of odds ratios for the levels of the derived categorical variable can be estimated with a logistic regression model.

Figure 31-4 provides an example of how this type of recoding is done. If the original categorical variable has *n* possible responses, then *n – 1* dummy variables are required to capture all the responses to the original question. In the example in the figure, there were four possible responses, so three dummy variables are required. If there were nine categorical responses instead, then eight dummy variables would be required. In some modeling approaches, all of the *n – 1* variables are routinely included in a regression model, even if some would otherwise be eliminated during a stepwise selection process. A statistician can provide guidance on appropriate use.

31.5 Confounding and Effect Modification

One of the main reasons researchers use multivariable statistical models including three or more variables is to explore the interactions that may occur among variables. A **third variable** is a variable that is associated with an exposure variable and an outcome variable but is not part of the causal pathway from an exposure to an outcome. Some third variables conceal or muddle the true relationship between the independent and dependent variables of interest. There are several types of third variable effects, including confounding and effect modification. They may make the association between an exposure variable and

FIGURE 31-4 Dummy Variables

If the response to the original question was . . .	Then the values of the dummy variables are . . .			
	B_Dummy	C_Dummy	D_Dummy	
	(Was B the response to the original question?)	(Was C the response to the original question?)	(Was D the response to the original question?)	Conclusion based on the dummy variables
A	0 (no)	0	0	The response was not B, C, or D, so it was A.
B	1 (yes)	0	0	The response was B.
C	0	1	0	The response was C.
D	0	0	1	The response was D.

an outcome variable appear more or less significant than it truly is.

A **confounder** is a third variable that is associated with both the exposure variable and the outcome variable and distorts the apparent relationship between the exposure and outcome. A confounder is not a causal factor for the disease, and it is not part of the causal pathway. However, a confounding variable may make the association between an exposure variable and an outcome variable appear more or less significant than the relationship truly is.

For example, suppose that the crude (unadjusted) odds ratio for the relationship between sedentariness and a first heart attack shows that the odds of physical inactivity in the past year are four times higher (OR = 4) among adults who have had a recent heart attack than among adults who have no history of heart disease. Age may confound this association. Older adults are more likely than younger adults to be inactive, and older adults are also more likely to have a heart attack. Age-specific analysis may show that the odds ratio for this association is OR = 2 among younger adults and it is also OR = 2 among older adults. The discrepancy between the crude association (OR = 4) and the age-specific associations (both OR = 2) is a sign that age is confounding the association between sedentary behavior and myocardial infarctions.

When a third variable like age is shown to be a confounder, an adjusted measure of association, such as an age-adjusted odds ratio, should be reported for the association between the exposure and the outcome. In the preceding example, the confounder is hiding the true association between physical inactivity and heart attacks. Instead of reporting a crude odds ratio of $OR_{cr} = 4$, it would be more accurate to report an age-adjusted odds ratio of $OR_{adj} = 2$. A **Mantel-Haenszel** weighting method can adjust odds ratios or other measures of association for an exposure variable and an outcome variable after using a third variable to stratify the data.

An **effect modifier** is a third variable that defines groups of individuals who experience different biological responses to various exposures. For example, menopausal status may be an effect modifier for some studies about women's health. Suppose that heavier weight is associated with a decreased risk of breast cancer in premenopausal women but has the opposite effect, an increased risk of breast cancer, in postmenopausal women. Grouping all women together without accounting for menopausal status might make it look like weight is not a risk factor for breast cancer, because the experiences of premenopausal and postmenopausal women would be averaged together. Reporting that there was no association between weight and breast cancer would hide a potentially important biological difference in breast cancer risk that may be related to hormonal status.

If a third variable is shown to be an effect modifier—one for which the stratified measures of association are different in populations with different biological characteristics—it is best to report separate stratum-specific measures of association for each level of the effect modifier, such as separate results for premenopausal and postmenopausal women. Pooling the results for the biologically different groups would hide meaningful differences, so an adjusted or crude measure of association should not be reported when effect modification is occurring.

Figure 31-5 summarizes the steps required to identify confounders and effect modifiers. To be a confounder or effect modifier, the third variable must be independently associated with both an exposure (or predictor) variable and an outcome variable. These two relationships should be confirmed. Then, a crude odds ratio (or other measure of association) for the relationship between the exposure and the outcome should be calculated, along with a separate measure of association for each level of the third variable, such as separate odds ratios for males and females. The stratum-specific measures of association are then compared using a

Breslow-Day test, which assesses the homogeneity of stratum-specific measures of association. Alternatively, a –2 log likelihood test or another statistical test can compare the strata. After running a suitable test, the interpretation of the third variables are as follows:

- If the stratum-specific measures of association are not different but they are different from the crude measure of association, the third variable is a confounder. Report an adjusted measure.
- If the stratum-specific measures of association are different from one another and they are also different from the crude measure of association, the third variable is likely an effect modifier. Report stratum-specific measures.
- If the crude and stratum-specific odds ratios are all similar, then neither confounding nor effect modification is occurring. Report a crude (unadjusted) measure.

Interaction occurs when the effect of one predictor variable on an outcome variable depends on the presence or absence of a second predictor variable. The two exposure variables together have an additive or multiplicative impact on the outcome variable. Synergistic interactions increase disease risk beyond the expected level. Antagonistic interactions decrease disease risk to below the expected level. Interaction terms added to regression models can help clarify the relationships between sets of variables, but they must be interpreted very carefully.

Several other types of third variable effects might also affect the interpretation and reporting of multivariable analyses. An **extraneous variable** is a third variable that produces an apparent but false association between two other variables that are not causally related. The term **spurious** describes results that are false or invalid, such as the spurious associations that are generated by extraneous variables. A **lurking variable** is a third variable that was not measured in a study but is affecting the apparent association between an exposure variable and an outcome variable.

FIGURE 31-5 Confounding and Effect Modification

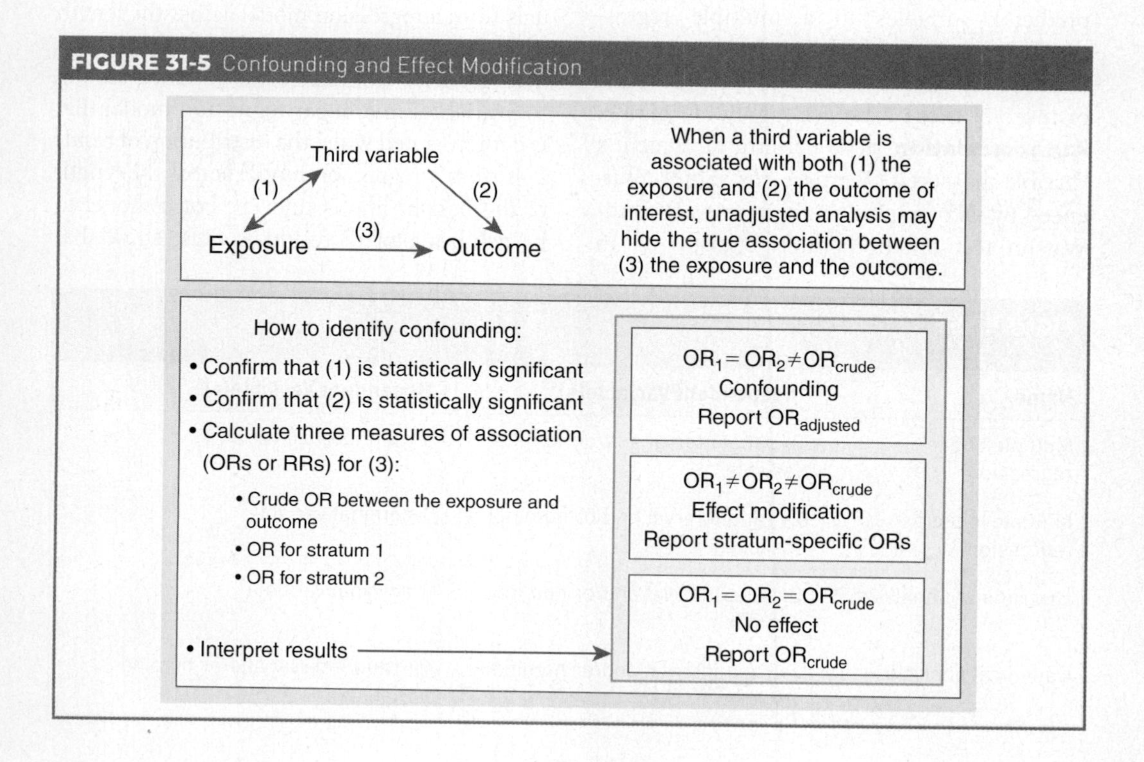

Third variable
(1)
(2)
Exposure
(3)
Outcome
When a third variable is associated with both (1) the exposure and (2) the outcome of interest, unadjusted analysis may hide the true association between (3) the exposure and the outcome.
How to identify confounding:
• Confirm that (1) is statistically significant
• Confirm that (2) is statistically significant
• Calculate three measures of association (ORs or RRs) for (3):
• Crude OR between the exposure and outcome
• OR for stratum 1
• OR for stratum 2
• Interpret results
$OR_1 = OR_2 \neq OR_{crude}$
Confounding
Report $OR_{adjusted}$
$OR_1 \neq OR_2 \neq OR_{crude}$
Effect modification
Report stratum-specific ORs
$OR_1 = OR_2 = OR_{crude}$
No effect
Report OR_{crude}

31.6 Multiple Regression

A variety of analytic approaches can test relationships among three or more variables while adjusting for possible confounders (**Figure 31-6**), including multiple linear regression and multiple logistic regression.

A **multiple linear regression** model examines the relationships between several ratio/interval and/or nominal predictor variables and one ratio/interval outcome variable when there is a linear relationship between the independent and dependent variables. Multiple linear regression models can have both continuous and categorical predictor variables, as long as the responses to categorical variables are expressed by numbers and all of the key assumptions of this type of model are met.

The independent variables in a multiple linear regression model must have reasonably independent errors. **Multicollinearity** is a problem that occurs when two or more predictor variables in a multiple regression model are highly correlated, and that redundancy makes the coefficients for one or more of those variables highly inaccurate. **Autocorrelation** is a pattern in which a variable measured over time has values influenced by its own past values as per a Durbin-Watson test or another test statistic or, in spatial analysis, a measurement of how similar one location is to nearby places. The presence of multicollinearity or autocorrelation might indicate that the model outcomes are inaccurate. The **variance inflation factor (VIF)** should be small enough (typically VIF < 10) to show that the independent variables in a regression model have reasonably independent errors and are not intercorrelated. (VIF values cannot be lower than VIF = 1.) Similarly, the **tolerance**, which is the inverse of the variance inflation factor (that is, 1/VIF), should not be too close to 0.

Another requirement for a multiple linear regression model is that the residuals for all variables must be normally distributed, as per a Kolmogorov-Smirnof test or another goodness-of-fit test. (Log transformations can sometimes be used to change nonlinear variables into ones that will result in a normal distribution.) **Homoscedasticity** is the homogeneity (similarity) of variance among the variables in a linear regression model that is demonstrated by the even distribution of residuals from a regression model across the length of the best-fit line. **Heteroscedasticity** is the heterogeneity (difference) of variance among the variables in a linear regression model that is demonstrated when the distribution of residuals from a regression model across the length of the best-fit line is uneven. For a model to be valid, a plot of residuals must show that

FIGURE 31-6 Examples of Multivariable Analysis Approaches for Testing Relationships Among Three or More Variables

Name	Independent Variable(s)	Dependent Variable(s)
Multiple linear regression	2+ ratio/interval and/or nominal variables	1 ratio/interval variable
Multiple logistic regression	2+ ratio/interval and/or nominal variables	1 nominal variable
Discriminant analysis	2+ ratio/interval and/or nominal variables	1 nominal variable
Canonical analysis	2+ ratio/interval and/or nominal variables	2+ ratio/interval and/or nominal variables

the error terms demonstrate homoscedasticity rather than heteroscedasticity.

Although it is easy to use a statistical software program to generate multiple regression models, researchers must carefully check all of the related output before concluding that the model is a valid one. **Figure 31-7** provides an example of how to interpret the output for a multiple linear regression model with two continuous predictor variables. The constant and the coefficients (the betas) for the predictor variables can be used to write an equation for a best-fit line. That equation can be used to examine the individual effect of each predictor variable on the outcome variable. To make this assessment, the value of one of the two predictor variables is kept constant so that the effect of a 1-unit change in the value of the other predictive variable on the expected outcome value can be ascertained.

Figure 31-8 shows how to interpret models with multiple types of predictor variables that do not interact. In the example, a 1-unit increase in the value of the "Predictor_2" variable is associated with a 2-unit increase in the value of the outcome (because the coefficient for "Predictor_2" is $\beta = 2.0$). This relationship between Predictor_2 and the outcome is the same for both males and females, even though males have an 18.7-unit higher value for the outcome than females (because the coefficient for sex is $\beta = 18.7$).

The predictor variables in multiple linear regression models may interact. For example, interaction may be occurring when the best-fit regression lines for males and females have significantly different slopes. **Figure 31-9** illustrates how to interpret models when interaction is occurring between some of the predictor variables. In the example, a 1-unit increase in the value of Predictor_2 is associated with a 2.4-unit increase in the value of the outcome for females but only a 1.2-unit increase for males. The equation for the regression model expresses this interaction through the use of a special interaction term. A **hierarchical model**, also called a **multilevel model**, is a multivariable regression model that adjusts for different levels of exposure, such as for both census tract and county.

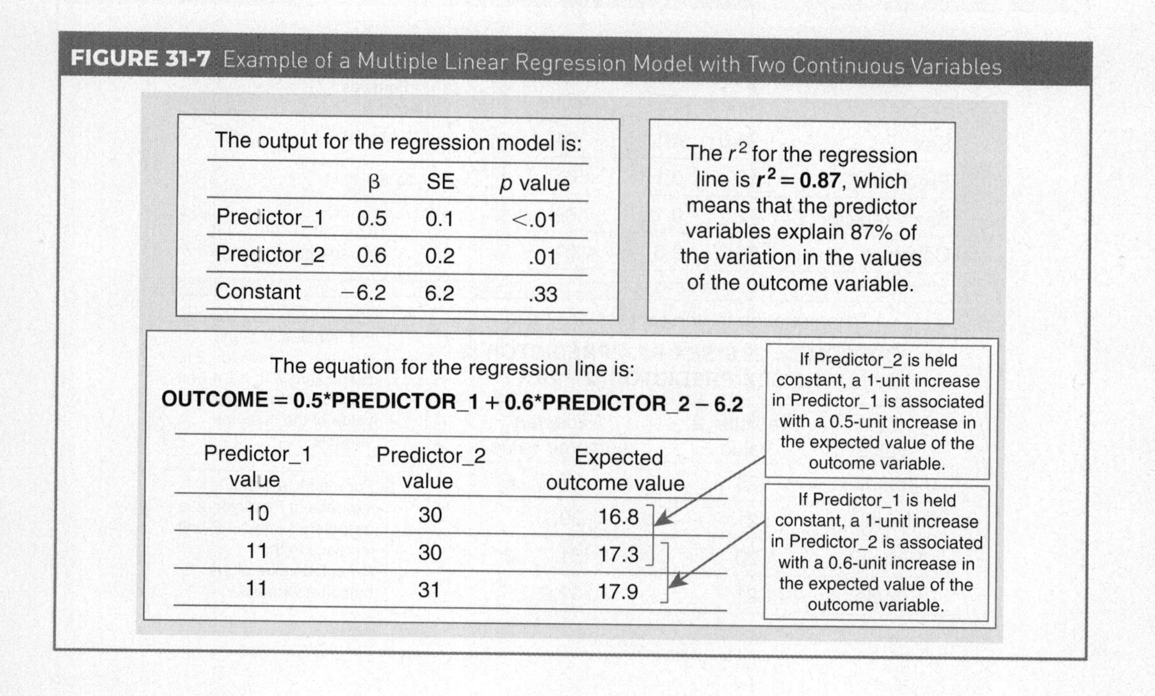
The output for the regression model is:
β SE p value
Predictor_1 0.5 0.1 <.01
Predictor_2 0.6 0.2 .01
Constant −6.2 6.2 .33
The r² for the regression line is r² = 0.87, which means that the predictor variables explain 87% of the variation in the values of the outcome variable.
The equation for the regression line is:
OUTCOME = 0.5*PREDICTOR_1 + 0.6*PREDICTOR_2 − 6.2
Predictor_1 value
Predictor_2 value
Expected outcome value
10 30 16.8
11 30 17.3
11 31 17.9
If Predictor_2 is held constant, a 1-unit increase in Predictor_1 is associated with a 0.5-unit increase in the expected value of the outcome variable.
If Predictor_1 is held constant, a 1-unit increase in Predictor_2 is associated with a 0.6-unit increase in the expected value of the outcome variable.

FIGURE 31-7 Example of a Multiple Linear Regression Model with Two Continuous Variables

FIGURE 31-8 Example of a Multiple Linear Regression Model with One Continuous and One Categorical Variable with No Interaction

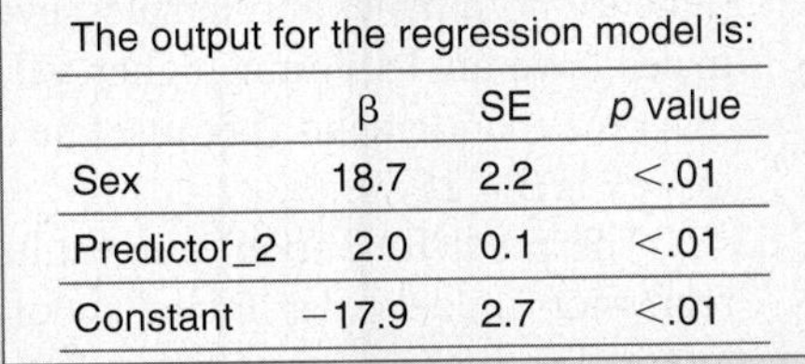

The output for the regression model is:

	β	SE	*p* value
Sex	18.7	2.2	<.01
Predictor_2	2.0	0.1	<.01
Constant	−17.9	2.7	<.01

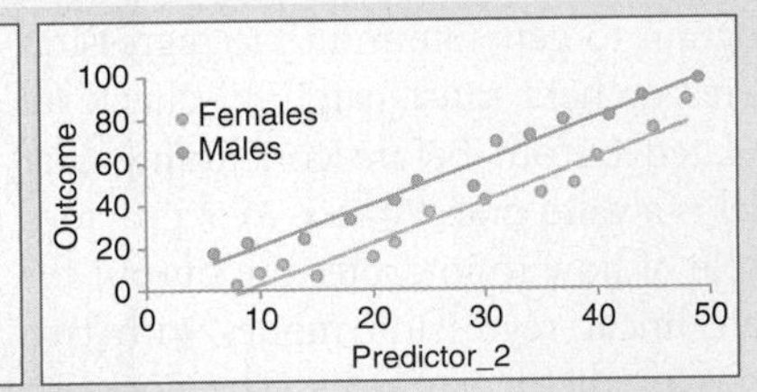

The equation for the regression line is:

OUTCOME = 18.7*SEX + 0.2*PREDICTOR_2 − 17.2

Sex value	Predictor_2 value	Expected outcome value
0	20	22.1
1	20	40.8
1	21	42.8

If Predictor_2 is held constant, a male has an 18.7-unit higher expected outcome value than a female.

For males, a 1-unit increase in Predictor_2 is associated with a 2.0-unit increase in the expected outcome variable.

FIGURE 31-9 Example of a Multiple Linear Regression Model with One Continuous and One Categorical Variable with Interaction

The output for the regression model is:

	β	SE	*p* value
Sex	28.0	4.1	<.01
Predictor_2	2.4	0.1	<.01
Sex*Predictor_2	−1.2	0.1	<.01
Constant	−20.3	2.8	<.01

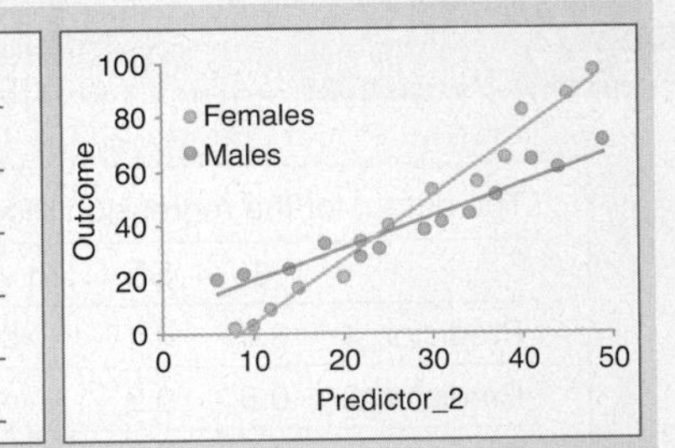

The equation for the regression line is:

OUTCOME = 28.0*SEX + 2.4*PREDICTOR_2 − 1.2*SEX*PREDICTOR_2 − 20.3

Sex value	Predictor_2 value	*Expected* outcome value
0 (female)	20	27.7
0 (female)	21	30.1
1 (male)	20	31.7
1 (male)	21	32.9

For females, a 1-unit increase in Predictor_2 is associated with a 2.4-unit increase in the expected value of the outcome variable.

For males, a 1-unit increase in Predictor_2 is associated with a 1.2-unit increase in the expected value of the outcome variable.

FIGURE 31-10 Example of a Multiple Logistic Regression Model

The output for the regression model predicting being a case (not a control) is:

	β	SE	OR (95% CI)	*p* value
Sex	0.59	0.49	1.8 (0.7, 4.7)	.23
Ate_food	1.44	0.52	4.2 (1.5, 11.7)	.01
Constant	−0.82	0.37		.03

OR for Sex = exp(β) = exp(0.59) = 1.8

Lower bound of 95% CI: exp(β − 1.96*SE) = exp(0.59 − 1.96*0.49) = 0.7

Upper bound of 95% CI: exp(β + 1.96*SE) = exp(0.59 + 1.96*0.49) = 4.7

OR for Ate_food = exp(β) = exp(1.44) = 4.2

Lower bound of 95% CI: exp(β − 1.96*SE) = exp(1.44 − 1.96*0.52) = 1.5

Upper bound of 95% CI: exp(β + 1.96*SE) = exp(1.44 + 1.96*0.52) = 11.7

Controlling for Ate_food (a yes/no variable for whether participants ate a certain food), there is no difference by sex in the odds of being a case: OR = 1.8 (0.7, 4.7).

The r^2 for the model is $r^2 = 0.14$, which means that the predictor variable explains only 14% of the variation in the values of the outcome variable.

Controlling for Sex, those who ate the suspected food item had significantly higher odds of being a case than those who did not eat the item: OR = 4.2 (1.5, 11.7).

A **multiple logistic regression** model examines the relationships between several ratio/interval and/or nominal predictor variables and the value of one nominal outcome variable. **Figure 31-10** shows how to interpret a multiple logistic regression model. Multiple logistic regression models can have both continuous and categorical predictor variables, and the predictor variables do not have to have a normal distribution, be linearly related, or have equal variances. A model with two predictor variables generates adjusted odds ratios for both variables. In the example, the food-adjusted odds ratio for sex has $p = .23$, which indicates that after adjusting for consumption of the suspected culprit food the people with gastroenteritis and the healthy controls did not have different likelihoods of being male rather than female. The sex-adjusted odds ratio for eating the suspected culprit food has $p = .01$ and an odds ratio of OR = 4.2 (1.5, 11.7), which indicates that after adjusting for sex the people with gastroenteritis had four times greater odds of eating the suspect food item than those who were not sick.

31.7 Causal Analysis

An **association** is a relationship between two variables. **Causation** is a relationship in which an exposure directly causes an outcome. A **causal factor** is an exposure that has been scientifically tested and shown to occur before the disease outcome and to contribute directly to its occurrence. The presence of a statistical association between two or more variables is not proof that a causal relationship is present. Correlation does not equal causation. The determination of whether an exposure is a causal factor requires statistical analysis plus additional considerations of causality. An apparent link between an exposure and an outcome may be causal, but it may be due to chance, confounding, or bias.

(Statistical associations deemed to be the result of error or bias rather than a true relationship are often said to be spurious associations or artifacts.) The presence of causality is usually determined with both quantitative analysis and a qualitative consideration of causal theory.

Temporality describes the timing of events. An exposure cannot cause an outcome if the exposure does not occur prior to the outcome. Showing that the exposure happens before the outcome is a necessary step toward demonstrating causality, but it is not sufficient evidence that the exposure is a causal factor. The **Bradford Hill criteria** (and more recent adaptations, such as the Hill-Doll criteria) are a set of conditions that provide support for the existence of a causal relationship between an exposure and an outcome. These criteria include temporality as well as the strength of the association, specificity, consistency, and other types of evidence (**Figure 31-11**). There is no requirement that all of these causal criteria must be met for an exposure to be considered the cause of an outcome, but the likelihood that a relationship is causal increases when more criteria are met. Most researchers are very cautious about using language that claims or implies causality, and the use of any words that suggest causal relationships must be carefully justified.

Etiology is the cause of a disease or other health disorder. **Multicausality** is a causal pathway in which many different risk factors or combinations of risk factors contribute to a disease occurring. A **logic model** is a visualization of the hypothesized causal pathways that lead to an outcome of interest. Logic models are typically illustrated with a series of arrows that show the assumed relationships among distal exposures, proximal exposures, and outcomes. (Logic models can also be used as part of program management when flowcharts or other graphics are used

FIGURE 31-11 Criteria for Causation

Temporality	Did the exposure happen before the onset of disease?
Strength of the association	Is the measure of association between the exposure and outcome strong (such a rate ratio or odds ratio having a value that is much greater than 1)?
Biological gradient (dose-response relationship)	Do people with a higher level of exposure have a higher risk of the outcome than people with a lower level of exposure?
Cessation	Does stopping the exposure reduce the risk of the outcome?
Specificity	Are the exposure and outcome both narrowly defined rather than general concepts?
Theoretical plausibility	Is there a reasonable biological explanation for why the exposure might cause the outcome?
Consistency	Has a potentially causal relationship between the exposure and outcome been observed in other studies and other populations?
Plausibility/coherence	Is a causal relationship between the exposure and outcome congruent with other knowledge about the variables?
Experimentation	If it is ethical to conduct an experimental study of the exposure and outcome, has experimental testing confirmed a causal relationship?
Consideration of alternate explanations	Are there reasons why what appears to be a causal relationship might not actually be causal?

to illustrate the key inputs, activities, and outcomes for the various phases of planning, implementation, and evaluation.) Once a logic model has been developed based on the existing scientific literature and the study hypotheses, mathematical analyses can be used to examine various aspects of the proposed etiologic pathways.

Multiple regression models cannot prove that an exposure caused an outcome, but they can provide insights about the etiology of a disease or other health disorder. Results of regression models can be used as part of qualitative considerations of causality. A **recursive model** assumes that all causal pathways are unidirectional. **Path analysis** is a recursive causal analysis strategy that uses regression models to examine causal patterns among variables. A **nonrecursive model** assumes that causal pathways can be bidirectional. **Structural equation modeling** is a nonrecursive causal analysis strategy that can be used to examine complexities in the directionalities of the path diagram.

31.8 Survival Analysis

Survival analysis is the statistical evaluation of the distribution of the durations of time that individuals in a study population experience from an initial time point (such as the time of enrollment in a study or the time of diagnosis of a particular condition) until some well-defined event, such as death, discharge from a hospital, or some other outcome. Some survival measures are based on **cumulative probability**, the probability of an event occurring by the end of a particular observation period. Others are based on **conditional probability**, the probability of an event occurring given that some prior event has already occurred. A cumulative survival study might ask what percentage of people born in the early 1900s lived to age 95 years, while a conditional survival study might determine the percentage of people who have already lived to age 90 years who survive 5 more years to age 95. This conditional probability can be written as P(B|A), where B represents the probability of surviving to age 95 and A represents the condition of having already survived to age 90. The vertical line between B and A can be read as "given"—the probability of B happening *given* that A has already happened.

The most common population-level measures of survival include the median survival time and cumulative survival at set times after enrollment or diagnosis, such as 1-year or 5-year survival rates. A **Kaplan-Meier plot** is a time graph that displays cumulative survival rates in a study population (**Figure 31-12**).

FIGURE 31-12 Example of a Kaplan-Meier Plot

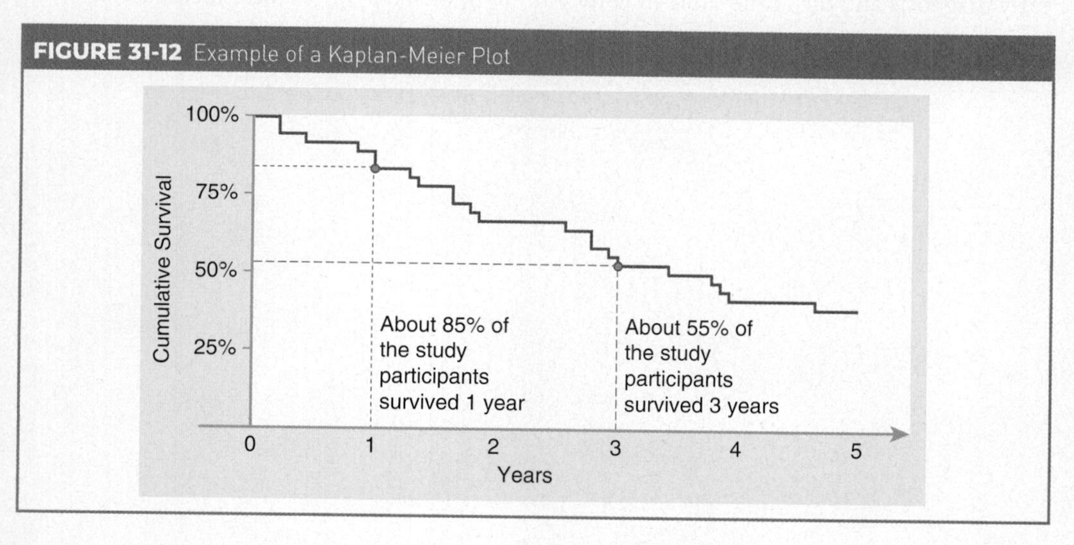

A **log-rank test** is a statistical test that determines whether survival rates are longer in one population than another.

A **life table** is an actuarial table that displays conditional and cumulative survival probabilities in a population. A **hazard function** is an equation describing the conditional probability of an individual having an event (such as death) at a particular time given that the person has survived to that time. A **hazard ratio** compares durations of time to an event in two populations. **Cox proportional hazards regression** is a type of regression model that estimates a hazard ratio.

Survival analyses require careful interpretation. For example, **lead-time bias** occurs when a screening test that enables early detection of an adverse health condition is incorrectly interpreted as prolonging survival with the condition. Suppose that a particular type of cancer has a median survival of 4 years after the onset of symptoms. A new screening test is developed that detects the cancer 2 years before onset of symptoms. If early diagnosis does not improve treatment outcomes, individuals diagnosed with the new test will not have a longer life expectancy than they did before the screening test was available. Instead of living 4 years after clinical diagnosis, they will live 6 years after early diagnosis and die at the same time they would have died if they had not been diagnosed early. The screening test may artificially increase the 5-year survival rate, because the survival rate is based on time from diagnosis, but that increase would not represent a real improvement in survival.

31.9 Cautions

Only a very limited number of studies require regression analysis or any of the other advanced statistics that are described in this chapter. User-friendly statistical software programs have made it possible for nearly everyone to run advanced statistical analyses, but these programs still require the user to select appropriate tests and decipher the outputs. Researchers should not use these tests without first knowing when to use them, what conditions have to be met to make their use appropriate, how to run them, and how to interpret them. They also need to make informed decisions about how to handle missing data and how to use sensitivity tests to confirm the robustness of study results. Advanced statistical tests should be used only when they are necessary for the research question. Specialty statistical references and experienced statisticians need to be consulted before attempting to implement any of these methods.

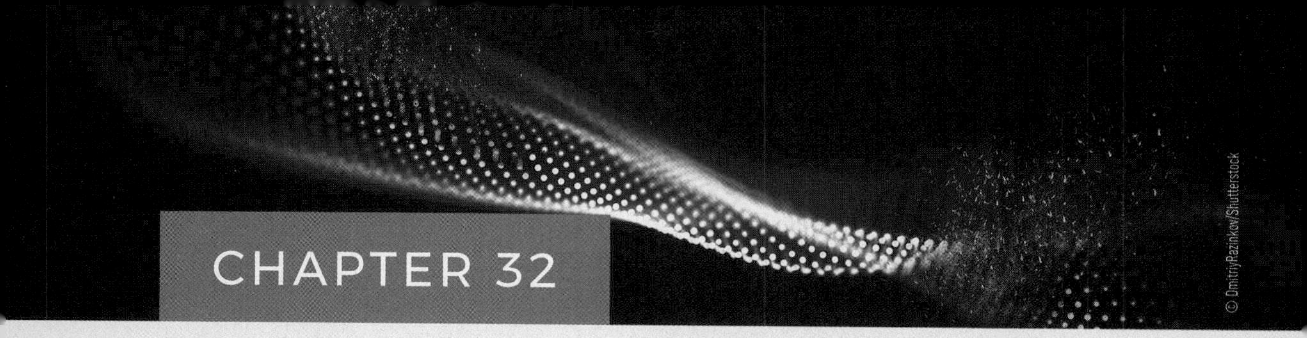

CHAPTER 32

Qualitative Analysis

Qualitative data analysis uses inductive methods to understand emergent themes and theories.

32.1 Overview

The goal of qualitative analysis is to understand the ways that people find meaning in their experiences and to develop themes and theories that explain phenomena (**Figure 32-1**). The naturalistic inquiry approach that is frequently used for qualitative research studies is a holistic process that might involve several rounds of data collection, analysis, and interpretation in which the preliminary themes and theories identified in one round inform the data that are collected in the next round. This cycle might continue until data saturation has been reached and new data are no longer providing new information.

There are three primary modes of reasoning: deduction, induction, and abduction. **Deduction** makes logical inferences based on facts or widely accepted premises, and the conclusions are assumed to be certain. **Induction** makes inferences based on observations, and

FIGURE 32-1 Comparing Typical Qualitative and Quantitative Analytic Approaches

	Quantitative Studies	Qualitative Studies
Goal	Predict	Understand
Data types	Observable phenomena (facts and numbers)	Attributed meanings (opinions and narratives)
Data collection approach	Formal, impersonal, and detached	Informal, personal, and reflexive
Data analysis approach	Deductive (theory-driven: use data to test theories) and reductionist	Inductive (data-driven: use data to generate theories) and holistic

the conclusions are assumed to be likely. **Abduction** makes inferences based on limited observations and minor premises, so the conclusions are assumed to be best guesses that are merely probable. Deductive reasoning moves from the general (theories or premises) to the specific (data), while inductive reasoning moves from the specific (observations) to the general (theories). Quantitative research typically uses a theory-driven deductive approach in which data are used to test preformulated hypotheses. Qualitative research typically uses a data-driven inductive process in which observations are used to identify patterns, generate hypotheses, and formulate theories.

32.2 Analytic and Interpretive Frameworks

There are a variety of approaches that can be used for the analysis and interpretation of qualitative data, including grounded theory, content analysis, narrative analysis, discourse analysis, and many others.

Analysis based on grounded theory uses inductive approaches to develop causal theories about a phenomenon. **Constant comparison** is a process in which qualitative data are collected and analyzed simultaneously, rather than waiting to begin analysis after all data have been gathered. The categories that emerge from one round of data analysis inform the next round of data collection.

Content analysis is the process of categorizing textual data. In this context, texts can be documents, speeches, photographs, videos, or other media. The analyst begins by systematically coding the text using labels and categories derived from the text or from existing theories or previous research findings. The analyst determines which codes occur most often and then uses that information to identify the most prominent patterns and themes in the text.

Narrative analysis is a qualitative analysis method that seeks to understand personal stories. Narrative analysis may focus on the content of stories, the structure of stories, the themes of stories, or the communication goals of stories. Postmodernism, feminism, or other established philosophies may be applied to help with the interpretation of the stories. **Discourse analysis** uses the tools of linguistics to evaluate the ordinary use of written and spoken language. The goal is to understand natural language use.

A variety of other philosophical orientations and frameworks can also be applied to the qualitative analysis process. For example, **hermeneutics** is the study of the interpretation of texts. The hermeneutic process seeks to understand the layers of meaning that are embedded within texts, images, and other artifacts. Researchers applying hermeneutics to qualitative data try to understand how people from different cultures and social groups interpret texts. **Semiotics** is the study of signs and symbols. Semiology seeks to understand the literal meaning of phrases and sentences and also to understand their underlying meanings within the context in which they are used.

32.3 Codes, Categories, and Themes

After data have been collected, they must be prepared for analysis. This process might involve generating and validating transcripts of interviews and focus groups or cleaning other files. Once the data files are complete and clean, several levels of coding are used to understand the data. The analytic process is a flexible one that must align with the theoretical paradigm and methodologies selected for the project. For example, projects applying grounded theory typically move through a process of collecting data, transcribing interviews, assigning initial codes, identifying categories, identifying themes, and developing

FIGURE 32-2 Qualitative Analysis Process

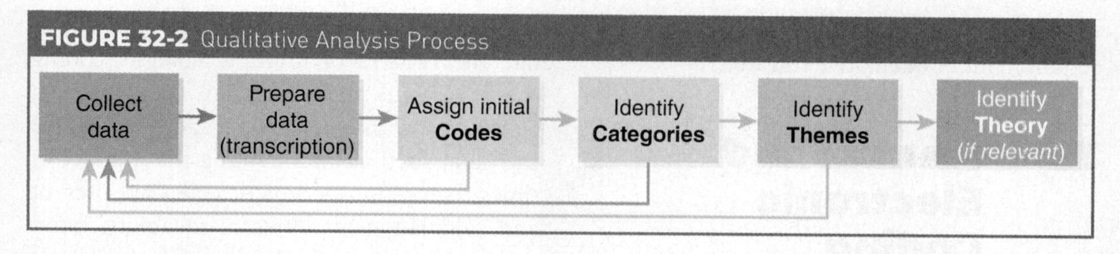

a theory (**Figure 32-2**). Many projects will include several cycles of data collection, analysis, and additional data collection and analysis rather than a single linear process of data collection, analysis, and reporting.

The first level of coding is often called open coding or initial coding. In qualitative analysis, **coding** (or **indexing**) is the use of words or short phrases to briefly summarize the contents, attitudes, processes, or other aspects of each item in a transcript or other qualitative document. A **code** is a label attached to a word or phrase. Several types of codes might be applied during the first and subsequent readings. Some codes might pertain to descriptions of participant characteristics, such as gender or socioeconomic status, or to place and time details. Some might capture emotions such as happiness, sadness, surprise, boredom, pride, confusion, relief, discomfort, hope, fear, admiration, jealousy, gratitude, or anger. Some might indicate the direction or magnitude of participants' perspectives, such as making a distinction between agreement and disagreement, positivity and negativity, presence and absence, or high and low interest. Some codes might denote participants' values, evaluations, or judgments. Some codes might describe processes or actions. ***A priori* codes** (or preset codes) are developed before the start of data analysis. The preliminary codebook may be based on the interview guide and previous publications about related topics. **Emergent codes** (also called emerging themes) are concepts that are identified during the early stages of qualitative analysis and assigned a label or code that describes them.

The second level of coding sorts the codes and then groups the codes into categories. A **category** is a group of related codes. The goal is to identify trends and patterns, look for relationships between codes, and begin to understand multiple layers of meaning. The analyst seeks to identify similarities and differences among codes, observe which codes occur frequently or infrequently, note the sequence of codes, and recognize other types of patterns and relationships. **Pattern coding** seeks to group codes into a limited number of categories. The analytic process may reveal that some of the initial codes express similar concepts and should be lumped together, or it may reveal that some of the initial codes should be split into separate codes. **Focused coding** seeks to identify the most frequent and important categories. The process of categorizing may reveal new meanings that were not explored with the initial coding, so several rounds of recoding and reclassifying may be necessary. **Iteration** describes a repetitive process. Iterative research processes are intended to generate new insights with each round of analysis and interpretation.

A third level of coding, typically called thematic coding, synthesizes the categories in order to identify the concepts, meanings, and themes that answer the study question. A **theme** is a concept that encompasses one or several categories. **Axial coding** identifies one core category or core phenomenon and several related categories that express the major and minor themes of the analysis. The core and related categories are then used to craft a narrative that explains the phenomenon. For some studies, a fourth level of coding generates a new theory about the phenomenon. A **theory** is a construct that provides a systematic explanation about a phenomenon. However, the development of a new abstract theory is not a requirement of qualitative research.

32.4 Manual and Electronic Coding

The process of coding and categorizing qualitative data can be completed by hand or facilitated by computer-assisted qualitative data analysis software (CAQDAS) programs such as ATLAS.ti and NVivo. Manual coding of qualitative data is usually done only when the data sets are relatively small, because this type of coding is a time-consuming process. Electronic coding is typically used for large data sets.

Coding begins with a line-by-line read of a document that has been segmented into meaningful units that can be coded. (For audio and visual files, coding begins with listening to and/or looking through the items.) If the coding is done using printed sheets of paper, key content can be highlighted with a pen or marker. Different colors of ink can be used to visually distinguish between different sets of observations. Words jotted in the margins can be used to add descriptions and other labels. Lines, arrows, symbols, and circles can mark important words and phrases and can show relationships. If the coding is done electronically in a word processor or spreadsheet file, highlighting and the comment function can be used to mark and annotate text. If computer-assisted coding is used, the codes are assigned within the software program. A codebook containing a master list of all the codes is generated and revised with subsequent rounds of coding. The analyst may also engage in **memoing**, documenting personal reflections and impressions about observations, participants, experiences, codes, categories, and themes.

Qualitative analysis software can facilitate the process of multiple analysts coding the same data files. Software programs can also assist with quantifying some aspects of the data, such as enumerating the number of times that various words or codes are present and calculating measures of reliability. Intercoder reliability is present when multiple coders code the data consistently. Intracoder reliability is present when one coder applies codes consistently.

32.5 Quality Assurance

Quality assurance for qualitative analysis includes considerations of credibility, transferability, dependability, and confirmability.

Credibility is present when the interpretation of the data accurately reflects the studied groups or texts. Credibility in qualitative research is an indicator of trustworthiness that is similar to internal validity in quantitative research. **Transferability** is present when the interpretation of qualitative data is likely to be applicable in other circumstances. Transferability is an indicator of applicability that is similar to external validity or generalizability in quantitative research.

Dependability is an indicator of consistency that is demonstrated through transparency about data collection, analysis, and interpretation methods. A dependable study is one that could be replicated. **Confirmability** is an indicator of neutrality that is present when the results of a study are shown not to be due to researcher bias. Confirmability may be enhanced through triangulation in which multiple data sources, methods, and theories are used to study a phenomenon from different perspectives.

CHAPTER 33

Additional Analysis Tools

A variety of spatial, mathematical, computational, and economic modeling techniques can enhance population health analyses.

33.1 Spatial Analysis

If global positioning system (GPS) coordinates or other geographic data have been collected as part of primary research or are available in secondary data sets, then special software programs may be useful for conducting spatial analyses. Once the geographic data have been incorporated into a geographic information system (GIS), it is possible to map the locations of events, examine the spatial distribution of events, search for patterns like disease clusters, and test for possible associations between various social and physical environmental characteristics and health status. For example, **Moran's *I* coefficient** tests for spatial autocorrelation, which is a measurement of how similar one location is to nearby places. A medical geography or health geography reference should be consulted for assistance with spatial analysis.

33.2 Bayesian Statistics

Statistics are often described as being frequentist or Bayesian. These terms refer to the two most popular ways of interpreting the meaning of probabilities.

A **frequentist** approach to probability is based on the expected frequency of an event occurring over a long time period or if an experiment is repeated many times. Inferential statistics assume that parameters are fixed and data vary, and they seek to determine whether the null hypothesis should be rejected. Suppose that a researcher is trying to estimate the mean age of a large population. A frequentist would say that the true mean age in the total population does not have a distribution, because it is a fixed value. A 95% confidence interval can be calculated for a sample mean age if a random sample of people is drawn from the total population. In frequentist analysis, confidence intervals are not probability distributions. The 95% confidence interval will either include the true mean age or not include the true mean age. The frequentist would interpret a 95% confidence interval as indicating that 95% of sample mean ages generated from random samples could be expected to contain the true mean age within their 95% confidence intervals. In other words, frequentists expect to be wrong 5% of the time.

A **Bayesian** approach uses data and prior beliefs (the priors) to predict the likelihood of a particular outcome (the posterior). Bayesian statistics assume that data are fixed and parameters vary, and they seek to examine whether the null hypothesis is likely to be better than the alternative. A researcher using a Bayesian approach to estimate a population mean age would say that the sample data are real and would interpret the 95% confidence interval (or credibility interval) as indicating that, given the available data, there is a 95% likelihood that the true mean age of the total population falls within the 95% interval. Complex computational methods can be used to calculate posterior distributions based on prior distributions. For example, **Markov chain Monte Carlo (MCMC)** methods are stochastic processes that use algorithms to take samples from simulated probability distributions.

Most statistical tests use a frequentist statistical approach, including inferential tests that are based on calculations of point estimates and their 95% confidence intervals and models that use maximum likelihood estimation methods. One limitation of frequentist analysis is that confidence intervals and *p* values are tied to sample sizes. Very large sample sizes will result in very narrow confidence intervals, and small sample sizes will result in wide confidence intervals. Bayesian analysis is useful when there are few data points, the researcher has strong intuitions about the priors for a model, or there is a lot of uncertainty about a particular model.

33.3 Mathematical Modeling

Mathematical modeling explores epidemiological trends or other changes over time in selected real-life or theoretical populations. For example, a **compartmental model** is a mathematical model in which each "individual" in the simulated population exists in only one of several states at one time, but over time these individuals can move between states. A compartmental model of aging would have one compartment for each age group. Suppose that 5-year age compartments are used. During one simulated year, approximately one-fifth of the members of the compartment for people ages 20 to 24 years would move to the compartment for people ages 25 to 29 years. The model could also incorporate population dynamics that allow individuals to be added to the population to simulate births and immigration and removed from the population to represent deaths and emigration.

An **SIR model** is a compartmental mathematical model of infection transmission that describes how the susceptible (S) individuals in a population may become infected (I) and then eventually recover (R) with immunity (**Figure 33-1**). In an SIR model, each member of a population is classified as susceptible, infectious, or recovered (sometimes called "removed"). Equations define the rate at which members move between these three compartments over time. The model can simulate new infections by moving members from the S to the I compartment. The impact of immunization can be modeled by moving vaccinated individuals directly from the S to the R compartment. Complexity can be added to the model by creating additional compartments, such as separate S, I, and R compartments for each age group or for different exposure groups. SIR models can also be modified to include additional disease states. For example, an SEIR model has four compartments—susceptible, exposed, infectious, and recovered—and

FIGURE 33-1 A Simple SIR Model

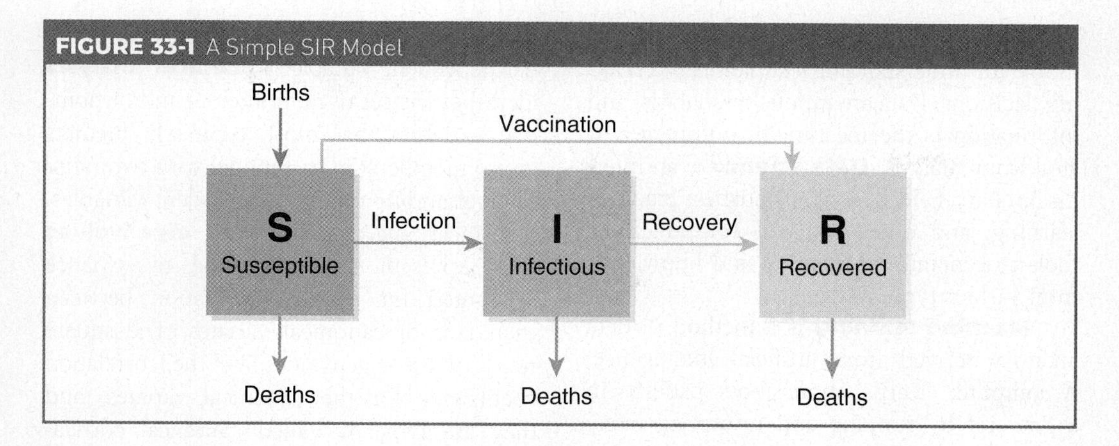

the E represents a latent stage in which individuals are infected but not yet contagious.

An **ordinary differential equation (ODE)** is an equation that includes one or more functions of one independent variable along with the derivatives of those functions. (ODEs are "ordinary" because they do not include the partial derivatives used in partial differential equations.) ODEs or other types of equations can describe the flows between compartments in a mathematical model over time. For example, one equation might describe the infection rate, which is the rate at which individuals move from the S compartment to the I compartment. Another equation might describe the rate at which individuals age from one age group's R compartment to the R compartment for the next oldest age group. To add more realism to a model, the distribution of population members across the compartments and the rates of flow between compartments usually are based on data from field studies.

A **deterministic model** is one in which the outcomes of the model are the same every time the model is run with the same inputs. A **stochastic model** is one in which the inputs vary according to a probability distribution, so the outcomes differ slightly every time the model is run. A distribution of the outcomes can be generated by rerunning the model hundreds or thousands of times.

Sensitivity analysis is the process of examining the robustness of statistical methods and the results of models. A variety of sensitivity tests can help ensure that a mathematical model has reasonable validity and provides insights into how the real world works.

33.4 Agent-Based Modeling

Agent-based modeling, sometimes called agent-based simulation or **individual-based modeling**, uses computers to simulate the actions and interactions of various individuals (agents) in a population. Once a set of assumptions about how the agents in the model behave and how they relate to one another is identified, those assumptions are written into the model's code. Specialized software is then used to run the simulation. Agent-based models can assist with developing and testing new theories as well as with understanding complex data.

33.5 Machine Learning

The word **data** (the plural of the word datum) refers to raw or unprocessed facts, figures, symbols, or signs. **Information** refers to data that have been processed and presented in a format usable for understanding a situation and making decisions. Data are inputs for analysis, and information is the meaningful output generated from analysis. **Data science** is an interdisciplinary field that uses statistics, machine learning, and other types of computational tools to generate information and knowledge from various types of data.

Machine learning is a method of data analysis derived from artificial intelligence. A computer "learns" more about patterns in a data set by running and rerunning many rounds of analysis. Machine learning is used to create and evaluate neural networks, decision trees, and a host of other emerging applications. For example, **social media analytics** is the process of compiling and analyzing data from social networking services like Instagram and Twitter. Machine learning can assist with **natural language processing**, a machine learning algorithm that is used in the analysis of qualitative and social media data to examine how people speak and write in real-life situations. Machine learning is also used for **cluster analysis** (sometimes called segmentation analysis or taxonomy analysis), which identifies groups of similar observations using an algorithm that seeks to minimize the variations among observations within each group.

The machine learning algorithms generated by iterative processes are often used for predictive analyses. Explanatory and causal models seek to explain observed associations by examining the strengths of the associations between variables. Predictive models have a different goal, aiming to determine which variables best predict group membership. In modeling, **discrimination** is the ability of a model to distinguish between independent groups. **Discriminant analysis** (or **discriminant function analysis**) is a statistical method that identifies the set of ratio/interval and/or nominal variables that most accurately predicts group membership in a model with a nominal dependent variable. **Canonical analysis** identifies the set of ratio/interval and/or nominal variables that most accurately predicts group membership in a model with two ratio/interval and/or nominal dependent variables. Canonical analysis calculates **eigenvalues** that represent the proportion of variance accounted for by the correlation between each pair of canonical variates. The square roots of the eigenvalues are the correlation coefficients for the canonical variates, and they are typically called canonical correlations. **Propensity score matching** predicts the probability of group membership while adjusting for covariates.

33.6 Cost-Effectiveness Analysis

A diversity of health economics methods are useful tools for health research. **Cost-effectiveness analysis** (CEA) is an economic analysis that compares the health gains from an intervention to the financial costs of that intervention. Cost-effectiveness studies typically calculate cost-effectiveness indicators as a ratio of health gains (in the numerator) to financial costs (in the denominator). Cost-effectiveness studies can be conducted for preventive, diagnostic, therapeutic, and curative interventions.

Health gains associated with various types of health-related interventions are often quantified in terms of the **quality-adjusted life year (QALY)**, a metric used in health economics to estimate the additional duration of life and quality of life conferred to populations by successful public health interventions. One QALY is equivalent to 1 year in perfect health. Diseases and disabilities reduce health status to less than perfect. A premature death is a death at any age that is younger than the target life expectancy in the population. Premature death reduces the QALY for that individual to 0. Health interventions such as prevention campaigns, screening tests that enable early diagnosis and treatment, clinical procedures, and rehabilitative therapies can restore the QALYs that in the absence of intervention would have been lost to illness and can prevent the QALYs that would otherwise have been lost to premature death. The average cost per QALY can be used to evaluate the cost-effectiveness of an intervention.

Quality of life (QOL) is a construct that captures an individual's perceived position in life in the context of that person's expectations, goals, values, and concerns. Various measures of QOL and health-related quality of life (HRQOL) can be used for economic analyses. Specialty references from health economics, health services research, and related disciplines provide information about how to estimate these measures.

33.7 Burden of Disease Metrics

The term **burden of disease** describes the adverse impact of a particular health condition (or group of conditions) on a population. Burden of disease studies often use a diversity of demographic and health measures to create metrics that summarize current health status in populations and describe changes in health status over time. An **indicator** is a variable used to measure performance, achievement, or change. A **metric** is a composite indicator derived from two or more other measures. The validity of a burden of disease metric is dependent on the quality of the component indicators and the rigor of the modeling methods used to generate the metric.

Years lived with disability (YLDs) are a burden of disease metric used to quantify the population-level reductions in health status attributable to nonfatal conditions. In burden of disease studies, disability is defined as any temporary or permanent reduction in health status. Any adverse health condition can cause disability: an acute infection, a pregnancy complication, a neonatal disorder, a nutritional deficiency, a chronic noncommunicable disease, a mental health disorder, physical impairment stemming from an injury, or any other cause of diminished health. YLDs are quantified using disability weights that assign higher burdens to health conditions that cause greater losses of productivity.

Years of life lost (YLLs) are a burden of disease metric used to quantify the population-level reductions in health status due to premature mortality. All deaths among people who are younger than a selected target life expectancy for a population are classified as premature deaths. If a population is defined as having a target life expectancy of 80 years, a toddler who dies on her 2nd birthday will contribute 78 years to the population YLL total and a man who dies on his 75th birthday will contribute 5 years to the population YLL total. The major contributors to YLLs in a population provide valuable information about the diseases and subpopulations that might be important targets for public health interventions.

The **disability-adjusted life year (DALY)** is a burden of disease metric that is quantified as the sum of YLDs and YLLs in a population. DALYs are similar to the QALYs that are frequently used for health impact assessments, but DALYs measure years of perfect health lost while QALYs measure years of perfect health gained. DALYs, QALYs, and other burden of disease statistics are useful when identifying population health priorities and making decisions about which health interventions to fund and support.

STEP 5

Reporting Findings

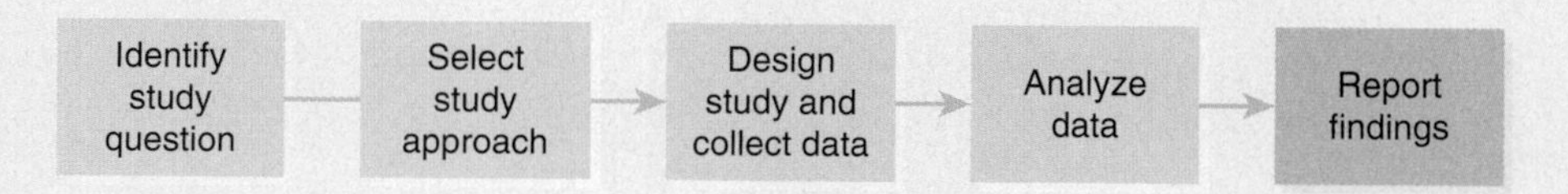

The fifth and final step in the research process is writing a research report and disseminating the results through presentation and publication. This section provides tips for writing, revising, presenting, and publishing findings.

- Posters and presentations
- Article structure
- Critically
- Critical revising
- Writing success strategies
- Reasons to publish

CHAPTER 34

Posters and Presentations

Research results are often publicly shared for the first time during an oral presentation or a poster session at an academic or professional conference.

34.1 Purpose of Conferences

The primary benefit of most professional and academic conferences is the networking that occurs during the gathering: meeting new people working in the same field of interest, catching up with former classmates and colleagues, and making and nurturing professional connections that may be helpful in the future. Conferences are a place to exchange ideas: to be inspired by the discoveries others in the field are making, to learn new methods and techniques in a discipline, and to share current work with others and receive advice from experts. Presenting new research in the form of a poster or an oral presentation can be a particularly useful way to get feedback on a project before submitting the work for review by a journal. Sharing findings is a way to gauge the strengths and weaknesses in the initial presentation that can be enhanced or corrected in the subsequent manuscript.

34.2 Structure of Conferences

Some conferences are annual events sponsored by professional organizations that draw thousands of attendees. Others are small gatherings of a few dozen scholars working in a narrow field of study. Most conferences include a mix of:

- Plenary sessions for all attendees, which often feature keynote addresses
- Concurrent sessions in which multiple panels of oral presentations are held at the same time in different rooms
- Poster sessions in which attendees can mingle while viewing research posters
- Exhibitions where attendees can visit informational displays set up by partner organizations, vendors, and other sponsors
- Business meetings run by the officers of the host organization

Presenters are usually assigned to deliver either an oral presentation or a poster presentation. An **oral presentation** involves one individual speaking to a group, typically for about 15 minutes. Oral presentations are generally considered to be more prestigious than posters, in part because there are usually more slots for poster presentations than for oral presentations. Oral presentations usually require delivering a prepared presentation and then participating in a question-and-answer period with the audience and other panelists. This interaction can be so helpful that some presenters are disappointed when no one in the audience points out a weakness in their work that they should address before submitting a research manuscript to a journal. However, oral presentations can be very stressful for those who are not experienced and confident public speakers.

A **poster session** is a designated time during a conference when selected researchers display printed placards and are expected to be available to talk about their posters with other attendees. Poster sessions usually do not require the presenters to make a formal speech. Instead, the sessions are designed to facilitate one-on-one and small group conversations. Posters may be taped along the walls of a room or displayed on long rows of easels, and attendees can browse through the posters at their own pace and interact with presenters if they want more information about a project. These relatively private conversations may allow for a more fruitful exchange of ideas than is possible during the question-and-answer time at the end of a concurrent session. Another benefit of posters is that they can be displayed in the hallway of an academic department or workplace for several months after the conference. However, posters often require more preparation time than oral presentations, and they may be expensive to print and a hassle to transport.

34.3 Writing an Abstract

Researchers who want to present at a conference usually are required to submit an abstract for consideration by the organizers. The scientific committee and other volunteer reviewers evaluate the submitted abstracts, decide which researchers will be invited to present, and then select who will give an oral presentation as part of a panel and who will be assigned to a poster session. All selected abstracts are usually printed in a bulletin that conference attendees use to decide which sessions to attend and which posters to seek out.

Abstracts often are due many months before a conference, yet it is not uncommon for conference guidelines to prohibit the submission of abstracts for studies that will be published in a journal prior to the conference. Thus, the ideal timing is to have preliminary results ready to include in the abstract, to prepare the final results for presentation at the conference, and then to use the feedback from the conference to finish the full-length manuscript that will be submitted for publication.

A good health research abstract includes key methods and results while also conveying one clear message that is appropriate to the audience expected at the conference. If the conference focuses on clinical practice, the abstract's applied message should be readily translatable into improved patient care. If the conference focuses on research theories and methods, the abstract should emphasize the novelty of the approaches used and their applicability to other research topics. If the conference focuses on health policy, the abstract should have a clear policy implication.

In a few scientific subdisciplines, selected conference presenters are invited to submit a **conference paper**, an article-length research report that will be published in the proceedings of the conference. These conference proceedings are not collections of abstracts; they are book-length volumes of research akin to an issue of a journal. In the fields in which these types of proceedings are commonly published, conference papers are considered equivalent to peer-reviewed journal articles. However, in most health science disciplines, the only written outputs from a conference are abstracts that are not complete enough to be considered part of the formal scientific literature. When only an abstract from a conference is published, researchers are encouraged to consider the conference presentation to be an intermediate step toward publication and not an end product.

34.4 Submitting an Abstract

Many conferences use online systems to collect applications, although some still ask for abstracts to be submitted via email. In addition to providing a title, an abstract, and the names of any coauthors, applicants may be invited to indicate their preferred presentation format. People who indicate a willingness to make either an oral presentation or present a poster may increase the likelihood that their abstracts will be accepted for the conference.

While most presentations at a conference are made by one individual, most conference organizers allow or expect abstracts to name multiple coauthors who contributed to the work. The general principle is that if an individual will be a coauthor of a journal article that reports the findings being presented at the conference, that individual should also be listed as a coauthor on the abstract for the conference. Every coauthor must approve the text of the abstract before it is submitted to the conference.

Submitting an abstract implies a commitment to attend the conference if the abstract is accepted. The sponsoring organization may keep track of dropouts and absentees and not allow them to present at future conferences. The fine-print instructions for the conference often specify the other expectations of applicants. Most conferences require presenters to pay a registration fee (often several hundred dollars) as well as cover all of their own travel expenses. Some schools and employers may reimburse some or all of these expenses for researchers who will present their work at a conference, but if funds are not available, the researcher will be responsible for these costs.

34.5 Preparing a Poster

Conference attendees are drawn to visually appealing, symmetric posters. Researchers preparing a poster must give attention to both the content and the design of the poster (**Figure 34-1**). Posters should be well organized and have a focused message. They should have a pleasing balance among text, images, and "white space" (background of any color that is not covered with words or images) and an inviting color palette.

Posters can be created by using either specialized graphic design software or a

FIGURE 34-1 Checklist for Poster Content and Design

Content	■ Keep the content focused on one core message. ■ Choose a descriptive title. ■ Include the names of all coauthors, brief author affiliations, and contact details for at least one author. ■ Do not list particulars about the conference (such as name, dates, or location) on the poster. ■ Consider skipping the abstract to save space. ■ Clearly state the main goal, the specific objectives or hypotheses, and the importance of the study. ■ Use a structured format, with introduction/background, methods, results, and conclusion/discussion sections (and a reference list, in small font, if previous studies are cited). ■ Be concise. Use short sentences and bulleted lists when possible. ■ Images like graphs, tables, flowcharts, photographs, and maps are more effective than words at conveying information.

(continues)

FIGURE 34-1 Checklist for Poster Content and Design *(continued)*

Design	■ Find out the size and shape (horizontal or vertical) of the display area that the conference organizers will provide and create a poster to fill the space. ■ Decide whether to print one large poster (preferred) or smaller panels that can be joined together at the conference venue. ■ Organize content into three or four columns or another structure with a logical flow. ■ Use boxes, color, and/or lines to group the information. ■ Select a visually pleasing color palette. ■ Ensure adequate contrast between the background (usually light) and content (usually dark). ■ Use large fonts and consistent typefaces that can easily be read several steps back from the poster. ■ Simplify graphs and make sure they can be read from a distance (which may require adding a title and directly labeling lines or bars rather than using a key). ■ Use high-resolution images that are owned by the author, licensed for use, or in the public domain (and remember that enlarged photographs become fuzzy).

presentation software program (like Microsoft PowerPoint). The size of a slide or page can be adjusted so that the dimensions match those required by the conference. A sample layout is shown in **Figure 34-2**, and the Internet has many examples of other poster designs. Asking several people to check both the content and the design of the poster before it is printed will improve the product.

Prior to having the poster printed, the researcher should inquire about printing costs (which will vary significantly depending on

FIGURE 34-2 Sample Poster Layout

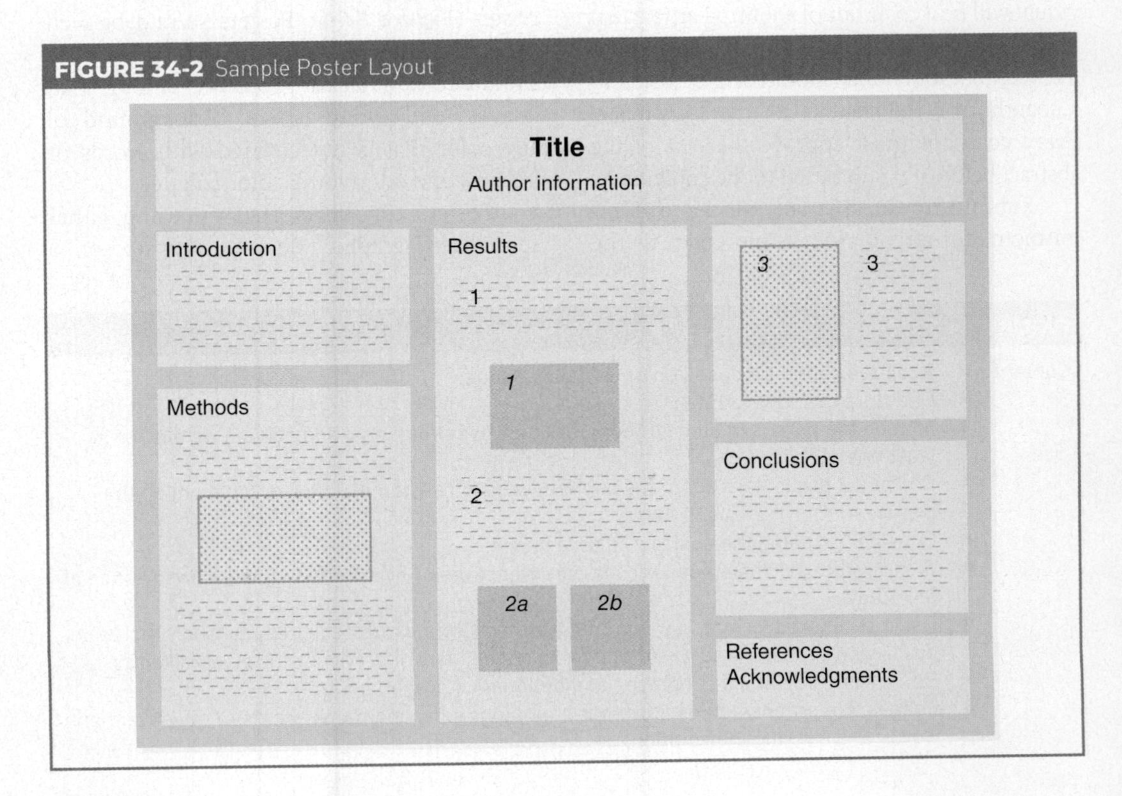

the size of the poster, the amount of color, the type of paper or fabric, and any special options like laminating or mounting) and the amount of time required for printing. If the presenter does not want to travel with a poster, it may be possible to arrange to have the poster printed by local printshop near the conference venue.

34.6 Using Images

Only images that are owned by the presenter (or a coauthor), are used with the written permission of the artist, are licensed for use (with or without attribution), or are in the public domain should be printed on a poster or projected during a talk. Images that are under copyright protection are usually not acceptable to use unless a license has been purchased. (Librarians can often offer expert advice about legal use of particular images.) Only images suitable for the professional setting should be selected for use; clipart and cartoons should generally be avoided.

Photo banks with images that are in the public domain or licensed for free use with attribution are available from Internet sites such as Pixabay and Wikimedia. Some search engines also offer image searching functionality that can locate relevant photographs, logos, and other types of graphics. However, most of the images found during online searches will not be licensed for reuse or in the public domain. For example, Google Images has a "Labeled for reuse" filter that can help searchers find appropriate images, but a sizable proportion of the images that make it through the filter are ones that are still under copyright protection. Authors should not assume that an image is in the public domain just because it is commonly used on websites without attribution or because it was created by a government agency or a nonprofit organization.

In some situations, the best option is for authors to take their own photographs. Be sure that the lighting is good, the images are nicely framed, the background is not distracting, and other guidelines for good photography are followed. For example, consider the "rule of thirds" (which suggests that the balance of a photograph is optimized when the primary point of interest is not centered) and other composition elements. After taking a photograph, be sure that it is not fuzzy and that it can be cropped appropriately to fit a screen or poster without distortion. Do not take photographs of other people without their permission. In medical and public health reports, identifiable images (such as ones showing a patient's face) and sensitive images (such as photographs of severely injured people, even if identifiable features are not shown) generally cannot be used without the written permission of the individual (or his or her legally appointed representative). It may be fine to include photographs of hands, feet, backs, and so on without written permission, but authors must follow the rules specified by their workplaces or universities.

A credit must be provided for all images not created by the coauthors and not in the public domain. One good option is to place the image credit—the name of the artist along with any other details required by the license, such as a year, a company name, or a license type—in small font so that the text overlaps the image or is located just next to it. (Image credits do not belong in a reference list.)

34.7 Presenting a Poster

At most conferences, the poster presenter is responsible for setting up the poster at an assigned time. Although some conference organizers provide all the necessary supplies, this is not always the case. Because a variety of display setups may be used, poster presenters should come prepared with binder clips (for clipping a poster to a stiff board set on an easel), pushpins (for pinning a poster to a corkboard), and tape (for taping a poster to a wall). The presenter is also responsible for taking down the poster at an appointed time. It is considered bad form to take down a poster early or to leave it up after the assigned time, when another researcher may be waiting to set up a poster for the next session.

Some conferences designate poster session times when presenters are expected to stand by their posters and interact with attendees for an hour or two. These sessions provide valuable time for one-on-one conversations with interested individuals. It is appropriate to greet each person who stops to view the poster, and it is acceptable not to interact with those who are merely passing by. Becoming so engaged with one person that all others with questions or comments are ignored—or, conversely, allowing only superficial banter—is a missed opportunity to network. Some presenters prepare a handout that is either a page-sized printout of the full poster or a sheet with highlights. Most presenters have business cards with contact information available for distribution.

FIGURE 34-3 Sample Distribution of Slides for a 10- to 12-Minute Talk

Content Area	Number of Slides
Title slide with author names and contact information for the presenter	1
Research goal	1–2
Background	2–4
Methods	2–4
Results	4–8
Strengths and limitations	1
Future directions	0–1
Conclusions/implications	1
Acknowledgments and/or invitation for questions	0–1
Total	12–20

34.8 Preparing for an Oral Presentation

A typical oral presentation time slot is about 15 minutes long. Because a couple of minutes are required for setup at the beginning and questions at the end, about 10 to 12 minutes of this time slot are available for the actual presentation. Most presenters at health science conferences prepare a set of computerized slides (typically using PowerPoint, unless the conference guidelines specify another format) that will guide their talks and provide visual information to the audience. Because most presenters can describe 1 or 2 slides per minute, about 12 to 20 slides are appropriate for a 10- to 12-minute talk (**Figure 34-3**). The slides should not attempt to reproduce a paper on the screen. They should highlight the key message of the presentation using images in place of words as often as is appropriate. Figures and tables of statistical results usually need to be very simple to be readable on a projection screen. References for any previous publications mentioned in the slide show can be listed in small font at the bottom of the relevant slides. **Figure 34-4** provides a checklist for the content and design of slides for a presentation slide show.

Preparing the slide show is only the first step in preparing for an oral presentation. **Figure 34-5** provides a list of content-, voice-, and performance-related items to practice extensively in the weeks before a presentation. A presenter can video-record a practice performance, review it, and identify areas for improvement. Colleagues and mentors may be willing to provide honest feedback. No one can plan for everything that might be encountered at the conference, including nerves, but practice makes a positive experience more likely.

A few weeks before the conference, check on the equipment that will be provided in the

FIGURE 34-4 Checklist for Presentation Slide Show

Content	■ Graphs, tables, photographs, maps, and other types of visualizations are used in place of words as often as is appropriate. ■ Key words and phrases are used instead of full sentences. ■ The number of slides is appropriate for the scheduled presentation duration (about 1–2 slides per minute, excluding time set aside for questions). ■ There are no more than about six lines of text per slide. ■ All bulleted phrases on one slide use a consistent voice (for example, all start with the word "to" or all start with an "-ing" word). ■ All words are spelled correctly, and all phrases are grammatically correct. ■ The content of each slide is accurate. ■ Every slide is relevant. ■ The slides are in a logical order. ■ Citations, references, and image credits are provided (if applicable).
Design	■ The background is simple and not distracting. ■ All tables and figures are easy to interpret. ■ A consistent, readable, and adequately large font is used for text, tables, and figures (which may require simplifying images and enlarging the text and other components). ■ There is an adequate contrast between the background and the text (either dark letters on a light background or light letters on a dark background) even under different lighting conditions (for example, when overhead lights are on or off). ■ A consistent and pleasant color scheme is used throughout. ■ The slides are not cluttered. ■ Unnecessary effects like sounds, animated components, and distracting slide transitions are avoided.

FIGURE 34-5 Items to Practice Before the Presentation

Content	Opening lines	Practice the exact opening sentences that will capture the attention of the audience.
	Message	Master the content of each slide enough to describe each one without referring to notes.
	Phrasing	Use relatively short, precise sentences with active verbs.
	Flow	Practice transitions from one slide to the next.
	Closing lines	Practice exact closing sentences about key conclusions.
Voice	Pace	Speak at a moderate to slow rate.
	Volume	Speak relatively loudly.
	Pitch	Vary your voice inflection.
	Enunciation	Speak clearly.
	Pronunciation	Check on the pronunciation of technical words and names.
	Fillers	Try to avoid fillers (such as "um," "ah," "like," and "you know").

(continues)

FIGURE 34-5 Items to Practice Before the Presentation *(continued)*

Performance	Engagement	Smile and make eye contact with members of the audience.
	Posture	Stand tall or sit straight.
	Delivery	Do not just read the slides or read from a script.
	Movement	Try not to fidget, sway, pace, or make other distracting gestures or movements.
	Technology	Become comfortable with advancing slides (using a mouse, keyboard, and/or clicker) and using a pointer, if applicable; face the audience when using these tools, if possible.

presentation room (such as a computer and an LCD projector).

- Some conferences expect presenters to bring their own laptop computers.
- Some conferences require presenters to upload their presentation files to a website in advance of the conference.
- Some ask presenters to email their files to the session moderator.
- Some expect presenters to have the file on a flash drive.

No matter what format is preferred, always bring a backup copy of the presentation file in an accessible format. Ideally, have versions of the presentation ready in different formats (such as saving a PowerPoint presentation as a PDF that can be projected if the primary file will not run on the device in the presentation room).

34.9 Giving an Oral Presentation

Figure 34-6 summarizes the key tasks for the day of the presentation. Conference organizers often advise presenters to:

- Arrive at the presentation room at least 15 minutes before the panel begins (not 15 minutes before an individual presentation time).
- Check in with the moderator.
- Set up the computer and projector or confirm that slides are ready to be projected.

Expect that some aspect of the session will not go as planned. There might be a technology glitch, or the order of presentations within the session might need to be adjusted. Arriving early and being prepared will minimize the stress of those last-minute changes.

Presenters must remember to be considerate of other presenters in their session by strictly adhering to their assigned time limits. If practice sessions before the conference consistently reach or exceed the time allotted for the talk, trim content before the conference so that the risk of talking too long is minimized.

At most conferences, time is allotted for questions from the audience, either after each presentation or after all the panelists in the session have spoken. If a microphone is not available for those asking questions, the respondent should repeat the question to ensure that members of the audience hear the question before it is answered. The appropriate etiquette is usually to:

- Keep responses short.
- Thank those who offer suggestions for improving the work.
- Acknowledge the limitations of the project while highlighting its strengths.
- Be respectful to everyone.

At the end of the session, one-on-one or small group conversation about the research may continue. Presenters can share business cards with attendees who have overlapping interests. When contact information is exchanged, it is appropriate for the presenter to send a follow-up email after the conference that expresses an interest in continued communication and possible collaborations.

FIGURE 34-6 Checklist of Tasks on the Day of the Presentation

Time	Tasks	
Fifteen minutes before the assigned presentation panel is scheduled to begin	Moderator	Check in with the session moderator or chair, if there is one.
	Q&A	Ask the moderator whether the question-and-answer time will take place after each presenter or after all of the presenters are finished.
	Time	Confirm the amount of time for the presentation, and ask the moderator whether there is a timekeeper and what sort of warning signs will be given when the allotted time is nearly finished. If there is no timekeeper, ask a friendly person in the front row to serve as one.
	Computer	If using a computer and/or projector, check that the devices are set up and the presentation is loaded on the computer and ready to use.
	Pointer	If using a pointer and/or clicker, check that they are working.
	Microphone	If using a microphone, conduct a sound check.
	Water	Bring a bottle of water and have it easily accessible during your presentation.
	Co-presenters	Greet other presenters in the session.
During other presenters' talks in the session	Listen	Pay attention to the other talks; do not focus on personal notes or preparation during this time.
	Connect	Listen for points of connection between the research talks being presented, especially if the question-and-answer period comes at the end of the session.
During the talk	Relax	Trust that practice will result in a proficient presentation.
	Be calm	Be alert to nervous behaviors, such as adding fillers to speech or swaying the body.
	Keep time	Do not exceed the allotted time period.
After the talk	Thanks	Thank the moderator, timekeeper, technology support person, and fellow presenters.
	Belongings	Check that personal items are not forgotten.
	Conversations	Wait in the room for at least a few minutes in case anyone has follow-up questions; move the discussion into the hallway as soon as the presenters for the next session begin setting up their talks.

© DmitriyRazinkov/Shutterstock

CHAPTER 35

Article Structure

Research manuscripts usually follow the same outline: abstract, introduction, methods, results, and discussion.

35.1 Outlining a Manuscript

Most scientific research manuscripts follow the same outline—introduction, methods, results, and discussion—often represented to by the acronym IMRaD. The most common information included in each section is summarized in **Figure 35-1**. Outlining a paper down to the paragraph level before writing allows authors to track progress toward a complete manuscript and ensure that no critical information

FIGURE 35-1 Key Content for Primary Research Manuscripts

Section	Content
Abstract (or Summary)	■ Summarize the article.
Introduction (or Background)	■ Provide essential background information. ■ State the objectives of the study (or, for experimental studies, the hypotheses tested).
Methods	■ Identify the study design. ■ Describe the person, place, and time characteristics of the study, explaining how the desired number of participants was estimated, how potential participants were selected and recruited, what the eligibility criteria were, and where and when data were collected. ■ Explain how data were collected and how potential sources of bias were minimized. ■ Describe the statistical or other methods used for analysis (including providing definitions for key variables in quantitative studies). ■ Discuss ethical considerations (such as which research ethics committee approved the project, whether an inducement was offered, and how informed consent was documented).

(continues)

FIGURE 35-1 Key Content for Primary Research Manuscripts (continued)

Section	Content
Results	■ Describe the study population, including the sample size (using a flow diagram to show the number of individual participants at each stage of the study, if that will be helpful to readers). ■ Report relevant results (using tables and figures when possible).
Discussion	■ Summarize (briefly) the key findings and state how they achieved the goals of the study. ■ Discuss the limitations of the study. ■ Describe the key implications of the study for practice, policy, and/or future research.
References	List all of the sources cited in the manuscript (and no sources that are not cited in the main text).
Title page or end matter	Provide the information requested by the target journal, such as a description of each coauthor's contributions, acknowledgments of the contributions of people who did not meet the authorship criteria, funding sources, and disclosure of possible conflicts of interest.

is inadvertently omitted. A sample outline for an 18-paragraph manuscript about a primary study is shown in **Figure 35-2**.

35.2 Abstract

The abstract is a paragraph-length summary of the report. The most important function of the abstract is to serve as an advertisement for the paper, catching the eye of potential readers. Even when researchers have access to the full text of an article, they will be unlikely to read past the abstract if the summary does not capture their attention. An abstract must convey the key message of the paper in a compelling way while also including critical information about the person, place, and time characteristics of the study as well as the key findings. Writing an accurate and reasonably complete synopsis can be a challenge when most journals limit abstracts to a maximum of 150 to 250 words.

A **structured abstract** is a research summary that uses subheadings like Objective, Methods, Results, and Conclusion. An **unstructured abstract** is a narrative research summary that does not use section titles to divide the content of the paragraph. Most journals' author instructions and most calls for abstracts for conferences will specify whether a structured or unstructured abstract is preferred. Many authors find it easiest to write the abstract after the rest of the paper has already been written and the focus, key results, and conclusions are clear. Other authors find it helpful to write the abstract first, so that it can guide the way they present their key message in the full manuscript.

Most research databases and Internet search engines have access only to abstracts. A carefully constructed abstract will include a diversity of relevant search terms in order to maximize hits from computerized searches. For example, although the MeSH (medical subject header) dictionary considers the terms "hypertension" and "high blood pressure" to be synonyms, other abstract databases might not. If an abstract about hypertension includes only the word "hypertension," someone searching for "high blood pressure" might not find the article. A stronger abstract will include both "hypertension" and "high blood pressure." Similarly, if a study about a particular country has relevance to a wider region, it is advantageous to include both the country name and

FIGURE 35-2 Sample Outline for an 18-Paragraph Primary Research Manuscript

	Section	Paragraph
1	Abstract	Summary
2	Introduction	Background/context (setting the stage)
3		Justification of the study's importance
4		The main study question and three specific aims
5	Methods	Study design and person, place, and time characteristics
6		Data collection methods
7		Analytic methods
8		Ethical considerations
9	Results	Description of participants (Table 1)
10		Key finding 1 (Table/Figure 2)
11		Key finding 2 (Table/Figure 3)
12		Key finding 3 (Table/Figure 4)
13	Discussion	Explanation of what the results show about the answer to the main study question
14		Commentary on key finding 1
15		Commentary on key finding 2
16		Commentary on key finding 3
17		Study strengths and limitations
18		Implications and conclusions
	References	
	Figures/Tables	

regional terms. For example, an abstract about a study conducted in Jordan might benefit from inserting terms like "the Middle East" and "the Eastern Mediterranean" (the name of the World Health Organization region that encompasses North Africa and the Middle East) into its objectives or conclusion statements.

Some journals now invite authors to submit a **graphical abstract**, a single visual representation that displays the most important finding of a study in a format that can be easily disseminated through social media. Graphical abstracts are supplementary material, and they must be submitted in addition to the standard written abstract.

35.3 Introduction

The **introduction section** (or **background section**) is the first section of a scientific report. It provides critical information a reader must know to understand the methods and results of the article. This section often presents the foundational theories that informed the study, defines critical terms, and provides contextual information about the study's key exposures, diseases, populations, and/or location. The section may also include a paragraph justifying the importance, significance, and novelty of the new study. The introduction section usually ends with a paragraph spelling out the overall goal and the specific aims, objectives, or hypotheses that the paper will address.

The length of the introduction section compared to the discussion section varies according to the target publication venue. For some journals, a typical introduction might consist of only one or two paragraphs, but a lengthy discussion is expected. For other journals, the introduction might be several pages long, but the discussion is relatively short. For example, a long introduction section might include a comparison to previous studies as part of explaining what is novel about the new study, but that content often appears in the discussion section instead.

35.4 Methods

The **methods section** is usually the second section of a scientific report, and it presents details about the processes used for data collection and analysis. The methods section of a paper can often be written even before data collection begins, because most of the methods are finalized before data collection starts.

The methods section typically begins by identifying the study design that was used. If person, place, and time characteristics are not provided in the introduction, they are typically listed at the start of the methods section. For primary studies, the methods used to identify, sample, and recruit potential participants are described, and the inclusion and exclusion criteria are listed.

Next, the methods for collecting data are described. For quantitative studies, the methods section describes measurement procedures related to surveys, interviews, laboratory tests, and other assessments. For qualitative studies, the research paradigm being applied is presented and any researcher characteristics that might influence the research are disclosed. For secondary analyses, this section explains who collected the data originally, how the data were collected, and how the data files were acquired for secondary analysis. For a full-length research report, enough methodological detail should be provided so that the study procedures could be replicated reasonably well by another researcher. The previous publications that informed the study's methods may be cited so that readers can easily track down additional details about the methodology. The methods section may also explain the steps taken during the study's design, implementation, and analysis stages to minimize bias.

If definitions for the key exposures, outcomes, and other variables are not part of the background section, they are presented in detail in the methods section. For case–control and cohort studies, the case definition is spelled out. For experimental studies, the intervention, control, and outcomes are described in detail. Some manuscripts provide the exact phrasing and order of questionnaire items and describe the steps taken to validate the survey instrument.

The methods section then describes how data were cleaned, coded, and analyzed. Quantitative research reports list the various statistical tests deployed, sometimes including an explanation about how to interpret key tests. They may also explain how the required number of participants was estimated before the study and how missing data were handled during analysis. Qualitative research reports describe how data were processed and how themes were analyzed.

The methods section also provides information about ethical considerations, such as which research ethics committees reviewed the project, whether community groups were consulted, how informed consent was documented, and whether inducements were offered.

(Details about ethics-related methods may also be included in the end matter, depending on the preference of the journal.)

A well-written methods section exhibits coherence and transparency. **Coherence** is the quality of being logical and consistent. A coherent research report demonstrates the alignment of the study goals, the selected methodologies, and the featured results and conclusions. **Transparency** is the quality of being open and clear about the methods and results of a research study. For example, a transparent report about a qualitative study explicitly describes what methods were used, how the study processes were implemented, and why the researcher decided to use those approaches.

35.5 Results

The **results section** is typically the third section in a four-section scientific report, and it contains key findings presented via text as well as tables and figures. A results section typically starts with a description of the study population that identifies the number of participants and the demographics of the participants (such as their distribution by age and sex). Additional results of quantitative and/or qualitative analyses are then provided, using tables and figures when possible. Most studies do not require multivariate statistics or other types of advanced quantitative analyses. The results of a statistical test should not be reported unless the researchers have confirmed that the test is an appropriate one for the data and the research question.

One common organizational strategy for the results section is to match results paragraphs to the specific aims of the study. For example, if there are three specific aims, then the results section might have four paragraphs: one that describes the characteristics of the study population, one that presents the results most relevant to the first objective, one with results for the second objective, and one for the third objective (**Figure 35-3**). Another organizational approach is to write one paragraph about each table and figure. The first table in a quantitative research report typically describes the study population and is the first paragraph of the results section. The remaining tables and figures are presented in an order that best aligns with the specific aims or hypotheses.

35.6 Discussion

The **discussion section** is typically the final section of a four-part scientific paper. It usually begins with a brief summary of the key findings of the new study, then compares the new findings to the prior literature on the topic, acknowledges the limitations of the study, and

FIGURE 35-3 A "Follow the Threes" Approach to Storytelling in a Scientific Paper

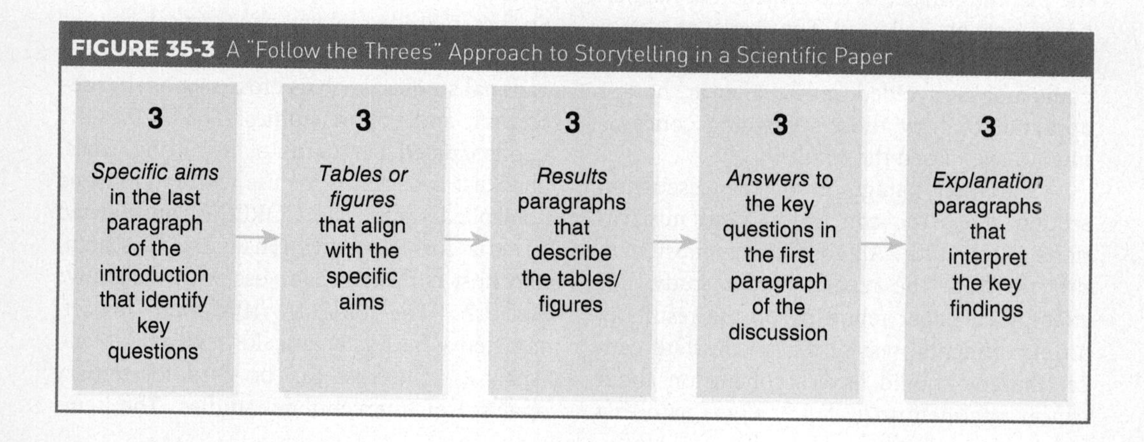
3
Specific aims in the last paragraph of the introduction that identify key questions
3
Tables or figures that align with the specific aims
3
Results paragraphs that describe the tables/figures
3
Answers to the key questions in the first paragraph of the discussion
3
Explanation paragraphs that interpret the key findings

summarizes the implications and conclusions of the study.

The key findings presented at the start of a discussion section should align with the aims, objectives, or hypotheses spelled out in the last paragraph of the introduction section. Ideally, the answer to the main research question posed at the end of the introduction section is answered in the first sentences of the discussion section. The subsequent paragraphs compare the new study to previous studies and include a thorough discussion of the existing literature that cites the most relevant publications related to the new work. The goal is not to show that the new study matches previous findings, but to show how the new study builds on previous research. A weak comparison section takes the form of "This study found X. Other studies also found X." A stronger comparison section uses prior publications to establish the context for the new study and explain the originality of the new results. See Chapter 3 for a review of what makes research original.

Every paper must include at least one paragraph about the strengths and limitations of the study. The limitations paragraph should identify potential types of bias and other problems that could make the study results inaccurate, invalid, or not generalizable beyond the study population. Most fixable problems should have been corrected long before the discussion section is written, ideally during the planning stages of the project that happen before data are collected. The limitations paragraph is the place to describe the issues that could not be avoided and to offer an honest appraisal of how those remaining concerns might have biased the results.

The final paragraph of the discussion section states the conclusions and implications of the study. All conclusions must stem directly from the results of the study. For example, a paper reporting on the results of an experimental test of a new prostate cancer therapy should have a conclusion about cancer treatment that closely aligns with that study's results. It should not have a conclusion about screening or diagnosis or about a therapy the authors did not test. A study about risk factors for sports-related injuries should have a conclusion about sports injury prevention that is closely related to that study's findings and not one that makes grand proposals about a diversity of prevention opportunities not examined during the study.

The appropriate types of conclusions vary by discipline and journal, but they might include a summary of the new theories that emerge from the analysis or recommendations for new preventive, diagnostic, or therapeutic practices and policies. A suggestion about directions for future research on the topic is generally the weakest conclusion that can be made. It is better to end with a specific key message directed at improving clinical or public health policy and practice, especially if the recommended action is a cost-effective one that could reasonably be implemented in the target population.

35.7 Writing Checklists

Several checklists have been developed for the specific content that reports about particular types of research studies should present. Some of the most frequently used checklists are listed in **Figure 35-4**. For example, the STROBE (Strengthening the Reporting of Observational Studies in Epidemiology) checklist can be used for primary and secondary reports about observational studies, such as cross-sectional, case–control, and cohort studies. The CONSORT (Consolidated Standards of Reporting Trials) checklist is designed for use with randomized controlled trials. The COREQ (Consolidated Criteria for Reporting Qualitative Research) checklist is designed for use with qualitative research. The PRISMA (Preferred Reporting Items for Systematic Reviews and Meta-Analyses) checklist can be used for tertiary analyses of interventional studies. The items

FIGURE 35-4 Common Reporting Guidelines

Study Approach	Checklist	
Case series	CARE	Case Report
	STARD	Standards of Reporting of Diagnostic Accuracy
	TRIPOD	Transparent Reporting of a multivariable prediction model for Individual Prognosis or Diagnosis
Cross-sectional study	STROBE	Strengthening the Reporting of Observational Studies in Epidemiology
Case–control study		
Cohort study		
Experimental study	CONSORT	Consolidated Standards Of Reporting Trials (for randomized controlled trials)
	SPIRIT	Standard Protocol Items: Recommendations for Intervention Trials
	SQUIRE	Standards for Quality Improvement Reporting Excellence
	CHEERS	Consolidated Health Economic Evaluation Reporting Standards
	TREND	Transparent Reporting of Evaluations with Nonrandomized Designs
Qualitative study	SRQR	Standards for Reporting Qualitative Research
	COREQ	Consolidated Criteria for Reporting Qualitative Research
Tertiary study	PRISMA	Preferred Reporting Items for Systematic Reviews and Meta-Analyses (for evaluations of interventions)
	MOOSE	Meta-analysis Of Observational Studies in Epidemiology

from the relevant checklist can be allocated to the most relevant paragraphs of a manuscript early in the drafting process.

35.8 End Matter

End matter is the information that some journals list between the end of the main text of an article and the start of the reference list. This end matter may include some of the following components:

- The affiliations of the authors and their contact details (if these items are not listed on the title page)
- The specific contributions of each author to the paper
- Acknowledgments of people who assisted with the study but who did not meet authorship criteria
- Information about some ethical aspects of research, such as a declaration that each participant gave informed consent along with the names and locations of the committees that reviewed and approved the project
- A list of all funding sources
- Disclosures of the presence or absence of possible conflicts of interest, including

personal financial conflicts of interest as well as potential conflicts related to being employed by an organization having a financial interest in the study

Some journals provide this information in the final published version of the paper but request that it be removed from the submitted manuscript because they use a blind review process and this information could reveal the identities or affiliations of the authors. The author guidelines of each journal will indicate what information should be provided on the title page or in the end matter of submissions.

35.9 Tables and Figures

Many health journals limit the number of tables and figures allowed for each article, often to a maximum total of four tables and figures combined. This limit means that the content for tables and figures must be carefully selected to highlight the most important aspects of the study.

In a research report, a **table** is the concise presentation of key findings in a grid. Tables are used to organize and present statistical results that cannot easily be listed in the text in a sentence or two. A high-quality table provides enough information that the contents can be interpreted and understood without reading the main text of the manuscript.

- The title of the table provides a brief but complete description of the content.
- The rows and columns each have a descriptive label and, when applicable, provide units and/or sample sizes (which are often designated by *n* for the number of participants).
- For each statistic, a confidence interval, *p* value, and/or other measure of uncertainty is provided, such as a standard deviation or standard error for a mean or an interquartile range for a median.
- All abbreviations used should be defined. It is sometimes possible to spell out the term at first use and to note the abbreviation in parentheses, as is done in running text, but if the table contains multiple abbreviations, they may need to be listed at the bottom of the table.
- All abbreviations used in the table are introduced at first use (with the full term spelled out and the abbreviation presented in parentheses) or are listed at the bottom of the table.
- A note just below the table (or just after the title) explains the meaning of asterisks (*) and other symbols (such as †, ‡, and §) denoting statistical significance and other items of interest.
- Consistent fonts, spacing, and number of digits after a decimal point are used for all tables in the manuscript.

Significant figures are the number of digits in a number that are known to be accurate. Zeroes presented to the right of a decimal point are, by definition, significant figures. If a column in a table is reporting percentages to the tenth (one number after the decimal point), numbers with a 0 in that final position should be reported just like any other value would be, so values of 7.2%, 8.1%, and 9.0% should be reported as such, and the 9.0% should not be presented inconsistently as 9%. The number of significant figures reported should be appropriate for the statistical power of the study. A study with a small number of participants should report values like 30% and 50% rather than 30.00% and 50.00%, because the numbers after the decimal point imply a level of precision that a study with few participants does not have the statistical power to provide.

A **figure** is the visual presentation of key findings in the form of a diagram, flowchart, drawing, map, photograph, or other graphic. Figures are used when a visual presentation of the material is more effective than words or numbers at conveying a result. All images in a report should be meaningful, not merely decorative.

Graphs are among the most frequently presented figures in scientific reports. A **graph** is an illustration of quantitative results, such as a scatterplot or a line graph that shows the values of a numeric variable over time. A graph should provide enough information in the title, figure, and/or legend or key for a reader to be able to interpret the graph even without reading the related portion of the manuscript. **Figure 35-5** highlights some of the features that may make a graph easier to interpret correctly, including the selection of an appropriate type of visual representation, the use of appropriate scales and labels for axes, and the inclusion of other relevant information about the data being presented.

High-resolution photographs, maps, and other images provided by the authors can also be used as figures. Photographs of study participants are usually not allowed to be published without the written permission of the subject or subjects. This is true even when a black bar covers the eyes or other distinguishing features. Clinicians who are considering writing a case report or case study and are documenting the progression of a patient's disease with photographs should usually secure written permission to use those images prior to taking the first photo.

A callout, also referred to as a text reference, is a note in the text of a manuscript that points readers to an element such as a figure or table. A callout for each table and figure—a phrase like "Figure 1 shows. . ." or a notation in parentheses like "(Table 2)"—should be placed in the text to indicate when the reader should first refer to a table or figure. Numbers provided in a table or figure are generally not repeated in the text, because the visual element is more effective than text at presenting those values.

FIGURE 35-5 Examples of Correct and Problematic Graphs

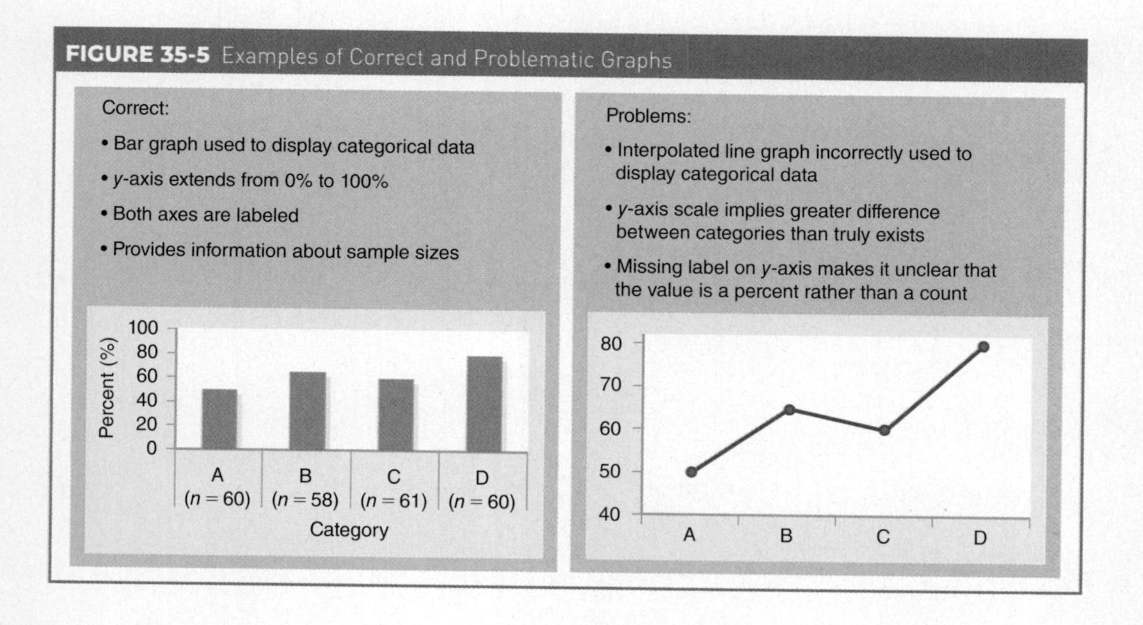

Correct:
• Bar graph used to display categorical data
• y-axis extends from 0% to 100%
• Both axes are labeled
• Provides information about sample sizes
Percent (%)
100
80
60
40
20
0
A (n = 60)
B (n = 58)
C (n = 61)
D (n = 60)
Category
Problems:
• Interpolated line graph incorrectly used to display categorical data
• y-axis scale implies greater difference between categories than truly exists
• Missing label on y-axis makes it unclear that the value is a percent rather than a count
80
70
60
50
40
A
B
C
D

CHAPTER 36

Citing

Research reports must contain accurate reference details for the sources that informed the methods, interpretations, and conclusions of a new study.

36.1 Referring to the Scientific Literature

The authors of every scientific paper need to explain how their new investigation fits with previous studies. The introduction section of a manuscript usually provides the background necessary to understand the importance of the new work. The discussion section typically provides an extensive comparison of the results of the new study and the results of previously published works. A typical article in the health sciences refers to about 25 or 30 other articles published in peer-reviewed journals, although some cite only a few and some (especially review articles) may cite hundreds.

Researchers find pertinent articles by searching electronic databases and by looking at the reference lists of articles already identified and determined to be helpful. (See Chapters 3 and 26 for more information about how to find relevant articles.) Most of the articles that are cited in the text of a manuscript and then included in the reference list at the end of the document provide evidence that supports the importance, validity, and conclusions of the new study. References can also be used to acknowledge alternative methodological approaches that could have been used, to identify areas where the new findings appear to contradict previous studies, and to provide varying perspectives on the policy and practice implications of the study.

The best articles to cite are ones that present results and key findings that are directly relevant to the new study. Authors should be cautious about citing commentary from the introductions and discussions of other papers, especially when the pertinent commentary is citing other sources. Suppose that "Paper 1" makes an interesting comment in its discussion section about the findings of "Paper 2" and "Paper 3." In that situation, the best option is to look up both "Paper 2" and "Paper 3" so that their methods and results can be examined and then cited if relevant. "Paper 1" does not need to be cited, because the supporting evidence for the new paper does not derive from the results of "Paper 1" itself.

Or suppose that "Paper 4" cites 8 articles at the end of a sentence as evidence that many previous studies have identified a particular exposure to be a risk factor for a particular disease. The best option in that situation may be to look for a review article about that association. The results section of most systematic review articles presents a summary and synthesis of the full body of literature on the selected topic. That analysis will clarify whether there is consensus about the effects of the exposure, or whether 8 studies have found a significantly risky association but 80 others have found no association. A systematic review article is a more appropriate source to cite than "Paper 4" or the 8 studies it cited.

Authors must read the full text of a report before citing it, because abstracts may incorrectly or incompletely summarize the methods and results of a study. For example, abstracts may omit critical information, like a very small sample size, a very low participation rate, or the use of data that are many decades old. Or they may report only the statistics that are most congruent with previous studies or the most shockingly different from them. Additionally, abstracts often state conclusions that the study's data do not support. Authors are responsible for confirming that the methods and conclusions of the reports they cite are sound.

It may also be helpful for authors to confirm that an article was not retracted after it was published. A **retraction** is the removal of a published article from the accepted scientific literature due to major errors or author misconduct. A retracted article has been withdrawn from the peer-reviewed scientific literature and should not be cited. A retraction is different from a correction. An **erratum** is a published correction to a minor error in an article that was introduced during the publishing process. A **corrigendum** is a published correction to a minor error in an article that was caused by the author rather than the publisher. If an erratum or corrigendum has been issued to correct an error in the article, the study's findings are still considered to be sound. The researcher should just be sure to read the updated version of the manuscript.

36.2 Formal and Informal Sources

Formal sources are scholarly works that were critically reviewed before being disseminated by a publishing group in a format that includes details such as author names, the name of the publisher, and the publication date. In the health sciences, peer-reviewed journal articles are typically the preferred source of evidentiary support. Books, book chapters, and scientific reports published by trusted governmental agencies and other organizations are also acceptable formal sources to cite. **Figure 36-1** summarizes the characteristics of formal reports.

A distinguishing feature of a formal report is that it does not change after it is published.

FIGURE 36-1 Characteristics of Formal Scientific Reports

Formal Scientific Reports . . .

- Are published in a peer-reviewed journal (or sometimes a peer-reviewed report or book), not in a magazine or as a page on a website
- Describe the study design and explain why it was appropriate for the objectives of the study
- Explain how the study population was selected (if relevant) and demonstrate that the sample size was sufficiently large
- Explain how exposures and outcomes were defined and assessed
- Describe the analytic approaches used and present results using easily interpreted tables and graphs
- Compare the new study to previous studies
- Discuss the limitations of the study
- Draw conclusions that are reasonable and that are derived directly from the study's data
- Follow a standard outline and other conventions for scientific writing

Formal reports are published on a particular date. An article assigned to a particular issue of a journal will typically show the month and year of the issue on its first page. A report issued by a government agency or an independent organization will usually show the date of publication on the title page or on a copyright page in the front matter of the report. After that publication date, the content of the report will remain unchanged. The same rules about the content being final as of the publication date apply whether a report is printed on paper or published online. Most formal reports that are available online are published as PDFs or other types of files that lock in the formatting and the content of the pages.

Informal sources like webpages, fact sheets, blogs, podcasts, and other types of information that are not peer reviewed and formally published should almost never be cited in formal research reports (**Figure 36-2**). These resources usually lack important information about data sources, the methods used to acquire and analyze information, authorship, and publication dates. Additionally, a website's content is rarely fixed in time. The content posted on Wikipedia might be updated or deleted at any moment. Even the information

FIGURE 36-2 Citable Sources

Source	Formal Source?	Citable?	Remarks
Webpage or fact sheet	No	Rarely	Webpages and fact sheets may be helpful starting places for informal research but should be cited in a formal manuscript only if they are from a trusted organization and no formal article or report provides similar information.
Newspaper or popular magazine	Maybe	Rarely	Popular media items should be referred to only when no formal scientific article or report provides similar information.
Statistical database	Maybe	Sometimes	Cite statistical databases and reports only when information is provided about how, when, and where the data were collected; there is a fixed date for the data; and no formal article or report provides the same information.
Official report	Yes	Yes	Reports are usually cited only when they are formal publications (with assigned publication years and/or other bibliographic information) from trusted organizations.
Book or book chapter	Yes	Yes	Although most scientific communication occurs through journals rather than books, scientific books are acceptable sources for formal manuscripts; general textbooks are rarely appropriate sources, but some highly techn cal textbooks are appropriate to cite.
Abstract	Yes	No	Cite only full-text articles; abstracts of conference presentations are rarely appropriate to cite.
Journal article	Yes	Yes	Articles from peer-reviewed journals are the preferred references for formal manuscripts.

on scientific websites, such as those of the U.S. Centers for Disease Control and Prevention and the National Institutes of Health, might disappear or change with no notice. Suppose that a webpage reports a statistic of interest, such as the prevalence of a particular disease. If that page mentions the original source of the information, like noting that the prevalence was determined by researchers at a particular university or published in a particular issue of a medical journal, it is easy to look up and cite the original source. If the page does not provide information about where the number came from, an Internet search for that number and the disease name might reveal the original source. If Internet searches do not yield a formal source reporting the details about the statistic, the number must be viewed with skepticism. Informal reports should be cited only when they are made available by a trusted organization and a more reliable and permanent source of information is not available.

36.3 Writing in One's Own Words

Almost no scientific articles quote directly from another source word for word. There are many reasons to avoid direct quotations. One of the most important reasons is that other writers' phrases and sentences may clash with the author's own writing style, and that mismatch will interrupt the flow of the document. Some people use a quotation when they feel that an original work was so perfectly written that it would be impossible to express the same thought equally well using different words. The reality is that communicating the same idea in one's own writing style usually is better for the new work because it allows the entire manuscript to have a consistent voice. Another benefit of paraphrasing is that it helps ensure that the article being cited has been understood. Using a quotation that is not fully understood is never a good idea. Accurate paraphrasing requires a level of comprehension that copying and pasting does not.

Paraphrasing does not remove the requirement to cite the original source; it just means that quotation marks do not have to be used. On the rare occasions when it is acceptable to include a direct quotation in a scientific report, the entire quotation must be in quotation marks (or indented from the left margin as a block quote, depending on the length of the quote and the publisher's formatting preferences) and an in-text citation must be provided. When the ideas or findings of other scholars are paraphrased, quotation marks are not used because the words are not being copied, but an in-text citation for the source of the original idea or information must still be provided. **Figure 36-3** illustrates the difference between a quotation and a paraphrase.

36.4 Common Knowledge and Specific Knowledge

Specific knowledge is information that is specific to a particular study, such as a particular statistic or a particular laboratory finding. All specific knowledge must be cited when it is mentioned in a scientific paper. By contrast, **common knowledge** (also called **general knowledge**) refers to information that should be familiar to a typical person working in that research area. Because this information is widely accepted to be true, it does not require citations and references in a research report. For example, health professionals generally agree that influenza is caused by a virus and that Germany is located in Europe. Both of these facts are well established, and a quick search for papers on influenza or about studies conducted in Germany would show that this information is not usually accompanied by a citation. However, a statistic about the proportion of Germans who seek clinical care

FIGURE 36-3 Examples of Quoting and Paraphrasing

Quotation (Almost Never Used in Journal Articles)	Paraphrase (Often Used)	Reference (Always Required for Quotations and Paraphrases)
A case-control study examining risk factors for ovarian cancer in Canadian women found that "age at first full-term pregnancy was not associated with risk of ovarian cancer."[1]	A case-control study of Canadian women found no association between ovarian cancer and the ages of participants at the time of their first full-term pregnancies.[1]	1. Risch HA, Marrett LD, Jain M, Howe GR. Differences in risk factors for epithelial ovarian cancer by histologic type: results of a case-control study. *Am J Epidemiol.* 1996;144:363–372.
The authors acknowledged that "since we did not adjust for depth of inhalation and age at smoking onset, the RR for women, compared with that for men, due to smoking was likely to have been underestimated by our results."[2]	The authors of the study acknowledged that they might have underestimated the magnitude of the increased risk of lung cancer in female smokers compared to male smokers because they had not statistically adjusted for smoking behaviors, such as the depth of inhalation.[2]	2. Zang EA, Wynder EL. Differences in lung cancer risk between men and women: examination of the evidence. *J Natl Cancer Inst.* 1996;88:183–192.
The investigators noted that "cholera is usually considered to be a water-borne disease, but, in this outbreak, the available evidence indicates that a food item served as part of a meal was the most likely vehicle of infection."[3]	The investigators concluded that the most likely cause of the cholera outbreak was food served to passengers on the airplane.[3]	3. Sutton RG. An outbreak of cholera in Australia due to food served in flight on an international aircraft. *J Hyg (London).* 1974;72:441–451.

for influenza in a typical year would be specific knowledge, and the source of that information would need to be cited. When in doubt about whether a bit of information is common knowledge, authors should err on the side of providing a citation. Any disputed fact should be well supported by one or more reliable sources.

36.5 Avoiding Plagiarism

Plagiarism is the use of someone else's ideas, words, images, or creative work without proper attribution. Copying the exact words of another person without using quotation marks and providing a full citation, engaging in "thesaurus plagiarism" that swaps in synonyms for words in an original source in order to avoid the need for quotation marks, paraphrasing a unique theory or observation without providing a citation, and using an image without permission and attribution are all forms of plagiarism. Failing to fully acknowledge the source of the original work deprives the author or creator of the recognition that person deserves, and it may result in the plagiarist getting credit for work that he or she did not do.

Plagiarism is a major violation of scholarly integrity, and it can have a damaging long-term impact on a professional career. A published article with extensive plagiarism must be retracted, and a retraction notice issued by the journal will remove the article from the accepted scientific literature and be a permanent and public record of wrongdoing. For students, plagiarism can result in expulsion from school. For employees, plagiarism can result in the loss of a job. The other possible consequences of plagiarism and other forms of research misconduct, such as redundant publication or the fabrication or falsification of data, are discussed in detail on the website of the Committee on Publication Ethics (COPE).

Several habits can be adopted to ensure that plagiarism does not occur. One helpful practice is never to cut and paste information from a website, article, or any other source into a document file that contains any draft material for an article. It is far too easy for those words, phrases, or even whole sentences or paragraphs to be unintentionally incorporated into the text of a manuscript. "Unintentional plagiarism" is still plagiarism, and it carries the same penalties. When browsing websites and other sources for background material, take the time to paraphrase the information instead of cutting and pasting it into a file for later review.

Another good habit is to always include a full reference alongside all research notes derived from particular sources. For example, if an article presents a theory that explains the findings of the new project, do not just make a note about the theory. Jot down the theory and put a bracket with the author and year next to it, as is typically done for in-text citations in journal manuscripts, and then add in the full bibliographic details for the article so that the source of the theory can be easily identified later on when writing is underway and a citation is required.

36.6 Citation Styles

Most of the citation styles used in the health sciences require two types of notations about each source of information:

- In-text citations where the sources are briefly identified in the text
- A reference list at the end of the document that provides full bibliographic details for each source

Every article included in an in-text citation requires a full reference. Every entry in the reference list must be cited at least one time in the main text.

No one citation style is used across the health sciences. The two most common ones are APA style and AMA style. **APA style** is the citation and reference style recommended by the American Psychological Association, and it is widely used by social science and nursing journals. **AMA style** is the citation and reference style recommended by the American Medical Association, and it is widely used by medical and health science journals. Variations of AMA style include Vancouver style, ICMJE (International Committee of Medical Journal Editors) style, and NLM (National Library of Medicine) style. The term **house style** describes a particular journal's or publisher's requirements for spelling, citation style, and other formatting details. Journals that require an AMA-type citation usually provide a guide to their house style on their website. Sometimes other styles are required, such as MLA (Modern Language Association) style, Turabian style, or Chicago style. Reference manuals and style guides are available for all of the widely used styles, and most journals provide instructions for authors on their websites that specify the journal's style preferences. Articles recently published in the target journal provide additional examples of the journal's required style. When authors have the flexibility to select a style for a manuscript, a consistent citation and reference style should be used throughout the document.

In-text citations are brief identifying details, such as a numeral or an author's last name and the year of publication, that allow readers to locate the source of information in the reference list at the end of the article. Examples of formats for in-text citations are shown in **Figure 36-4**. Some journals will

FIGURE 36-4 In-Text Citation Styles

Citation Style	One Source	Two Sources	Three Sources
First author's last name and publication year	. . . [Ruiz, 2014].	. . . [Ruiz, 2014; Yamamoto, 2001].	. . . [Ivanov, 2008; Ruiz, 2014; Yamamoto, 2001].
Author(s) and publication year	. . . (Ruiz, 2014).	. . . (Ruiz & Sanchez, 2014; Yamamoto et al., 2001).	. . . (Ivanov, 2008; Ruiz & Sanchez, 2014; Yamamoto et al., 2001).
Number in brackets (square brackets)	. . . [1].	. . . [1, 2].	. . . [1–3].
	. . . .[1]	. . . .[1, 2]	. . . .[1–3]
Number in parentheses (round brackets)	. . . (1).	. . . (1,2).	. . . (1–3).
	. . . .(1)	. . . .(1,2)	. . . .(1–3)
Superscript number	. . . .[1]	. . . .[1,2]	. . . .[1-3]

convert bracketed citation numbers in submitted manuscripts to superscript numbers during the editing and layout process, and the author guidelines will state which submission style is preferred.

The reference list at the end of the article presents cited works either alphabetically in order of the first authors' last names or in the order of first appearance of the cited work in the text of the article. Sources appear only one time in each reference list. In AMA style, the first article cited is referred to as reference 1, typically denoted by a superscript 1, any time it is cited in the manuscript. In APA style, the authors' names are listed in the in-text citation every time the article is cited. The only change that occurs when an article is cited more than one time is that an article with three, four, or five authors will list all of the authors in the first in-text citation but subsequent in-text citations will list only the first author's last name followed by "et al." (the abbreviation for the Latin phrase *et alia*, which means "and others"). If the article has more than five authors, all citations, even the first, include only the first author's last name and "et al."

When preparing a manuscript for submission to a journal, authors should check the document carefully for compliance with the journal's style specifications. Journals using AMA style or a variant typically list authors by last name and first initials (with no periods after them), then the title (with capital letters only for proper nouns), an abbreviated journal name (which uses a formal journal title abbreviation, as specified in *Index Medicus*), the publication year, the volume number, and page numbers. However, publishers may request minor adjustments to these components. Some publishers expect all authors to be listed no matter how many there are, while some use an abbreviated version for six or more authors, such as listing only the first three authors followed by "et al." Some use abbreviations for journal titles (such as shortening *Journal* to just *J*), while others use the full journal name. Some list journal issue numbers, but many do not. Some list the full page numbers (such as 202–209), and others use a slightly shorter elided version (such as 202–9). Some use italics or bold type for some parts of the bibliographic entry. Some request nonstandard use of periods (full stops), semicolons, and commas to separate components of reference entries.

Most reference styles cite formal reports accessed online rather than in print format as though they are print publications, but some include the web address for the report. Some publishers ask authors to provide a digital object identifier for all sources that have DOIs. A **digital object identifier (DOI)** is an alphanumeric code assigned to a document by a registration body to allow quick online access to the document or its abstract. The publisher's webpage for a document will open when the DOI is pasted after http://dx.doi.org/ in a browser's address bar. For example, if the DOI is 10.1000/1, the web link to the document will be http://dx.doi.org/10.1000/1.

Authors need to be careful to use a consistent style across all entries in the reference list. A sloppy reference list may cause reviewers of a submitted manuscript to worry that the authors were similarly careless in their data collection and analysis. It is worth taking the time to compile, check, and recheck a flawless reference list.

CHAPTER 37

Critically Revising

Editing improves the content and clarity of research manuscripts.

37.1 Clarifying the Storyline

Research writing is not just about providing facts to readers. A strong scientific report conveys one key message, and every section, every paragraph, and even every sentence of the report supports that "storyline." The first step in editing a manuscript draft is to confirm that one big picture message is being clearly communicated. A **précis** is a concise one- or two-sentence summary of a research study's key finding that is typically limited to 35 words or less. The goal of editing is to craft a précis that captures the essence of the story, an abstract that tells the entire story in one convincing paragraph, and a full manuscript that conveys a cohesive message with a strong plotline (**Figure 37-1**).

A compelling research manuscript has a well-structured "plot" that establishes a research question, presents a sequence of relevant information, and then answers that question. A captivating research manuscript follows the outline of a mystery story.

The introduction section of most papers in the health sciences spells out the core question—the "mystery"—that the paper will explore and answer. In some disciplines, it is common to start a manuscript by writing "In this paper, we will show that . . ." or "In this paper, I will show that . . ." and then revealing the key argument. Most papers in the health sciences take a different approach, posing a question in the introduction that will not be answered until later in the paper.

The methods and results sections provide the evidence that will allow the "mystery" to be solved. These sections explain what the researchers did and describe what they observed. They provide all the necessary "clues" for answering the main research question, and they demonstrate that those observations are valid, sufficiently comprehensive, and reasonably unbiased.

The discussion section ties all the parts of the "story" together, neatly presenting the solution to the "mystery." Many fictional mystery stories end with a detective revealing the culprit and succinctly explaining how this determination was made. A similar approach is often used in nonfiction science writing.

FIGURE 37-1 Does the Manuscript Tell a Compelling "Story"?

- Does the manuscript have a clear "storyline"? Can the "plot" be summarized in one sentence?
- Does the title of the manuscript reflect the key message of the study?
- Does the abstract accurately summarize the key parts of the story?
- Do the opening paragraphs draw the reader into the story?
- Does the introduction overtly ask the main research question?
- Does the methods section explain how the study design allows the main research question to be answered?
- Does the results section provide all the evidence necessary to answer the study question?
- Does the discussion section overtly answer the main research question?
- Are there any missing parts of the story that need to be added so that it is complete and compelling? Do any gaps in logic need to be addressed?
- Are any parts of the manuscript redundant or peripheral to the main story? Can these be removed to tighten the storyline?
- Are the conclusions fully supported by the results?

37.2 One Paper, One Story

Some manuscript drafts do not read well because they present many related observations but do not clearly identify one key message. These reports improve when the authors select one storyline, delete all of the sentences and paragraphs that are not central to that core theme, and then use the remaining components of the draft to write a new manuscript that aligns with the narrative.

Other manuscript drafts read poorly because there are two or more plotlines, and none of those stories is presented completely. One paper must tell one story, not several stories. Sometimes it is better to write two well-organized and persuasive papers on related but distinct topics than to try to fit all of the results of a research project into one unfocused paper.

Recognizing that a manuscript draft is incoherent because it is telling too many stories and simplifying it so that it presents one complete story is very different from parceling one story into several incomplete analyses. "**Salami publication**" (sometimes called fragmentary publication) occurs when authors inappropriately write two or more similar manuscripts about the same research finding rather than telling the complete story in one manuscript. Authors are sometimes tempted to thinly slice their results to maximize the number of journal articles they can publish from one research study. Salami publication is problematic for two reasons. First, it tells the same story more than one time, diluting the scientific literature. Second, it usually means that none of the papers tells the story completely. Rather than telling one story completely in one paper, the authors spread the relevant evidence across several manuscripts. Many editors consider salami publication to be a form of research misconduct because the authors are presenting only partial results and are therefore not being fully truthful about their findings.

37.3 Structure and Content

Once a manuscript's storyline is clear, the next step is to check the structure and content of the draft (**Figure 37-2**). The manuscript should be well organized. Each paragraph should have one clear theme, and that theme should be an essential part of telling the overall story. The text must accurately describe what the researchers did and what they observed, and it should be complete yet concise.

FIGURE 37-2 Checklist for Structure and Content of the Manuscript

- Is the manuscript well organized? Is the content focused?
- Does every paragraph have one theme? Does every sentence within a paragraph fit with that paragraph's theme?
- Does the order of paragraphs within each section support the plotline?
- Does the introduction section provide all essential background information, including person, place, and time details?
- Does the introduction support the importance of the research question and explain why the study is novel?
- Are the methods described in adequate detail?
- Does the results section present a complete set of findings related to the research question? Should any findings be removed because they are peripheral to the research question?
- Are the tables and figures well designed? (And are all statistics presented either in the figures/tables or in the text, but not in both?)
- Does the discussion section provide a concise summary of key findings and then place the new findings in the context of previous research? Does the discussion section avoid redundancy with the results section? (No statistical results should be reported in the discussion section.)
- Does the discussion section adequately address the potential limitations of the study?
- Is every claim in the introduction and discussion sections supported by citations of formal reports (and/or by the study's results)? Should additional references be added to further support the key message of the manuscript? Should any entries in the reference list be removed because they are not directly related to the key message?
- Has the manuscript been double-checked to ensure that no part of it is plagiarized or paraphrased without proper attribution?
- Is every part of the manuscript truthful? (For example, does the text report the methods that were actually used rather than an idealized version of them? Does the manuscript report the results of the most appropriate statistical tests rather than results from less appropriate tests that happened to produce statistically significant results?)
- Is the manuscript's word count satisfactory?

Scientific manuscripts written for publication in journals must comply with the word limits of target journals. Full-length journal articles in the health sciences are often capped at 3000 or 3500 words (excluding the abstract, references, tables, and figures). Short reports may be limited to 1500 or 2000 words, or as few as 800 or 1000 words, with only one table or figure allowed. Being aware of these restrictions prior to beginning a draft allows an appropriately focused narrative to be crafted.

37.4 Style, Clarity, and Consistency

In a final check, writers should confirm that each word, sentence, paragraph, and section has a clear meaning and a consistent style (**Figure 37-3**):

- Words must be used carefully.
- Sentences must be concise and clear.
- The voice must be consistent.

FIGURE 37-3 Checklist for Style and Clarity

- Are words used precisely? (For example, are terms like "associated," "correlated," and "caused" used appropriately? Are "incidence" and "prevalence" used correctly? Is the word "who" used with people, writing "people who" rather than "people that"?)
- Is unnecessary jargon avoided? Are definitions provided for all key terms?
- Are all abbreviations introduced at first use?
- Is the tone of the writing appropriate? Is the writing style fact-based rather than emotion-based?
- Does the article consistently use a third-person voice or consistently use a first-person ("I" or "we") voice? Is the voice correct for the target journal?
- Do all subjects (nouns or pronouns) agree with their associated verbs? (For example, since "data" is a plural word, is "data are" used rather than "data is"?) Are all other grammatical conventions followed?
- Is active voice rather than passive voice used whenever possible? (Avoid passive phrases like "It is found that . . . ," and simply report the finding.)
- Is the verb tense consistent? (For most papers, the past tense is used rather than the present tense because the data were collected in the past.)
- Is each sentence clear? Are phrases as concise as possible?
- Are all words spelled correctly? (Each report should consistently follow the spelling conventions of one country.)
- Is all punctuation correct? (For example, are there extra or missing commas?)
- Are all sections of the text formatted consistently? Are all in-text citations and reference list entries formatted consistently?

- Grammar, spelling, and punctuation must be correct.
- If applicable, the manuscript must match the style requirements of the target journal.

Any document submitted to a journal for review (or sent to a professor or supervisor) should have clean and consistent formatting. For example, the same typography should be used on all pages (rather than, for example, using Arial for some elements and Times New Roman for others). Line spacing and paragraph separation styles (extra lines between paragraphs and/or indentation at the start of new paragraphs) should be consistent throughout the main text. All entries in the reference list should be complete and use a consistent reference style, including consistent use of punctuation and any special fonts, such as italicization for book and journal titles. All tables should have the same design. Figures should have a high resolution so that they do not appear blurry. A carefully formatted manuscript signals to reviewers and readers that the research process was implemented carefully.

CHAPTER 38

Writing Success Strategies

A variety of tactics can help a researcher move successfully through the writing process.

© DmitriyRazinkov/Shutterstock

38.1 The Writing Process

By the time a researcher is ready to write a final report about a research study, the vast majority of the work on the project has been completed. A study question has been identified and refined, a study approach has been selected and a protocol developed, and data have been collected and analyzed. The end of the project is in sight, but the prospect of creating a report that is intended to be disseminated beyond those immediately involved in the project can be intimidating. Postponing the writing process is easy. The writing can drag on, and in some cases it is never completed.

Few writers have the ability to sit down and compose a complete manuscript in one burst of productivity. Most writers experience cycles of high motivation and productivity followed by periods of limited or no interest in their work. **Figure 38-1** illustrates a typical writer's productivity levels during the writing process. The durations of each stage vary among writers and for different projects, but most writers need strategies for motivation at three key times:

- First, writers must overcome the barriers to getting started.
- Second, writers must find ways to prolong the period of high productivity that often occurs at the start of a writing project.
- Finally, most writers become fatigued during the writing process and at some point lose all desire even to think about their projects. At such points, they must find the motivation to persevere and complete the manuscript.

38.2 Getting Started

The only way to get started on a writing project is to start writing. Because scientific papers follow a standard outline, an easy way to start filling pages is to:

- Put a working title for the paper at the beginning of the file, along with the names and affiliations of all the coauthors.

FIGURE 38-1 Typical Variations in Productivity During the Writing Process

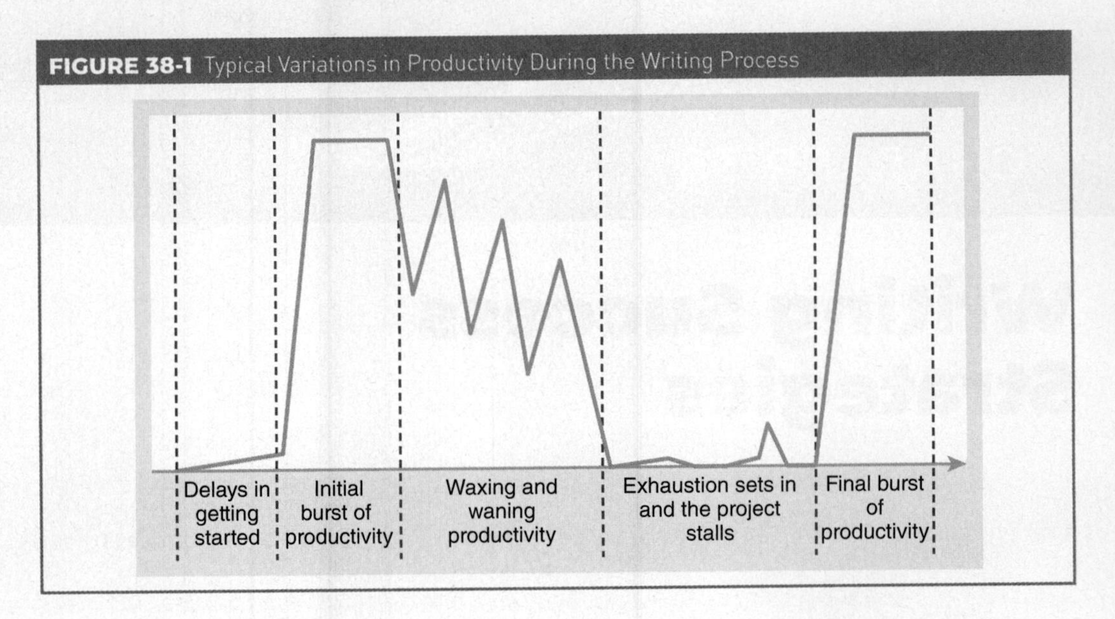

- Add in the headers for the Abstract, Introduction, Methods, Results, Discussion, and References.
- Paste in a table or figure that was created during the analysis process and will be included in the final report.
- Paste in some relevant lines about methods from the protocol.
- Add bibliographic information for several articles that will be cited in the report to the reference list.

Then start filling in the gaps.

Creating a detailed outline that specifies exactly what topics each paragraph in the paper will cover is often helpful for organizing the document and making progress toward a full draft. Articles from the target journal can provide a template for the outline. For example, if a model paper includes paragraphs on statistical methods and ethical considerations, headers for those paragraphs can be inserted at the end of the methods section in the new draft. A brief list of what to cover in each of those paragraphs can then be added based on what was reported in the model articles. For example, the methods section will likely include sentences about informed consent, ethics committee review, the significance level used for statistical tests, and so on. Be careful not to plagiarize any ideas or phrases from the model articles. Focus on the general outline of the topics covered and the order in which those topics are presented, not on the particular words used in the template articles.

After a manuscript has been outlined, the content of the manuscript does not need to be added in any particular order. Many authors of scientific manuscripts find it easiest to start with the methods, then to write the results, then the introduction and discussion, and finally the abstract, but that order is not required. Many authors skip around in the paper, adding a few sentences at a time here and there. Some authors find it helpful to write throughout the research process (**Figure 38-2**). They may draft the introduction as soon as the study question and approach have been selected, the methods as soon as the study protocol is finalized, and the results as soon as data have been analyzed. Then they draft the discussion section and edit the earlier sections of the manuscript to ensure that the paper tells a focused story. In short, when getting started on a paper, a good

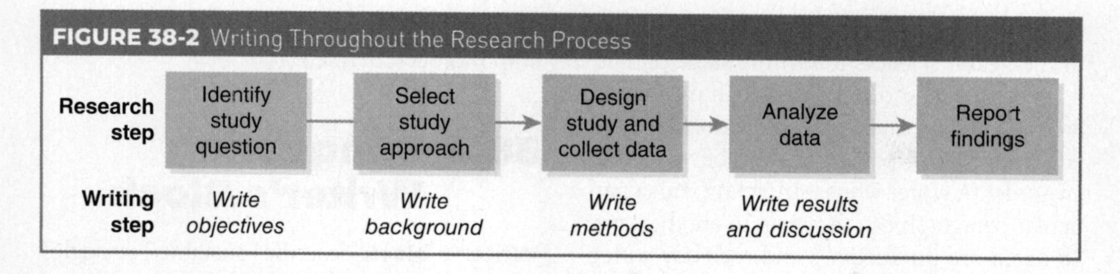

FIGURE 38-2 Writing Throughout the Research Process

plan is to first write whatever part of the paper is ready to be put into words. Then just keep on writing.

If the barrier to getting started is not having a clear sense of how best to tell the "story" of the paper, it may help to try an oral, visual, or kinesthetic method for moving toward writing productivity. Researchers who process their thoughts best by talking about them should seek out opportunities to have conversations about their research with coauthors, colleagues, and friends. Answering the questions these audiences ask will provide valuable practice describing and explaining the project. Try developing an "elevator pitch" that tells the main lesson learned from the project in 30 seconds. Try narrating the story of the paper aloud and recording it, then transcribe those words as a first step toward drafting a paper. It may be easier to edit spoken language into more formal written language than it is to start from scratch on formal writing.

Researchers who are visual processors may find it helpful to create a poster about the research project or to create a slideshow for a presentation about it. Organizing a poster or presentation helps clarify the flow of the storyline and the relationships among the study's objectives, methods, results, and conclusions. A presentation at a research seminar or conference can garner feedback that will improve the content and organization of the subsequent manuscript. An added benefit is that the visuals created for the presentation may become figures for the paper.

Researchers who are kinesthetic learners may find it helpful to take a long walk away from a computer and to use that time to think through the story that needs to be written. Walk, then write. Consider using a standing desk or walking desk (a desk with a treadmill) if movement improves writing productivity.

38.3 Staying Motivated

Most writers experience times when they have a strong desire not to write, but they can take action to regain motivation. Sometimes changing habits or scenery helps, such as writing in a new place or at a new time of day. However, a better practice may be to develop a writing routine, and to write at the same place at the same time every day. Writing daily, even when one is not in the mood to write, means making daily progress toward project completion. Remove distractions from the writing area, including banning music, videos, computer games, and email if those are barriers to productivity. Ensure that the writing space has a supportive chair and other provisions to make writing comfortable.

A manuscript can be completed relatively quickly when the author writes a small fraction of the draft daily. Writing 100 words each weekday will yield a complete draft of a 3000-word manuscript in 6 weeks. If the daily writing time during the following 2 weeks is used for editing, the manuscript can be completed and ready to submit in about 2 months. Writing just one paragraph each day (or making one table or figure during the daily writing time) will yield a complete manuscript in about 1 month. Because most journal articles in the health sciences follow the same outline, it is possible to know exactly which paragraphs are needed for a draft even before the first word is written. These paragraphs can be written in any order. A writer who is struggling with a particular paragraph can move on to another one. To maximize productivity, end each day with a plan for what paragraph to write the next day.

Setting a timeline for completing portions of a paper is often helpful. A timeline can include a schedule of events to celebrate intermediate successes on the way to a completed paper. Writers can select rewards that they will give themselves if they achieve their writing targets. It may be helpful to ask others to aid in enforcing those self-imposed deadlines. A supervisor can mandate a particular level of output, but coauthors and other motivators can also serve in this role. Writing groups can be excellent support systems for propelling their members to productivity (but writers should be cautious about support groups that validate excuses for not writing rather than equipping members to become productive). Mentors can also help with accountability, if that is the way the relationship has been defined.

38.4 Conquering Writer's Block

Writer's block describes sustained struggles with writing that an author might experience due to fear of failure or other barriers to productivity. Writer's block creates a negative thought cycle that can be difficult to break. The underlying issues leading to writer's block are often fear of being judged and fear of failure. Acknowledging these worries is an important step toward getting back to writing. Struggling writers also need to initiate new behaviors to facilitate success, and they need to stop engaging in writing avoidance behaviors. **Figure 38-3** lists various types of writer's block and the realities that counter them. Writer's block can be overcome when an aspiring writer makes writing a priority.

The completed manuscript will not be perfect. No paper is perfect. By the time a report is written, there are likely to be

FIGURE 38-3 Forms of Writer's Block and Writing Avoidance

Reason to Avoid Writing	Reality Check
"I don't know how to write a scholarly paper."	The best way to learn how to write is by writing. A writing support group, coauthors, and/or mentors can help with this process.
"I don't have time to write."	Almost everyone can find 15 or 30 minutes a day to write if that is a priority. Do not use "I'm too busy" as an excuse to avoid writing.
"I only write well when I'm under pressure from a deadline."	Most people do not do their best work when they are stressed. Fear of missing a deadline may motivate a person to "just get the thing done," which can feel like sweet success, but the work will be less thoughtful. Pressure to finish a product does not allow time for thoughtful writing and careful editing.
"I don't know how to get started."	Coauthors and mentors will be happy to offer advice about how to move forward.
"I don't know what to do next."	

Reason to Avoid Writing	Reality Check
"This project was not interesting, so it is not publishable."	If the topic was interesting enough to merit designing a study and collecting data, then it is probably interesting enough to present and publish. Check with a mentor about options for appropriately disseminating the findings.
"This research project had some flaws."	Every study has flaws, but few are fatally flawed. Ask a mentor about how to address the limitations of the study. Write a paragraph about the strengths and weaknesses of the project for the discussion section, then move on to writing the rest of the paper.
"This study is not going to change the world."	Most studies make only minor contributions to moving a field forward, but the only way to make any contribution is to publish.
"I'm stuck on this one section, and I can't work on anything else until I finish this part."	Writing and rewriting the same section over again is a waste of time. Work on another section of the paper. Ask a coauthor or mentor for assistance with the difficult section.
"I need to read some more articles and run some more tests before I start writing."	These preparations can become stall tactics. There is always one more article that could be read and one more test that could be conducted, but these are not good reasons not to write.
"I don't want to disappoint or be criticized by my supervisor/professor/mentor."	Supervisors want a paper to be as good as it can be, and they are obligated to make suggestions about critical revisions if they are coauthors. A writer who is nervous about sharing drafts can ask a writing support group member or a trusted friend to critically review manuscript drafts before they are shared with a supervisor. Procrastination will only increase anxiety about being evaluated.
"If I submit this manuscript and it is rejected, I will be embarrassed."	Comments about a manuscript are not criticisms of the person who wrote it. The only people who will know about the status of a manuscript are those whom the authors choose to tell about it. Research supervisors know that many papers are submitted to several journals before they are accepted for publication. Procrastination will only delay the start of the review process and the possibility of acceptance and publication.
"If this manuscript is published, someone might discover a flaw in it, and that would be embarrassing."	Coauthors, reviewers, and editors will not let an obviously flawed or badly written manuscript proceed to publication. No paper is perfect, and at some point the authors need to stop revising and finish the manuscript.
"I'm not a good writer." "I'm not good at writing in English."	Coauthors, colleagues, and friends can help edit the manuscript, but only after it has been drafted.

several imperfections in the study design and implementation that cannot be fixed. These flaws are normal and expected. Authors cannot remedy or hide those issues, but they can ensure that they:

- Fully explain the actual methods used
- Conduct all the appropriate analyses
- Honestly identify the limitations of the study and explain what was done to address them
- Include a helpful set of references that support the methods and results
- Polish the prose
- Ask coauthors, mentors, and others to provide feedback on drafts

Writing does not have to be a solitary activity. Most research projects in the health sciences are team efforts. While the lead author may have the primary responsibility for drafting the research report, that person does not have to work in isolation. Collaborators who are aware that a coauthor is struggling with writer's block will often be eager to help that individual get back on track with productive writing.

38.5 Finishing a Manuscript

Most people will always be able to find something they would rather do than write. It is easy to allow distractions to crowd out writing time. Researchers who want to disseminate their work must force themselves to stop planning, stop working on other tasks, and just write. Being a consistently productive writer often is easiest when authors:

- Have a regular writing routine
- Set deadlines for making progress toward a complete manuscript
- Identify and address the excuses they use to avoid writing
- Have mentors, coauthors, and others who support their writing goals
- Focus on the story they want to tell and the population that will be served by that story being shared

Figure 38-4 summarizes a diversity of strategies for getting started on a writing project, staying motivated, and seeing a paper through to completion.

FIGURE 38-4 Thirty Tactics for Writing Success in the Health Sciences

Focus on the story.	1. Identify the "mystery" and the "plot" of the paper.
	2. Identify the most important practice or policy implication supported by the results of the project. Write a paper that justifies this call to action.
	3. Develop an "elevator pitch." Be able to tell the story of the project in 30 seconds. Use that as the starting point for writing.
	4. Tell the story of the project in images. Start by making tables and figures.
	5. Write a précis (one sentence) and an abstract (one paragraph) that summarize the storyline before starting on the full paper.
	6. Make a poster that tells the story of the project.
	7. Make a slideshow that tells the story of the project, focusing on the order in which different parts of the story are presented.
	8. Tell the full story of the project to others, focusing on the mystery, the evidence, and the solution. Talk, then write.
	9. Record yourself telling your "story," transcribe the recording, and then edit the transcript into formal scientific writing.

Be organized.	10. Use one or more recent articles from the target journal to create an outline for the new manuscript.
	11. Use a writing checklist (like CONSORT or STROBE) to add details to the outline of the manuscript.
	12. Create an outline that shows the alignment of the objectives, methods, results, and discussion sections.
	13. Start by outlining the paragraphs within each section, and then outline the sentences within each paragraph.
Make steady progress.	14. Write throughout the research process.
	15. Write a set number of minutes every day.
	16. Write a set number of words (or one paragraph) every day.
	17. Set target dates for completing drafts of each part of the paper, and reward yourself for meeting those deadlines.
	18. Write whatever is easy to write on a particular day rather than writing the paper in order from the first paragraph to the last paragraph. To maintain writing momentum, end each writing session by deciding which paragraph to write during the next writing session.
Learn from others.	19. Read other people's papers. Pay attention to how they frame their arguments, present their evidence, and support their conclusions.
	20. Seek mentorship.
	21. Join a writing support group.
	22. Welcome feedback from coauthors, colleagues, supervisors, conference attendees, reviewers, editors, and others.
Be persistent.	23. Have a writing routine. Write at the same time and place regularly.
	24. Try something new when stuck. Try a new writing place or a different approach to writing, and make that part of your revised writing routine.
	25. Walk, then write. Step away from the computer to compose your thoughts, and then return to type up the ideas developed while being physically active.
	26. Remove distractions when writing. Turn off music, videos, games, and phones.
	27. Recognize waning motivation and take action to prolong periods of writing productivity.
	28. Stop writing avoidance behaviors. Stop planning, reading, and analyzing, and start writing.
	29. Acknowledge fears about writing and publishing. Do your best work, but take comfort in knowing that no paper is perfect.
	30. Remember why you are writing the paper. A publication may accomplish personal goals, fulfill educational requirements, and contribute to professional advancement while also making a positive contribution to advancing the health and well-being of others.

© DmitriyRazinkov/Shutterstock

CHAPTER 39

Reasons to Publish

Many professional and personal benefits accrue from publishing research findings in peer-reviewed journals.

39.1 Scientific Dialogue

Publishing in peer-reviewed journals is the way scientists publicly communicate with one another. Submitting a manuscript to a journal for review is a first step in a series of conversations about a research report. The initial discourse occurs among authors, editors, and reviewers. After the article is published, the conversation continues as other researchers read, discuss, cite, and apply the work. It is typical for experienced authors writing new papers to cite the papers that cited their previous publications. Maps of these citation networks often illuminate vigorous written exchanges between research groups. Having one's work cited by others is a permanent record of participation in the dialogue about a particular health issue.

Presenting research findings at conferences is a helpful part of the scientific conversation, but presentations are not entered into the permanent record of scholarly discourse. The abstracts published from conferences are generally not cited in future publications because they are so incomplete. (An exception to this occurs in fields that publish full-length conference papers in book-like conference proceedings.) Conference abstracts are generally considered to be previews of works in progress, and not final products like published journal articles.

If the results of a research study are not published, then for all practical purposes, it is as if the research was never done. The findings do not become part of the conversation among scientists because there is no formal record of the project. Although the researcher may have learned from the project even if it is never formally written up, an unfinished report does not further scientific knowledge or improve clinical or public health practices and policies.

39.2 Critical Feedback

The peer-review process is an opportunity to receive expert constructive feedback about a manuscript. Reviewers are usually quite adept at identifying weaknesses in a manuscript and asking authors to carefully think through the problem areas and fix to them. Responding to suggestions from reviewers and editors requires authors to:

- Understand and appreciate different perspectives.
- Balance conflicting sets of advice about what would strengthen a paper.
- Rewrite the parts of the paper that were confusing to reviewers.
- Recover from negative comments and demonstrate resiliency by moving forward.

All of these are skills that make authors better researchers and better health professionals, not just better writers. Gaining alternative perspectives about a research area may improve the design and implementation of subsequent research studies while also providing new insights about how to serve diverse patients, clients, and communities. The ability to weigh competing points of view and chart an acceptable path forward is a valuable skill for primary investigators and for leaders in any sector. Learning how to explain a procedure or decision better improves communication in any workplace. Building resilience and compassion in response to criticism and rejections can be beneficial for professional and personal growth.

The most valuable part of critical feedback is that it improves science. Reviewers challenge authors to tell their stories better. They identify weaknesses in a manuscript and propose solutions for them. They refer authors to helpful resources. The detailed and specific feedback provided by journal reviewers is usually not available after a conference presentation. Subjecting a manuscript to criticism and possible rejection can be intimidating and unpleasant, but the peer-review process produces better scientists and stronger manuscripts.

39.3 Respect for Participants and Collaborators

When participants donate their time to a project, the researcher has an ethical obligation to make sure those people's time was not wasted. One way to fulfill this responsibility and show respect for the contributions of volunteers is to share the results of a study with appropriate audiences. If a project generates a meaningful discovery, that finding should become part of the scientific literature. It can be more challenging to publish a null results study than it is to publish a paper showing a strong statistical association, but statistically insignificant papers can still be quite meaningful. If a well-designed and high-powered study fails to reject the null hypothesis, the researchers should seek to publish those results. A complete record of research results improves scientific knowledge and allows other scientists not to waste time and resources on a redundant project.

Completing the dissemination phase of a research project also shows respect for collaborators. Being a coauthor on a published article is often the only "compensation" that supervisors, professors, and mentors receive for the time they invest in research projects. A lead author who fails to see a research project through to publication is denying all of the coauthors public recognition of their involvement in the project and an additional line on their CVs or résumés. A publication in a respected journal is an achievement shared by the entire research team. Mutual respect and appreciation result from a successfully concluded project.

39.4 Personal Benefits

A published article proves that a researcher is part of the scholarly community, has the ability to handle constructive criticism, and can see

FIGURE 39-1 Completing the Research Cycle

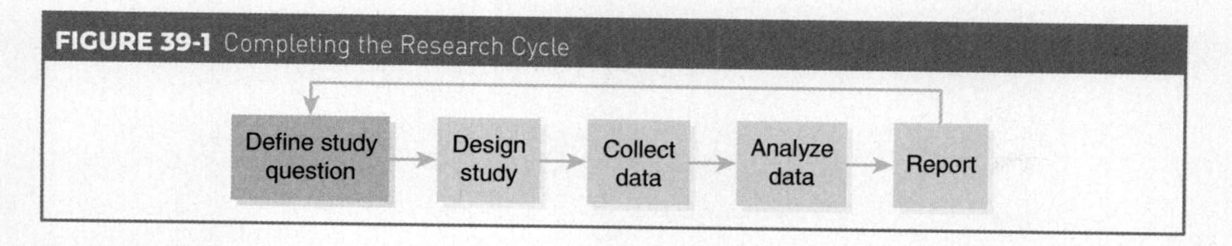

a project through to completion. Publishing enhances the author's CV and résumé. A published article becomes a part of each coauthor's permanent record, because the paper will be indexed in abstract databases for decades and certainly for the length of each author's career. And although authors of scholarly journal articles are not paid for their writing—and, in fact, are often happy when they do not have to pay publication fees—the payoff often comes in terms of improved job opportunities and promotions. Scientific publishing is unlikely to bring a person fame and fortune, but it does provide a tangible product after all the many hours that the researcher spent reading, planning, collecting data, conducting analysis, and writing. A published paper is evidence of the author's professional expertise and commitment to improving health for individuals and communities.

For those seeking to build and grow focused research portfolios, publications provide opportunities to gain expertise and recognition in a particular area of research. The research process does not necessarily end when a first report is published. The research process is a cycle in which data analysis and reporting naturally feed back into the formation of new study questions (**Figure 39-1**). Publishing marks an important step in this cycle. The next goal is not to publish the same results again (especially since redundant publication is a violation of professional standards and may result in the retraction of both the original article and the duplicate article), but to expand the research in a new but related direction. Some aspects of the data set that were not covered in the first publication might be worth exploring. Some newly identified gaps in the literature could be investigated with new data. A publication provides the momentum to launch into examining new study questions raised by the published paper.

© DmitryRazinkov/Shutterstock

CHAPTER 40

Selecting Target Journals

The culmination of a well-designed and carefully conducted health research project is often the dissemination of results through publication.

40.1 Choosing a Target Journal

Researchers who want to publish their findings must identify one or more journals that could reasonably be expected to disseminate their reports. Selecting a **target journal**—the journal a researcher intends to submit a manuscript to first—early in the writing process makes it easier to hone the paper's message for that journal's audience. An examination of recent articles published in the target journal provides guidance about the best outline to follow, including how to divide commentary between the introduction and discussion sections and which subsections to include in the methods section. Recently published articles will also provide insight about the appropriate voice and writing style, the amount of technical detail to include, and the reference and citation style. A first step toward identifying journals likely to consider a manuscript for publication is inspecting the items in the manuscript's reference list, because the journals cited most often in the manuscript are likely to be suitable target journals. Internet searches for relevant journals can expand the list of potential publication venues. That list can then be refined by examining abstract databases, journal rankings, and library holdings. Identifying several candidate journals and then choosing one target journal is a process that entails many considerations, including the aim, scope, and audience of the journals; the journals' impact factors and other bibliometrics; and online access options and the possible costs of publication.

40.2 Aim, Scope, and Audience

The most important consideration when identifying potential target journals is the fit of the research topic with the aims, scope, and audience of the journal. The aim is the overall goal of a journal. The **scope** of a journal describes the subject areas the publication covers. Some journals are very broad in focus, while others are very narrow and publish in only one subspecialty area. Some are international journals that publish research from around the world, while others have a very specific local or regional focus and publish only articles pertaining to that geographic area. The **audience** of a journal describes the readership that the publication intends to reach. Some journals are purely academic in nature, some are written for the members of clinical or professional organizations, and some have other types of readers as priorities.

Determining whether an article is a good match to a specialty or regional journal is often straightforward. A journal focused on liver disease in Argentina will not be interested in a paper about osteoporosis in Mongolia, but it will review a manuscript on cirrhosis in Buenos Aires. A journal focused on nutrition in Southeast Asia will not review a manuscript on vision disorders in Sweden, but it will consider a paper on iodine deficiency in Cambodia. Editors of journals with broader scopes use a more complex set of criteria to evaluate the fit of a manuscript with their journals. Some prestigious general journals will publish only articles expected to have a significant and nearly immediate impact on clinical practice. Some general journals in medicine, nursing, public health, and other health science fields will consider articles on just about any topic that could be considered relevant to the journal's scope. Reviewing the tables of contents for recent issues might provide insight about a journal's current thematic priorities.

The intended audience for a manuscript can also help authors determine whether a journal might be a good match to their work. If the manuscript's key message is targeted toward clinicians working in a focused geographic area, a journal sponsored by a regional professional society that provides a copy of each issue to all members of the organization might be the best venue. Publishing in this type of journal will ensure that the paper reaches those who will most benefit from reading it. If the study has conclusions that are relevant to an international audience, then a journal with a global readership might be more appropriate. However, the expansion of the Internet is making regional and international journals less distinct. Libraries and researchers nearly anywhere in the world are able to acquire copies of even relatively obscure publications.

The author guidelines of potential target journals provide additional information pertinent to the journal selection process. Authors of a systematic review manuscript should confirm that the target journal will accept reviews. Authors of a case report, a small case series, or a replication study might want to confirm that the target journal publishes short reports. Authors of comprehensive reports that far exceed the usual 3000- or 3500-word limit or the standard maximum of four tables and/or figures combined will require a journal that has flexible word limits, which will typically be an online-only journal.

40.3 Impact Factors

A secondary consideration when selecting the target journal may be the impact factor, ranking, or reputation of the journal. The **impact factor** of a journal is an annual determination by the Clarivate Analytics company (previously published by Thomson Reuters) about the average number of times an article published in a particular journal is cited by other articles during its first 2 years after publication. A few of the most prominent journals (like *Science*, *Nature*, *JAMA*, *The Lancet*, and the *New England Journal of Medicine*) have impact factors of 10 or greater, but most journals in the health sciences have an impact factor closer to 1 or 2. Specialty journals may have an impact factor less than 1, but they can still be important within the specialty area. Impact factors are often listed on journal websites, and many university libraries subscribe to the annual *Journal Citation Reports*, published by Clarivate Analytics, which presents detailed bibliometrics for indexed journals.

A growing number of other groups publish variations of the impact factor. Some of these are derived from rigorous evaluations of citation counts and other types of impact metrics, and they are transparent about their methods. However, there also many groups that publish purported bibliometrics that are completely fictitious or are derived from invalid data. Be alert to fake impact factors and other misleading bibliometrics placed on the websites of some low-quality journals.

40.4 Other Journal Characteristics

Many other factors may influence the selection of the target journals:

- The expected duration of the review process
- The acceptance rate
- Whether the journal is indexed in preferred databases
- Whether an online submission system is available
- The print and online availability of the journal
- The type of peer review used
- The publisher
- Data-sharing availability or requirements

Preferences related to these factors should be discussed with coauthors as part of the journal selection process.

When selecting the target journal, one set of factors relates to the journal's performance metrics. Some journals provide information about their average time from submission to first decision, their average time from submission to publication of accepted articles, and their overall acceptance rates. Many high-profile journals with very low acceptance rates have a turnaround time of only a few days because they send very few manuscripts out for external peer review. Specialty journals with higher acceptance rates may have a turnaround time of many months because three or more external referees review every manuscript. Some authors prefer to choose a target journal with a rapid time to a first decision. Some prefer to submit to competitive journals with low acceptance rates, while others prefer to submit to less competitive journals with high acceptance rates.

Some researchers prioritize journals indexed in MEDLINE or in other disciplinary collections because being indexed in those databases increases the likelihood that a published article will be read and cited by other scholars. The list of abstract databases that index the journal may also provide additional insight about the scope of the journal and its target readership.

Another consideration is the method of submission. Most journals have moved to online submission systems that allow authors to upload manuscripts to a website and then track the progress of their manuscripts through the review process. Some authors prefer online tracking systems over other modes of submission (such as email) because they like to be able to monitor the status of their manuscripts. Similarly, some authors have a preference for published articles being printed on paper and/or being available online. Although the vast majority of print publications now also offer online access to subscribers (usually libraries), not all do. A growing number of journals are online-only and do not print their issues. Although most online journals are likely to remain available on the Internet for many years to come, some researchers remain wary about publishing in online journals that do not leave a paper trail, especially if those journals are new and unproven.

Some authors have a preference for a particular type of peer review. In publishing, a **double-blind** peer-review process is one in which the reviewers do not know the identity of the authors and the authors do not learn the identity of the reviewers. A **single-blind** peer-review process is one in which the reviewers are provided with the authors' names, but authors are not given reviewers' names. Some journals use an open review process in which the names of the authors and reviewers are disclosed. Some even post reviewer comments and names on their websites alongside published articles.

Some researchers have strong opinions about particular publishers. For example, some scholars preferentially publish in journals that are owned by large publishing companies because they consider these presses to have high standards for editorial quality. Others preferentially avoid journals published by companies that generate hefty annual revenue while charging subscription fees the researchers perceive to be exorbitant and profiting from the free labor provided by authors, editors, and reviewers. Some researchers preferentially publish in open-access journals, while others are unwilling to pay author fees and consider many open-access journals to have low editorial standards. There are no set rules about how to make this type of decision, other than the caution to avoid journals that might be engaging in unethical publishing practices.

Data sharing is the willingness of a research team to make their data and methods freely available to other researchers. Most data collected by scientists are the property of those individuals or their institution, and the owners are not obligated to share their data with others. However, some funding agencies require data collected with their support to be made available to other research teams, either by publishing the data as a supplement to a journal article or by making the data files available to others upon request. Additionally, some journals have mandatory data sharing policies. Authors who are obligated to share data may prefer to select a journal with an open data policy. Authors who do not wish to share their data as a supplement to a paper should select journals that do not mandate data sharing.

40.5 Open Access and Copyright

The publication costs for many journals are covered through subscriptions, advertising, and/or the support of a professional society, but an increasing number of publishers mandate that authors cover some or all of the costs of publishing. While many publishers operate under a non-profit model, there are a growing number of for-profit publishers that are seeking to maximize income from their enterprises. As a result of these developments, there are now three types of journals in the health sciences: subscription journals, open-access journals, and hybrid journals (**Figure 40-1**).

A **subscription journal** is a journal that covers its costs from library and/or individual subscriptions and advertising and does not charge any author fees. Most articles published in a subscription journal are placed behind a paywall. People who want to access an article on the journal website must log in using credentials from a subscribing library or must pay to purchase access to the article.

An **open-access fee** is a charge that authors pay to a journal in order to make their articles available without a paywall. An **open-access journal** is a journal that mandates that authors pay a publication fee before their manuscripts are published. The content of these journals is freely available to readers on the Internet, and no subscriptions to the journal are sold to libraries. However, authors who are not able or willing to pay for an article to be published cannot publish their work in an open-access journal. For many journals in the health sciences, the open-access fee is several thousand dollars.

A **hybrid journal** is a subscription journal that gives authors the option of paying to

FIGURE 40-1 Publishing Models

Publishing Model	Cost to Authors	Cost to Readers
Subscription	Free.	Readers must pay for a journal subscription (individually or through an institution) or pay a per-article fee to access the article.
Open access	Authors must pay a fee of several hundred to several thousand dollars.	Free.
Hybrid	Authors choose whether they want to pay for open access or not; if they do not pay an open-access fee, the article will be published under a subscription model.	Depends on the authors' decision about whether to pay an open-access fee.

make an article freely available to all. Authors with funding may opt to pay for open access under this model. Authors without funding may publish at no cost, but their articles will be behind a paywall.

Gold open access is a publishing model in which authors or their funders pay to make a journal article freely available to readers on the Internet as soon as it is published. Some funding agencies require articles written with their support to be publicly available immediately upon publication. In those situations, authors must publish in an open-access or hybrid journal. Some funders mandate that a version of papers written with their support be made open access, but they allow a grace period of 1 or 2 years after publication. In this situation, authors have the option of publishing in a subscription journal that supports **green open access**, a publishing model in which authors are allowed to post a version of a published article on their personal websites or in institutional repositories, usually after an embargo period of 1 year or longer. The authors typically are allowed to archive only an unformatted version of their manuscripts, because the journal owns the formatting style of the published version. Authors may not post any version of the article on a public website until after the embargo period is over. Some subscription journals do not allow any open-access options for authors.

Copyright defines the legal rights assigned to the owners of intellectual property such as written and artistic works. Subscription publishers typically require authors to assign the copyright for a manuscript to the publisher before their articles are published. Open-access publishers typically allow authors to retain the copyright for their articles. Authors should carefully examine the copyright rules for journals before submitting their work.

A variety of copyright licenses can be used when an author or publisher makes a copyrighted work available to others. A **Creative Commons (CC) license** is a public copyright license that enables the free distribution of a copyrighted work. CC designations specify whether users of the work must give attribution to the author (indicated by "BY" or "CC-BY"), if users can distribute the work (indicated by "SA," for share-alike), if the work can be used only for noncommercial purposes (NC), and if no derivative products can be made from the original work (ND). Many open-access publishers mandate that authors select a CC license option that makes their articles freely available to others as long as the authors are credited for their work.

40.6 Publication Fees

Open-access charges and publication fees are usually disclosed in a journal's author guidelines or elsewhere on the journal's website. Authors are advised to look carefully for this information when considering publishing options. When authors are unable or unwilling to pay fees, they should not submit to a journal until after they have confirmed that no fees will be assessed. A few journals that charge open-access and/or other publication fees may allow authors to request waivers of some costs if the authors are from low-income countries and/or if the project was not supported by a contract or grant. These requests usually must be made before the paper is reviewed.

In addition to open-access fees, there are several other fees that authors might be asked to pay. A few journals require authors to pay a small **submission fee** prior to review. The journal will not review an article until this payment is received. Paying the submission fee is not a guarantee of acceptance, and the submission fee is not refunded if a manuscript is rejected. Some journals charge a small or large publication fee that does not make the article open access. A per-article charge that some subscription journals mandate prior to an accepted article being published is often called a **processing fee** or **processing charge**. A per-page fee assessed prior to publication in a subscription journal is usually called a **page fee** or **page charge**. The number of pages is determined by the final typeset article, not by the number of pages in the submitted manuscript. Some journals that are run by professional societies require the corresponding author of an accepted paper to become a member of the sponsoring society. In this situation, publication requires payment of a membership fee if the author is not already a society member.

40.7 Predatory Journals

Authors must be aware of the growing number of dubious journals being launched by publishing businesses that accept every submission (or nearly every submission) upon receipt of payment by the author. Many open-access journals are well respected and regarded as having strong peer-review systems. However, a subset of open-access journals have a reputation for being deceptive pay-to-publish schemes. A **predatory open-access journal** is an exploitative journal that does not provide the quality editorial and peer-review services associated with legitimate journals. Predatory journals do not follow good practices for editorial and peer review. Many are not transparent about their policies and fees. They may charge authors hidden publication fees that are not disclosed in advertising emails or on the journals' websites. These journals should be avoided. Before submitting to any journal, confirm that the journal is legitimate and respected (**Figure 40-2**). Beware of journals that send unsolicited spam to email addresses, promise a very quick time to decision and publication, have just launched or have published very few articles, and have poorly written web content and author guidelines.

FIGURE 40-2 Signs of Journal Quality

Low-Quality Journals	High-Quality Journals
The journal's website is disorganized, has low-quality writing, and is missing critical information.	The journal has a well-designed, professional, and up-to-date website.
The articles published in the journal are of poor scientific quality and/or have not been copyedited and professionally formatted.	The articles published in the journal present strong science and are well written and professionally formatted.
It is not clear who owns the journal or whether it is affiliated with a legitimate professional organization.	The name and contact information for the publisher are provided on the journal's website.
The website advertises a very quick time from submission to acceptance and boasts of a high acceptance rate rather than emphasizing quality and merit.	The journal website provides clear instructions to authors and reviewers.
The journal is not indexed in respected disciplinary databases.	The journal is indexed in relevant databases.
The journal is vague about fees and copyright.	The journal discloses all fees or clearly states that no fees are assessed.
The journal does not have an editorial board.	The journal's editorial board includes known experts in the field.
The journal's website features deceptive bibliometrics (such as listing an impact factor even when the journal has not yet published the 2 years of issues required for a Clarivate Analytics impact factor to be calculated).	The journal is a member of appropriate publishing groups, such as the Committee on Publication Ethics (COPE).
The journal solicits submissions by spamming academic email accounts with poorly written calls for papers promising to publish manuscripts within days.	The journal does not send spam emails; if it has an email list, it is easy for recipients to unsubscribe from it.

© DmitriyRazinkov/Shutterstock

CHAPTER 41

Manuscript Submission

Manuscripts should be formatted and submitted to one peer-reviewed journal as soon as the coauthors agree that the document is ready for external peer review.

41.1 Submission Timing

Publication is a priority for many health researchers because results that have not been published are not contributing to advancing knowledge and improving practice and policy. From the perspective of the broader scientific community, an unpublished project never happened. Submitting to a journal as soon as a revised and polished manuscript has been crafted and all the coauthors have signed off on it is critical. Procrastination can render the study useless, because data in the health sciences may quickly become obsolete and no longer be publishable. Submission does not mean that a manuscript is perfect, just that it is ready to receive comments from external reviewers. Submission is not the end of the writing process. Additional revisions will almost certainly be required, even if the first journal to which a manuscript is submitted eventually accepts the paper. Another incentive to submit as soon as possible is that revising a manuscript is easiest when the project is fresh in the minds of the coauthors.

41.2 Journal Selection

Once all of the coauthors are satisfied that the manuscript is ready to be submitted for peer review, *one* journal must be selected as the first journal for submission. A preliminary target journal may have been identified early in the research or writing process to serve as a guide. After a manuscript has been drafted and revised, a variety of journals should again be considered. Only one can be selected as the first place to submit the completed manuscript.

Duplicate submission, submitting the same manuscript to two or more journals at the same time, is not permitted in the health sciences. Although editors of some popular magazines may compete for manuscripts from paid freelance authors, nearly all of the labor in the academic journal system is voluntary. Editors may receive little or no compensation for their time, and reviewers and authors are unpaid volunteers. It would be a major strain on the editorial and peer-review system if every manuscript was sent to several journals at the same time. Most journals therefore require a statement with each submitted manuscript affirming that the manuscript is under consideration only by that one journal. This rule should be assumed to be true for all journals. After a manuscript has been submitted to a journal, it cannot be submitted elsewhere until the authors are notified that it has been rejected or the authors formally withdraw the manuscript from consideration. Authors should expect to wait several months for a decision after submission of a manuscript to a journal.

Journals in the health sciences typically publish only findings that have not already been published elsewhere. **Duplicate publication** occurs when authors publish a second paper that is identical or very similar to a paper they have already had published. Duplicate publication is usually considered to be a form of research misconduct. Any possible overlap with previously published material must be disclosed to the editor at the time of submission of a new manuscript. For example, some journal editors are willing to publish full-length articles even when an abstract or short report from the same project has already been published elsewhere, but the authors are obligated to disclose the prior dissemination of the study findings so that the editors can make an informed decision about whether the new submission is sufficiently different from the prior publication to merit consideration by the journal.

Authors should also avoid "**text recycling**," or "self-plagiarism," which occurs when one's own words from one publication are copied into a new manuscript. It can be tempting to copy and paste background and methodological information from one document into another one when preparing multiple manuscripts based on one research project. This should not be done, because text recycling can lead to charges of duplicate publication. Reusing material is especially problematic when the author has assigned the copyright for the original publication to the publisher of that work.

The **Committee on Publication Ethics (COPE)** is an organization that provides guidance on how to avoid research misconduct during the dissemination phase of a research project. The COPE website provides helpful information about appropriate conduct for authors and the repercussions for those who violate publishing norms. The editors of several thousand biomedical journals are COPE members who have agreed to follow the standards put forth by the organization. Violating the guidelines posted on the COPE website may result in serious consequences for a researcher.

41.3 Manuscript Formatting

Author guidelines, also called **instructions for authors**, are detailed directives from a journal about how manuscripts should be formatted prior to submission. The directions must be carefully followed. See **Figure 41-1** for examples of formatting preferences, which vary by journal.

FIGURE 41-1 Manuscript Formatting Requirements Addressed by Journals' Author Guidelines

Title page	Should only the title be listed on the title page? Or should author names and affiliations, word counts, keywords, running headers (abbreviated versions of the title), and other information also be listed? Should the title page be submitted as a separate file from the rest of the document?
Blinding	Should authors' names be removed from the manuscript? Should other identifying information be blacked out in the manuscript, possibly including the citation information for references to previous works by the research team?

Abstract	Should the abstract be structured, with subheadings for each section, or unstructured? If a structured abstract is expected, are there preferred subheadings (such as Objective, Methods, Results, and Conclusion)? What is the word limit for the abstract? Should the abstract appear on its own page in the manuscript file? Is an additional one-sentence summary (a précis) required? Is a separate set of statements about the contributions the paper will make to the literature required?
Keywords	How many keywords (search terms that will be linked to the article) should be provided? Must these be MeSH (Medical Subject Heading) terms? Where should these keywords be listed in the manuscript (such as on the title page or immediately below the abstract)?
Sections	Is there a preference for how sections within the document are labeled and formatted?
Acknowledgments and end matter	Should acknowledgments of funding sources or personal assistance be included on the title page or at the end of the manuscript? Is any additional end matter to be included, such as details about the contributions of each coauthor, information about ethics committee review, declarations of potential conflicts of interest, or other disclosures?
In-text citation style	How should in-text citations of works listed in the reference section be shown: as superscript numbers, as numbers within brackets, by placing the last name of the first author and the publication year in brackets or parentheses, by listing the names of several authors along with the publication year in brackets or parentheses, or by some other method?
Reference list order	Should the entries in the reference list be in alphabetical order (by the first author's last name) or numbered in order of first appearance in the manuscript?
Reference style	What specific style does the journal require for the reference list? For example, how many authors should be listed for articles with more than six coauthors? Should the journal title be written in full, or as an *Index Medicus* abbreviation? Should the volume and issue be listed, or just the volume? Should any of the parts of the reference be in bold or italics? What types of punctuation should separate various components of the reference list entries?
Page formatting	What margins and line spacing are required? Do the lines on each page need to be numbered?
Page numbering	Should page numbers be shown at the bottom center of each page, the top right of each page, or elsewhere?
Fonts	Do particular typefaces and font sizes need to be used? (For example, is 12-point Times New Roman or 11-point Arial recommended?)
Word and page limits	What is the word limit or page limit? Do these limits apply only to the main text of the article, or do they also include the abstract, references, and tables?
Tables and figures	Is the number of tables and/or figures limited? Should tables and figures appear in the manuscript following the paragraph in which they are first mentioned, or should they all be placed at the end of the manuscript file after the references? Should each table and figure be saved as a separate file, or should tables be placed at the end of the manuscript file but figures submitted as separate files?

Special attention should be paid to figures and other graphics when formatting the manuscript. Most journals will reformat the tables of all accepted manuscripts into their house styles when they convert the text into the single-spaced, small font, two-column layout that is popular in health science journals. However, graphs, maps, and other illustrations are rarely redrawn by a journal's graphic designer prior to publication, so all figures should have a professional appearance prior to submission. Journals may require image files to be submitted in a specific electronic format, which may or may not be a standard file type. Because an image may be resized prior to publication, authors should confirm that each image can be enlarged and reduced without distortion. Most journals charge a fee for printing color images but not for grayscale images, so color should be used only when it is absolutely necessary. Alternatively, some journals charge for color in the print version but allow the online version of the manuscript to use color at no cost. In this situation, authors may submit a color version of the image but should confirm that the grayscale version of that color image has appropriate tones and adequate contrast.

41.4 Cover Letter

A **cover letter** usually accompanies a submitted manuscript in order to briefly explain the importance of the work to the editor of the journal. Even though most submissions are made via computer rather than by postal delivery, most online submission systems still expect a cover letter to be uploaded. **Figure 41-2** summarizes the typical content of a cover letter for manuscript submission. The letter should provide a brief description of the project and the major conclusions, and it should seek to convince the editor that the work is important, valid, original, and a good fit with the aims and scope of the journal. After a manuscript is submitted, the editor's decision about whether to consider the article for publication may be made solely on the basis of the abstract and the cover letter, so both of these items must be compelling. (Cover letters

FIGURE 41-2 Sample Cover Letter Content

Salutation	Address the letter to the editor(s) by name ("Dear Dr. ____") if names are available on the journal's website.
Basic details	Provide the title of the manuscript. If the journal publishes several categories of articles—such as original research, short reports, reviews, and commentaries—specify the type of submission. (In this context, "original research" typically refers to full-length manuscripts presenting new analysis of primary or secondary data.)
Summary	Provide a short summary of the study design and key findings. (Do not copy the abstract into the letter. Write a new summary that emphasizes the key findings and implications of the study.)
Importance and fit	Make the case for why the manuscript is important, significant, and original, and why the manuscript might be a good fit to the aims and scope of the journal.
Required declarations	Some journals require the cover letter to affirm that the manuscript is not under review elsewhere and has not been previously published; to declare that all listed coauthors meet the authorship criteria of the International Committee of Medical Journal Editors (ICMJE), including approving the submission of the manuscript to the journal; and/or to disclose any potential conflicts of interest. Some journals may additionally require information about the funders of the research project and/or the specific contributions of each coauthor.

Thanks	Thank the editors for considering the manuscript for review and possible publication.
Names and signatures	Some journals require the signatures of all authors to appear on the cover letter. When this is required, a signed letter can be scanned into a computer and uploaded on the journal's submission website, or it can be faxed to the journal office. If signatures are not required, it is acceptable to simply type in the name of the corresponding author as the author of the letter.

are not sent to external peer reviewers. They are seen only by the editorial staff.)

41.5 Online Submission

Once the manuscript files have been prepared and all the required supplemental information has been compiled, the manuscript is ready to be submitted. The authors may need to email the manuscript and cover letter to the editor or send paper copies by postal mail, but most journals require online submission via a website.

Creating an account with a journal's publication management system usually takes only a few minutes. In addition to facilitating submission of the manuscript, the online account enables the corresponding author to track the manuscript's progress through the review process. Most online systems will indicate when the editorial office is considering an article, when the article is undergoing external review, and when reviews have been submitted and a decision is pending.

Most of the time, only the corresponding author needs to register with the journal. The **corresponding author** is the coauthor who will communicate with the journal and take the lead on answering questions from readers after the paper is published. The corresponding author may be the first author, the senior author, or the coauthor with the most stable email address and affiliation. Some journals request or require corresponding authors to link their profiles to ORCID accounts, which can be created at https://orcid.org. An **ORCID**—a term derived from *o*pen *r*esearcher and *c*ontributor *id*entifier—is a 16-digit number that serves as a unique persistent identifier for a researcher. The ORCID is for individual researchers what a DOI is for individual journal articles, books, or other publications: It is an enduring link to an online publication profile. An ORCID allows bibliometric algorithms to distinguish between two or more researchers who happen to have the same name and to link records generated by one researcher who has had several institutional affiliations.

Online submission usually takes about half an hour, but it may require more time if a lot of information must be typed in and there are multiple steps in the uploading process. Most submission websites start by asking for basic details about the article, such as the title, abstract, and keywords. It may be possible to type or paste in the keywords, or the keywords may need to be selected from a list provided by the journal. Some journals will also ask for:

- The type of article (such as original research, review article, or letter)
- The word count
- The number of tables
- The number of figures (grayscale and color)
- Statements about ethics approval, funding, data sharing plans, possible conflicts of interest, and author contributions
- Confirmation that the article is being submitted to only one journal

A second step asks for information about all contributing authors. The corresponding author should check ahead of time with coauthors about the preferred forms of their names. Most authors in the health sciences choose to use a middle initial when publishing, since PubMed and several other abstract databases list authors by their last names and their first and middle initials. Some journals also request a job title and affiliation, degrees earned (or the one highest degree earned), and contact information (including email addresses, street addresses for the workplace, telephone numbers, and even fax numbers) for all authors. These details should be collected from all coauthors before beginning the submission process, just in case they might be required.

The term **affiliation** describes the institution where an author was employed or enrolled when the individual was conducting or contributing to a research project. If a research manuscript was produced as part of a coauthor's work responsibilities, the employer is typically listed as the primary affiliation. If work was produced as part of fulfilling educational requirements, the school where the student coauthor is enrolled is typically listed as the affiliation. Some journals allow multiple affiliations for each author, while others allow only one affiliation to be listed per author. When only one affiliation can be listed, many journals expect the listed organization to be the one where the coauthor was located when the bulk of the work on the project was being conducted, even if that is not the current institution. Other journals specifically request addresses and affiliations that are current at the time of submission.

There may be additional steps. For example, the journal may request the names and contact information for three or more potential reviewers. Some journals require a list of potential reviewers before a submission will be processed; some make this information optional. A senior author can usually offer guidance on how to select appropriate names to add to this list. They must not be people who have a conflict of interest that would prevent them from reviewing the manuscript fairly. For example, journals may specify that the listed individuals cannot have written a paper with any of the coauthors within the past 5 years or may put other stipulations in place. The named individuals should not be contacted by the authors. If the editors select listed individuals as reviewers, the editor will contact them directly. The editors are under no obligation to contact any of the suggested individuals. Some journals also allow the corresponding author to identify people who should not be reviewers because of a known conflict of interest, but it is not binding on the editor to respect this request.

The final step is uploading the manuscript files. The website will provide instructions about how various files should be attached. Some journals require the title page to be uploaded separately from the rest of the manuscript, especially if the journal uses double-blind review. Many journals require each table and figure to appear in a separate file. The file types acceptable for figures vary among journals. The journal may also request additional files, such as a publishing agreement signed by all authors or a checklist showing compliance with required content and formatting.

All of the manuscript files are typically combined into one PDF file by the online management system during the submission process. Prior to finalizing the submission, the corresponding author should carefully review this file for completeness, page numbering, line numbering (if applicable), and the legibility of tables and figures. Some systems automatically link references in the manuscript to abstract databases, so that reviewers can easily access the abstracts of the cited articles. References that the computer cannot link to an entry in an abstract database may be flagged as possible errors. These submission programs typically allow the author to review an HTML version of the uploaded paper, check the linked references for accuracy, and correct errors in the reference list. After the manuscript and supporting files have been uploaded and any required corrections to the materials have been made, the author is usually prompted to click a link to approve submission to the editorial office. When the submission is authorized by the corresponding author, the submission is complete.

© DmitriyRazinkov/Shutterstock

CHAPTER 42

Peer Review and Publication

Manuscripts submitted to peer-reviewed journals are evaluated by external reviewers who provide feedback about how to improve a manuscript and by editors who make a decision about whether the paper is suitable for publication in a particular journal.

42.1 Initial Review

After a manuscript is submitted to a journal, the journal's editorial staff does a preliminary review and decides whether to send the manuscript to external peer reviewers or to reject it without review. Although the organizational structures of journals vary, the editor-in-chief who oversees the journal often assigns new submissions to assistant editors for initial review. The assistant editors identify *ad hoc* reviewers to review manuscripts deemed worthy of further consideration. *Ad hoc* reviewers are not on the journal's editorial board but are asked to review submissions because of their methodological or subject-matter expertise. Some journals send nearly all manuscripts out to external reviewers; others select only a small fraction of submissions for peer review.

One of the advantages of the initial review process is that it allows authors whose submission is not a good fit to one journal to quickly submit their work to a more suitable alternative. Rejection without review (sometimes called a **desk rejection** or **bench rejection**) is often based solely on an editor's evaluation of whether the cover letter, title, and abstract show an alignment with the journal's current aim and scope. Some manuscripts are declined because they are poorly written and sloppily formatted, but most of the time a declination of a manuscript is about fit rather than quality. If a manuscript is rejected without review, the authors should identify a different journal that might be a better fit, update the manuscript to match the writing style and formatting requirements of the new target journal, and submit to the new journal as soon as possible.

Authors are often notified of a decision to reject without review within a few days or a few weeks of submission. When an article is selected for external review, notification of the first decision about the manuscript usually takes 3 months or longer (**Figure 42-1**). Authors should usually not contact editorial offices to inquire about the status of their

FIGURE 42-1 Sample Timeline from Submission to Final Decision

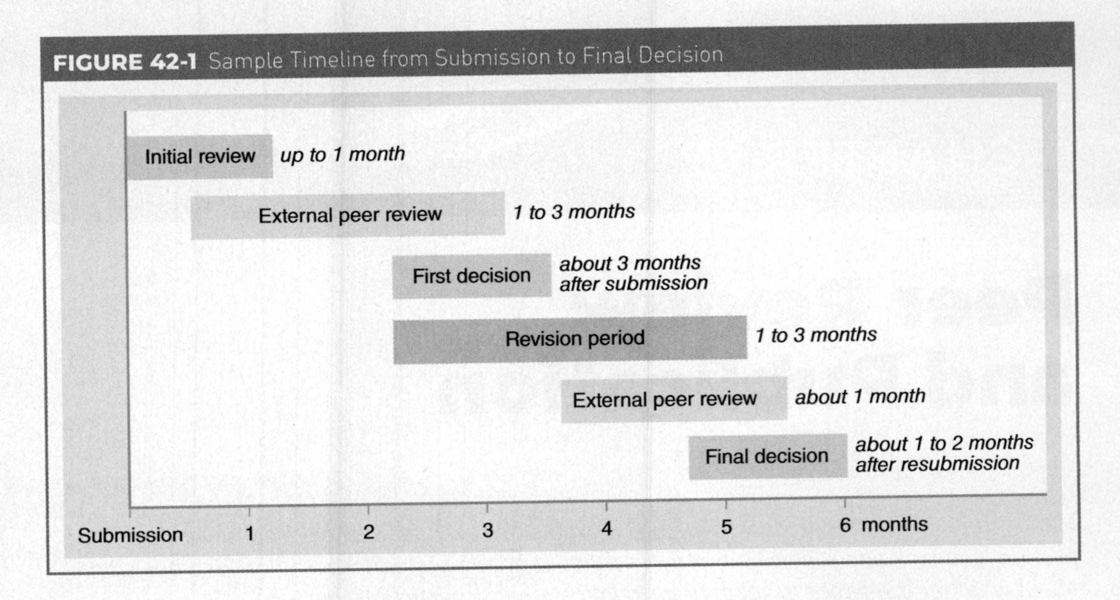

manuscript until at least 4 months after submission. Even then, a request for an update should be made only if the status of the paper has not recently been updated in the online submission management system.

42.2 External Review Results

Manuscripts are typically reviewed by at least two external reviewers. Each reviewer usually submits two sets of comments to the editor. First, the reviewer provides a list of major and minor issues in the manuscript that should be addressed by the authors prior to publication. These observations are addressed to the authors and will be sent to them along with the editor's decision letter. Second, each reviewer provides an assessment that only the editor will see. Reviewers are often asked to make a recommendation about whether the manuscript should be accepted with minor revisions, reconsidered after major revisions, or rejected. Reviewers may also be asked to rate the manuscript's novelty, importance, and fit with the journal in addition to assessing the quality of various aspects of the work.

An external peer review can lead to three possible results: rejection, an opportunity to revise and resubmit, or acceptance (**Figure 42-2**). An article determined to be methodologically sound and well written may receive low scores for importance and relevance to the journal, so it is possible for a manuscript to be rejected even if all the comments shared with authors are very positive. Alternatively, an article deemed to be lacking in writing quality may receive high scores for the originality of the topic and the apparent significance of the work, and this may result in an invitation to revise the manuscript and submit it to the same journal for another round of review. Often reviews are mixed, with one or more reviewers being very critical and one or more being quite positive. When the reviews are mixed, the editor may decide to reject the article or may choose to offer the authors the opportunity to revise their manuscript and resubmit it to the journal for further consideration.

FIGURE 42-2 The Journal Review Process

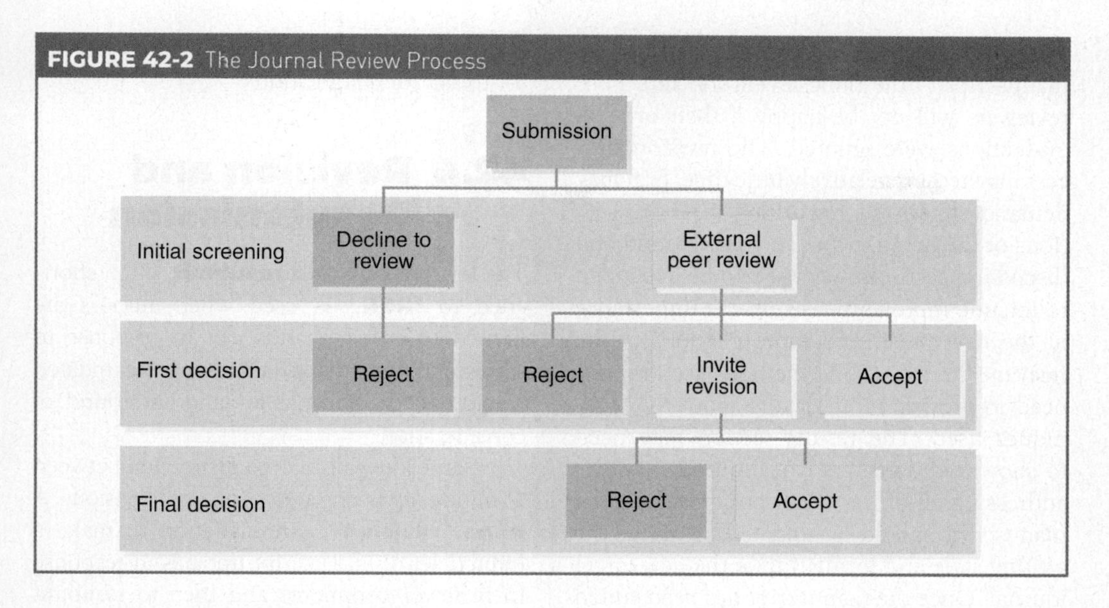

42.3 Rejection

Some manuscripts are rejected because they are poorly written, unsound, or of limited interest to those not directly involved in the project. However, many rejected manuscripts are well written, robust, and will be interesting to a wide audience. Many journals have low acceptance rates and routinely reject high-quality papers. An appeal to the editor to reconsider a rejected manuscript will almost never result in a different outcome. Rather than contesting a decision, authors should direct their energy into revising the manuscript for submission to another journal. A manuscript has a high likelihood of eventually being published if it is well written, if the study methods were valid and reasonably rigorous, and if the authors clearly link their results to broader applications or implications.

Rejection does not mean a manuscript has been rejected by all journals and will never be published anywhere. It simply means one journal has decided that the paper is not suitable for its audience. Many authors find it helpful to take a few days to be disappointed about the rejection, to vent about some of the reviewer comments, and to complain to coauthors about the editorial decision. But one rejection—or even several rejections—does not mean a manuscript is not publishable. Each set of reviewer comments can be used to strengthen a research report. Most studies are not so badly designed and conducted that they are fatally flawed and cannot contribute to the scientific literature. Most research manuscripts can be made suitable for publication somewhere with several weeks or several months of additional work. As long as researchers are willing to learn from each set of reviewer comments, the manuscript will continue to become stronger with each submission.

It is usually best to begin work on revisions soon after receipt of a rejection letter. As time elapses after the completion of data collection and analysis, remembering the original aims, methods, and results becomes increasingly difficult. All reviewer comments should be read and carefully considered, with appropriate edits made. Authors should never submit to a second journal without taking advantage of the input provided by the first set of reviewers. The most important reason to make these updates is that the feedback from the reviewers will improve the manuscript. A secondary reason to take revisions seriously is that the new journal may send the manuscript to the same reviewers, and those reviewers will not be happy if their original evaluations were ignored. The revision process may require relatively little time, or it may demand significant reworking of entire sections of the manuscript. The background and discussion sections may need to be expanded to include more emphasis on the importance of the new paper and more citations of the relevant literature. The methods section may need to provide more details about the techniques used. The results section may need to show additional statistical output. Besides addressing all of the reviewer comments, the manuscript should be updated to match the writing style and formatting of the new target journal. Once the manuscript has been edited to the satisfaction of all coauthors, it should be submitted to the new journal.

Some journals are now offering a "**reject and resubmit**" option, in which a rejection letter from a journal editor invites the authors to revise a manuscript and resubmit it to the same journal for consideration as a "new" article. This response is typically given when the editor wants the freedom to invite a different set of external peer reviewers to review the resubmitted manuscript. (This option is also sometimes preferred by editors who want to reduce the reported acceptance rates for their journals. A manuscript submitted as a revision and then accepted will be reported as one submission and one acceptance, whereas a manuscript submitted twice—as a "new" submission each time—is reported as two submissions and one acceptance. This strategy can reduce the reported acceptance rate for a noncompetitive journal from nearly 100% to about 50%.) "Reject and resubmit" decisions should usually be treated like a "revise and resubmit" opportunity. Most rejection letters specify that a different version of the same manuscript will not be considered by the journal. If the decision letter is not clear about whether a rejection is final or the authors are being invited to revise and resubmit the manuscript, the corresponding author should contact the editor to ask for clarification.

42.4 Revision and Resubmission

The term **revise and resubmit**, often shortened to "**R&R**," is used when authors are invited to edit a manuscript in response to reviewer comments and then send the updated version to the journal for another round of consideration.

Some journals make a distinction between a minor revision and a major revision. A **minor revision** is an invitation to make a limited set of manuscript updates in response to reviewer comments and then to resubmit the manuscript for editorial review and a final decision about acceptance. A **major revision** is an invitation to significantly update a manuscript in response to reviewer comments and then to submit the revised manuscript for another round of review by external peer reviewers. A minor revision might be reviewed solely by an editor after resubmission, but a major revision will almost certainly be sent back to the original reviewers and perhaps also to new reviewers.

Authors are usually assigned a deadline for resubmission, often 1 to 3 months from the time the decision letter is issued. A journal may allow only a very short time, often just a few weeks, for a minor revision to be returned. A major revision may be allocated a revision period of 3 months or longer. If this deadline is missed, the revised manuscript might be treated as a new submission or it might be rejected.

When authors accept a revise and resubmit opportunity, they need to prepare two documents. One is an edited version of the manuscript that marks each change made to the file. The other is a file providing a response to each and every reviewer comment. Reviewers who are asked to examine a revised manuscript are provided with a copy of both files. Since re-reviews are typically performed by the same reviewers who evaluated the original submission, every response to a reviewer comment needs to be carefully constructed and respectful. Examples of responses to reviewer comments are shown in **Figure 42-3**. Some reviewer suggestions—often marked within their comments as "minor"—will be easy to respond to, such as correcting typos, reformatting tables, or adding a few more citations. Others—often marked as "major" or "compulsory"—may require more thought and time.

Responding to comments that are complimentary and to points that the authors agree strengthen their papers is fairly easy. Responding to negative comments is much more difficult. Authors who disagree with the suggestion of a reviewer are not obligated to change their

FIGURE 42-3 Sample Responses to Reviewer Comments

Sample Comment	Sample Response(s)
The specific aims of this paper should be clearly stated early in the manuscript.	We have edited the final paragraph of the introduction section to make it clear that the three specific aims of the paper are (1) to . . . , (2) to . . . , and (3) to . . .
The paragraph on . . . is unclear.	We have rewritten this paragraph to improve clarity and to emphasize . . .
Did your survey include a question about . . . ?	The data set we analyzed did not include a variable for . . . . However, even without that information, our analysis shows that . . .
	This would have been a helpful question to ask, but, unfortunately, it was not included in our questionnaire.
	We did not ask this question in the baseline survey presented in this paper, but we do plan to ask a question about . . . in our follow-up study next year. We agree that this will be an interesting question to explore.
	We did ask this question, and found . . . . We have added this finding to the results section.
The sample size seems too low to have adequate power for this study design.	We used . . . software before initiating our study to estimate our required sample size. With expected inputs of . . . and a power of 80%, a sample size of . . . was estimated to be required. In total we recruited . . . participants. Based on the results of our study, and our power calculation during data analysis, which showed . . . , our sample size has sufficient power to yield significant results.
Table 3 seems incomplete. It should also report the results of the . . . test for each row.	We have done the additional analysis requested and have added a new column to Table 3 that shows the results of the . . . test. What we found was . . . , which is consistent with the results of our other statistical tests.
You used the . . . test to analyze . . . , but a . . . test would be more appropriate.	The . . . test that we used is the appropriate test because . . . . The alternate . . . test is not appropriate because . . .

(continues)

FIGURE 42-3 Sample Responses to Reviewer Comments *(continued)*

Sample Comment	Sample Response(s)
In the discussion section, the authors claim . . . , but is it possible that . . . is happening instead?	Our assertion that . . . is happening is based on . . . . This interpretation is supported by several recent publications, including . . . . We have expanded our rationale for this conclusion in the discussion section and added additional references to previous literature.
	The reviewer raises a very interesting point. We agree that both of these interpretations are possible and now discuss both perspectives in the discussion section.
The conclusion about . . . is not supported by the data.	We have removed this claim. Our primary conclusion, which is fully supported by our results, is . . .
You should include a discussion of . . .	Thank you for raising this interesting point. We have added commentary on . . . to the discussion section. We agree that this is an interesting topic, but since . . . is only tangentially related to our specific aims, we do not have space to discuss it in this paper.
You need to add a paragraph on the limitations of the study.	We have added a paragraph on limitations to the discussion section: ". . ."
Several recent publications have addressed the themes of your work and should be cited, including . . . , . . . , and . . .	Thank you for bringing these articles to our attention. The articles by . . . and . . . were helpful in supporting our implications section and are now included as references.
I am not convinced that the study is important enough for publication in an international journal. It may be a better fit for a regional journal.	We have added an additional paragraph to the introduction that highlights what is new and significant about our findings. We have also added an additional paragraph to the discussion section that discusses the implications of our findings for other settings. We believe that our paper is important because . . .
There are typos in lines . . . and . . . on page . . .	Thank you for catching these typos. We have corrected both of them.

paper to suit the reviewer, but they do need to write a thoughtful and courteous explanation of their point of view.

- Sometimes a reviewer's comments are hard to decipher or vague, such as "The entire manuscript is lacking focus and clarity." An appropriate response is to refer to exactly where and how the paper has been improved.
- Sometimes a reviewer's comments exhibit a lack of comprehension. Although it is tempting (and sometimes accurate) to assume that the reviewer was reading carelessly, the authors should consider how that part of the manuscript might be revised to promote clarity for future readers.
- Sometimes two reviewers offer conflicting advice. The responses to both of the comments should summarize both reviewers' perspectives and explain how the authors decided to address the underlying issue in the manuscript.

The responses to the reviewers' comments should be prepared in a separate file from the revised manuscript. Once all of the documents associated with the revision have been compiled—including a new cover letter, the revised manuscript that highlights or tracks all of the changes made to the originally submitted file, and the file with the responses to reviewer comments—these files can be uploaded to the journal submission website. The cover letter should thank the editors for the opportunity to revise and resubmit, thank the reviewers for their comments and their advice that improved the paper, and affirm that each reviewer comment has been addressed and responded to.

The time needed for a second round of review ranges from a few days to several months, depending on how many parties are involved in the re-review. Journal editors rarely promise authors that revisions will be accepted. However, the likelihood of acceptance is usually strong. The editors would not request a revision unless they were seriously considering accepting and publishing the edited version. Some journals may ask for third or even fourth or more revisions, with each round strengthening the paper's arguments. This can be frustrating to authors, but it is also evidence of the editor's continued interest in accepting the paper for publication. Unless there is a very good reason to move on to another journal, the best option is to revise and resubmit to any journal that offers an "R&R."

42.5 After Acceptance

A **provisional acceptance** is a notification from a journal that a manuscript will be accepted once minor adjustments are submitted. If a provisional acceptance is offered, editors will often ask that the required updates be made within a short period of time, sometimes in as little as a few days or 1 week. It is important for the authors to be responsive to those requests, because failure to submit the required updates in a timely fashion may convert a pending acceptance to a rejection. After the editor receives the corrected manuscript, a final acceptance letter will be sent to the corresponding author, usually by email.

After a paper is formally accepted, it may be sent to a copyeditor, who checks the paper carefully for grammar, spelling, and adherence to the journal's style. **Copyediting** is the process of correcting errors or inconsistencies in grammar, syntax, terminology, spelling, and punctuation in a manuscript prior to publication. Some journals have a style manual for copyeditors that specifies the preferred phrases, terms, abbreviations, and spellings for articles published in that journal, and the copyeditor will ensure compliance with those standards.

The manuscript is then sent to a layout specialist, who formats the document to look like all the other articles published in the journal. **Page proofs** (also called **galley proofs**) are the copyedited and formatted version of a manuscript that is sent to a corresponding author for review prior to publication, usually as a PDF file. **Proofreading** is the process of confirming the quality of the almost-final version of a soon-to-be-printed manuscript. Copyediting focuses on content, while proofreading typically focuses on the appearance of the file. Authors are usually given only 1 to 3 days to meticulously proofread the formatted document, respond to any queries from the editor, and make any final requests for corrections. This is not the time to request any substantive changes. Requests for modifications should be limited to new problems, such as errors in the way a table is being displayed, and easy-to-fix changes, such as adding a missing "a" or "the" or correcting a misspelled word. It is not appropriate to rewrite paragraphs or add new commentary at the proofing stage. However, authors should read every line carefully, confirm every statistic in the text, examine every figure for clarity and crispness, inspect every table for alignment, and check details like the spelling of authors' names, the contact information provided for the corresponding author, and the completeness, correctness, and consistency of the items in the reference list. This is the last opportunity to identify and fix errors.

After the authors return the page proofs, the time to publication of the article depends on the publisher. **Advance access** is the availability of an article on a journal's website prior to the assignment of that article to a particular issue of the journal. Some journals post a **preprint**, an unformatted version of the manuscript, in their advance access sections shortly after acceptance. Some post corrected page proofs in an advance access section of their websites. Some journals do not post the article online in any format until it has been assigned to an issue and published in print form. An article may be published in an issue mere weeks after acceptance or many months after the page proofs are approved. Soon after the article is posted online or published, the abstract will be added to the databases that index the journal. The article may be cited for the first time in another article about a year or so after publication. At this point, the full research cycle is complete.

© DmitriyRazinkov/Shutterstock

Glossary

A

Abduction A mode of reasoning in which inferences are based on limited observations and minor premises, so the conclusions are assumed to be best guesses that are merely probable.

Absolute risk A term for the incidence rate that emphasizes that the number is a measured value in one population rather than a comparison of several observed values.

Abstract A one-paragraph summary of an article, chapter, or book.

Abstract database An online collection of abstracts that allows researchers to search for articles using keywords or other search terms.

Accuracy In a survey instrument, diagnostic test, or other assessment tool, a condition that is established when the responses or measurements are shown to be correct; also called *validity*.

Action research Qualitative research in which participants work together to solve a problem.

Active surveillance The process of public health officials contacting healthcare providers in their jurisdictions to ask how often the clinicians are diagnosing particular types of disease.

Adjusted statistic A statistic that has been corrected to account for the effects of one or more other variables.

Advance access The availability of an article on a journal's website prior to the assignment of that article to a particular issue of the journal.

Adverse event A negative outcome that may be the direct result of a study-related exposure or may be a coincidental occurrence that is not directly related to the study but happens after an individual receives a study-related exposure.

Adverse reaction A negative side effect of a medication, vaccination, or other exposure, or another bad outcome related to a study.

Affiliation The institution where an author was employed or enrolled when he or she was conducting or contributing to a research project.

Age adjustment Methods that improve the validity of comparisons of two or more populations with different age distributions.

Agent A pathogen or a chemical or physical cause of disease or injury.

Agent-based modeling A type of modeling that uses computers to simulate the actions and interactions of various individuals (agents) in a population; sometimes called *individual-based modeling*.

Age-specific rate A rate for a particular age group.

Age standardization The application of age-specific rates from one or more study populations to a "standard population," or vice versa, to generate comparable statistics for populations with different age structures.

Age-standardized statistic A fictitious statistic for a study population that is created by applying age-specific rates to or from a standard population.

Aggregate study A study that analyzes population-level data and does not include any individual-level data; also called a *correlational study*.

Allocation bias A form of bias that occurs as a result of nonrandom assignment of participants to experimental study groups.

Allowable costs Expenses that are approved for a funded grant or contract as opposed to items that are not acceptable according to the terms of the grant or contract.

Alpha A Greek letter (α) used to indicate the probability of a type 1 error.

Alternative hypothesis A statement describing the expected result if there truly is a difference between the two or more values being compared.

AMA style The citation and reference style recommended by the American Medical Association, which is widely used by medical and health science journals.

Analytic epidemiology Studies that seek to identify the risk factors for various adverse health outcomes or to test the effectiveness of interventions intended to improve health status.

ANCOVA Analysis of covariance, a statistical test that compares the means of a ratio/interval variable in two or more independent groups while controlling for one or more additional ratio/interval or nominal variables.

Annotated bibliography A list of related publications that includes, at the minimum, a full reference for the document being reviewed, a brief summary of the article or report, and a note about the resource's potential relevance to a new study.

Anonymity The inability of the identity of a participant to be discerned from his or her responses to a survey instrument or records in a database.

Anonymized data set A data file that has been stripped of all potentially identifying information, such as names, street addresses, and personal identification numbers.

ANOVA Analysis of variance, a statistical test that compares the mean values of a continuous variable across independent populations.

Anthropometry The measurement of the human body, such as the measurement of height, weight, waist circumference, and hip circumference.

APA style The citation and reference style recommended by the American Psychological Association, which is widely used by social science and nursing journals.

***A priori* codes** In qualitative analysis, codes developed before the start of data analysis.

AR% An abbreviation for *attributable risk percentage*.

Area under the curve (AUC) The area under a receiver operating characteristic (ROC) curve that displays the diagnostic accuracy of a test; AUC values range from 0 to 1, with 1 indicating a perfect test.

Arms The treatment and nontreatment groups of an experimental study.

Ascertainment bias A form of bias that occurs when the individuals sampled for a study are not representative of the source population as a whole; also called *sampling bias*.

Assent The expressed willingness to participate in a study by a child or another person who is deemed not legally competent to provide his or her own consent.

Association A statistical relationship between two variables; this term does not indicate anything about whether the relationship is or is not causal.

Assumption A premise that is presumed to be true.

Attack rate The cumulative incidence of infection during the course of an epidemic.

Attributable risk (AR) The absolute difference between the incidence rates in two independent populations (often an exposed group and an unexposed group in a cohort study); also called *excess risk*.

Attributable risk percentage (AR%) The proportion of incident cases among the exposed population in a cohort study that is due to the exposure.

AUC An abbreviation for *area under the curve*.

Audience For a journal, the readership that the publication intends to reach.

Audit A systematic check of financial records and other actions and decisions that is conducted to confirm accuracy and compliance with standards of practice.

Author guidelines Detailed directives from a journal about how manuscripts should be formatted prior to submission; also called *instructions for authors*.

Autocorrelation A pattern in which a variable measured over time has values influenced by its own past values, as per a Durbin-Watson test or another test statistic, or, in spatial analysis, a measurement of how similar one location is to nearby places.

Autonomy An ethical principle requiring that only an individual (or his or her legal guardians) is authorized to decide whether to volunteer to participate in a research study.

Axial coding Qualitative analysis that identifies one core category or core phenomenon and several related categories that express the major and minor themes of the analysis.

Axiology The study of values.

B

Back translation A translation approach in which one person translates a questionnaire from the original language to a new language and a second person then translates the survey instrument in the new language back into the original language to ensure that the correct meanings were conveyed in the translation; also called *double translation*.

Background section The first section of a scientific report, which presents foundational theories, provides important definitions, and spells out the study goals; also called an *introduction section*.

Bar chart A graph that presents categorical data using equal-width rectangles with lengths that are proportional to the values they represent.

Baseline An initial measurement used as a benchmark for examining changes over time.

Basic medical research Studies of molecules, genes, cells, and other smaller biological components related to human function and health.

Bayesian An approach to statistics that uses data and prior beliefs (the priors) to predict the likelihood of a particular outcome (the posterior).

Before-and-after study A nonrandomized experimental study that measures the same individuals before and after an intervention so that each participant's "before" status can serve as that individual's control.

Belmont Report A report published by the U.S. National Commission for the Protection of Human Subjects of Biomedical and Behavioral Research in 1979 that defined the key research principles of beneficence, respect for persons, and distributive justice.

Bench rejection The rejection of a manuscript from a journal without external peer review; also called *desk rejection*.

Beneficence The ethical imperative for a research study to maximize possible benefits and minimize possible harms.

Berkson's bias A form of bias that can occur when cases and controls for a study are recruited from hospitals and therefore are more likely than the general population to have comorbid conditions.

Beta A Greek letter (β) used to represent statistical power or to indicate the coefficients of predictor variables in a regression model.

Bias A systematic flaw in the design, conduct, or analysis of a study that can cause the results of a study not to accurately reflect the truth about the source population.

Bibliometrics Quantitative analyses of publications and citations.

Big data A term used to describe data sets that are so large and complex that they must be analyzed using powerful hardware and special statistical software applications.

Bimodal A numeric variable with a two-peaked distribution.

Binomial test A statistical test that compares the proportion expressed by a binomial variable to a selected value.

Binomial variable A categorical variable that has only two possible responses and has been coded as having values of only "0" or "1."

Bioinformatics The use of computer technologies to manage biological information.

Biosketch A brief summary of a person's professional and educational accomplishments.

Biostatistics The science of analyzing data and interpreting the results so that they can be applied to solving problems related to biology, health, environmental science, or related fields.

Bivariable analysis Statistical analyses such as rate ratios, odds ratios, and other comparative statistical tests that examine the relationship between two variables.

Blinding An experimental design element that keeps participants (and sometimes some members of the research team) from knowing whether a participant is in the active intervention group or the control group; also called *masking*.

Block randomization An allocation method that randomly assigns some groups of people to an intervention group and other groups of people to a control group; randomization occurs at the group rather than individual level.

Boolean operators Conjunctions such as AND, OR, and NOT that define relationships between search terms.

Boxplot A graphical depiction of a numeric variable that displays the median, the interquartile range, and any outliers; also called a *box-and-whisker plot*.

Bracketing The process of a researcher intentionally setting aside any preconceived ideas about reality in order to be open to new meanings that might be expressed by participants.

Bradford Hill criteria A set of conditions that provide support for the existence of a causal relationship between an exposure and an outcome.

Brainstorming A process of generating long lists of spontaneous ideas about possible research questions.

Breslow-Day test A test for assessing the homogeneity of stratum-specific measures of association.

Burden of disease The adverse impact of a particular health condition or group of conditions on a population.

Canonical analysis A statistical method that identifies the set of ratio/interval and/or nominal variables that most accurately predicts group membership in a model with two ratio/interval and/or nominal dependent variables.

Carryover effects Residual effects from the first part of an experimental study that may bias the results of the second part of a crossover study if a sufficient washout period between the two arms of the study is not implemented.

Case A study participant with the infectious or parasitic disease, noncommunicable disease, neuropsychiatric condition, injury, or other disease, disability, or health condition of interest.

Case–control study A study that compares the exposure histories of people with disease (cases) and people without disease (controls).

Case definition A list of the inclusion and exclusion criteria that must be met in order for an individual to be classified as a person with the disease of interest in a case series, a case–control study, or another type of study.

Case detection rate (CDR) The proportion of people with a disease who are diagnosed as having that disease.

Case fatality rate (CFR) The proportion of people with a particular disease who die as a result of that condition.

Case report A report that describes one patient.

Case series A report that describes a group of individuals who have the same disease or disorder or who have undergone the same procedure.

Case study A qualitative research approach that uses multiple data sources to examine one person, group, event, or other situation in detail.

Categorical variable A variable for which the values have no inherent rank or order; also called a *nominal variable*.

Category In qualitative analysis, a group of related codes.

Causal factor An exposure that has been scientifically tested and shown to occur before the disease outcome and to contribute directly to its occurrence.

Causation A relationship in which an exposure directly causes an outcome; the presence of causality is usually determined with both quantitative analysis and a qualitative consideration of causal theory using the Bradford Hill criteria or other guidelines.

CBPR An abbreviation for *community-based participatory research*.

CEA An abbreviation for *cost-effectiveness analysis*.

Censoring Removing from further analysis participants in a prospective or longitudinal study who die, drop out, or are lost to follow-up for another reason.

Census A complete enumeration of a population.

Central tendency The average value for a numeric variable, such as a mean or median.

Certificate of confidentiality A legal document that protects the identity of participants in a study of sensitive topics from being subject to court orders and other legal demands for information.

Chance A random event that occurs by happenstance rather than design.

Chi-square goodness-of-fit test A statistical test that compares the proportion of responses to a nominal variable to a selected value.

Chi-square test A statistical test that compares the value of a nominal variable in two or more independent populations.

Clinical research Evaluations of the best ways to prevent, diagnose, and treat adverse health issues that affect individuals and families.

Closed-ended questions Survey or interview questions that allow a limited number of possible responses.

Closeout The process that determines that all applicable administrative actions and all required work for an award have been completed by the grantee.

Cluster analysis The identification of similar observations using an algorithm that seeks to minimize the variations among observations within each group.

Coauthorship The process of two or more collaborators working together to write a research report.

Cochran's Q statistic A statistic used to examine heterogeneity among the studies included in a meta-analysis.

Cochran's *Q* test A statistical test that compares the values of frequencies or proportions in three or more matched sets of binomial or nominal data.

Code In qualitative analysis, a label attached to a word or phrase.

Codebook A guide written for a particular study that describes each variable and specifies how the collected information will be entered into a computer file.

Coding In qualitative analysis, the use of words or short phrases to briefly summarize the contents, attitudes, processes, or other aspects of each item in a transcript or other qualitative document; also called *indexing*.

Coefficient of determination (r^2) A statistic that shows how strong a correlation is without indicating the direction of the association; r^2 values range from 0 to 1, with 1 indicating perfect correlation.

Coercion Compelling an individual to participate in a research study; this is a violation of the principles of autonomy and respect for persons.

Cohen's *d* A statistic used to evaluate whether the result of a *t* test comparing means is significant enough to be meaningful in applied practice.

Cohen's kappa A statistical measure that determines whether two assessors who evaluated the same study participants agreed more often than is expected by chance.

Coherence The quality of being logical and consistent.

Cohort A group of similar people followed through time together.

Cohort study An observational study that follows people forward in time so that the rate of incident (new) cases of disease can be measured.

COI An abbreviation for *conflict of interest*.

Committee on Publication Ethics (COPE) An organization that provides guidance on how to avoid research misconduct during the dissemination phase of a research project.

Common knowledge Information that should be familiar to a typical person working in that research area; also called *general knowledge*.

Common Rule The U.S. Federal Policy for the Protection of Human Subjects.

Community-Based Participatory Research (CBPR) Research partnerships in which academicians and community representatives work together to identify research priorities and conduct applied research in a community.

Comorbidity Two or more adverse health conditions occurring at the same time.

Comparative statistics Tests that compare the characteristics of two or more independent populations or compare the before-and-after characteristics of a study population being followed forward in time.

Compartmental model A mathematical model in which each individual in the simulated population exists in only one of several states at one time, but over time these individuals can move between states.

Concept A theory informed by observations.

Concept mapping A visual method for listing ideas and then grouping them to reveal relationships; this technique can be useful when identifying a study question and as part of narrative analysis of qualitative data.

Conceptual framework A model that a researcher sketches out using boxes and arrows to illustrate the various relationships that will be evaluated during a study.

Concordance Agreement.

Concrete validity Survey instrument validity demonstrated when an established test or outcome is used as a standard for confirming the utility of a new test that examines a similar theoretical construct; also called *criterion validity*.

Concurrent validity Survey instrument validity demonstrated when participants complete both an existing test and a new test and the correlation between the test results is strong.

Conditional probability The probability of an event occurring given that some prior event has already occurred.

Conference paper An article-length research report published in the proceedings of a conference.

Confidence interval A statistical estimate of the range of likely values of a statistic in a source population based on the value of that statistic in a study population; a narrow CI indicates more certainty about the value than a wide CI.

Confidentiality The protection of personal information provided to researchers.

Confirmability An indicator of neutrality that is present when the results of a qualitative study are shown not to be due to researcher bias.

Conflict of interest A financial or other relationship that could influence the design, conduct, analysis, or reporting of the study, or could appear to have caused bias.

Confounder A third variable that is associated with both the exposure variable and the outcome variable and distorts the apparent relationship between the exposure and outcome.

Constant comparison A process in which qualitative data are collected and analyzed simultaneously.

Construct A theory informed by complex abstractions and not merely by observations.

Construct validity Survey instrument validity demonstrated when a set of questions measures the theoretical construct the tool is intended to assess.

Constructivism A qualitative research paradigm in which researchers have a relativist perspective that considers each individual's reality to be a function of that person's lived experiences.

Content analysis The process of categorizing textual data.

Content validity Survey instrument validity demonstrated when subject-matter experts agree that a set of survey items captures the most relevant information about the study domain; also called *logical validity*.

Contingency question A survey question that determines whether the respondent is eligible to answer a subsequent question or set of questions; also called a *filter question*.

Contingency table A row-by-column table that displays the counts of how often various combinations of events happen; also called a *crosstab*.

Continuing education The completion of approved learning activities in order to maintain a professional licensure or credential.

Continuous variable A numeric variable that can take on any value within a range.

Contract Research funding that requires the researcher to deliver an agreed-upon product to the funder.

Control A participant in a case–control study who does not have the disease being examined or a participant in an experimental study assigned not to receive the active intervention.

Control definition A list of all of the eligibility criteria for inclusion in a comparison population.

Controlled observation A method of qualitative field observation in which a researcher observes study participants in a laboratory setting.

Controlled trial An experiment in which some of the participants are assigned to an intervention group and some are assigned to a nonactive comparison group.

Convenience population A nonprobability-based source population selected due to ease of access to those individuals, schools, workplaces, organizations, or communities.

Convergent validity Survey instrument validity demonstrated when two items that an underlying theory says should be related are shown to be correlated.

Copyediting The process of correcting errors or inconsistencies in grammar, syntax, terminology, spelling, and punctuation in a manuscript prior to publication.

Copyright The legal rights assigned to the owners of intellectual property such as written and artistic works.

Correlation A statistical measure of the degree to which changes in the value of one variable predict changes in the value of another.

Correlational study A study that uses population-level data to look for associations between two or more characteristics that have been measured in several groups.

Corresponding author The coauthor who will communicate with journal editors and take the lead on answering questions from readers after a paper is published.

Corrigendum A published correction to a minor error in an article that was caused by the author rather than the publisher.

Cost-effectiveness analysis An economic analysis that compares the health gains from an intervention to the financial costs of that intervention.

Count A number that enumerates the quantity of similar items.

Covariance A measure of the joint variability between two random variables.

Cover letter A letter that accompanies a submitted proposal or manuscript in order to briefly explain the importance of the work.

Covert observation Qualitative research in which the researcher does not inform study subjects that they are being investigated.

Cox proportional hazards regression A type of regression model used for survival analysis that estimates a hazard ratio comparing the duration of times to an event in two populations.

CPT codes Current Procedural Terminology codes published by the American Medical Association.

Cramér's V A statistical measure of the degree to which changes in the value of one categorical variable predict changes in the value of another categorical variable.

Creative Commons (CC) license A public copyright license that enables the free distribution of a copyrighted work.

Credibility An indicator of quality assurance that is present when the interpretation of qualitative data accurately reflects the studied groups or texts.

Criterion validity Survey instrument validity demonstrated when an established test or outcome is used as a standard for confirming the utility of a new test that examines a similar theoretical construct; also called *concrete validity*.

Critical theory A qualitative research paradigm that considers reality to be dependent on social and historical constructs and assumes that reality can be uncovered by identifying and challenging power structures.

Cronbach's alpha A measure of internal consistency that is used with variables that have ordered responses.

Crossover design An experimental study design in which each participant serves as his or her own control; some participants are assigned to receive the active intervention first and then the control, and others are assigned to receive the control first and then the active intervention.

Cross-sectional study A study that measures the proportion of members of a population who have a particular exposure or disease at a particular point in time; also called a *prevalence study*.

Crosstab A row-by-column table that displays the counts of how often various combinations of events happen; also called a *contingency table*.

Crude statistic A raw or unadjusted statistic.

Cultural competency The ability to communicate effectively with people from different cultures and backgrounds.

Culture A way of living, believing, behaving, communicating, and understanding the world that is shared by members of a social unit.

Cumulative incidence The percentage of people at risk in a population who develop new disease during a specified period of time; also called the *incidence proportion*.

Cumulative probability The probability of an event occurring by the end of a particular observation period.

Cutpoint A value that divides a numeric variable into separate categories; also called a *threshold*.

D

DALY An abbreviation for a *d*isability-*a*djusted *l*ife *y*ear.

Data Raw or unprocessed facts, figures, symbols, or signs.

Database A data management system that stores data in tables in which each row represents one record, and related records in different tables can be linked.

Data cleaning The process of correcting any typographical or other errors in data files.

Data management The entire process of record keeping before, during, and after a research study.

Data mining The process of examining big data sets to identify patterns and develop new knowledge.

Data saturation In qualitative research, the time in the research process in which no new information about a particular theory is emerging from additional data collection because variations across population members have already been captured.

Data science An interdisciplinary field that uses statistics, machine learning, and other types of computational tools to generate information and knowledge from various types of data.

Data security The process of protecting computer files with passwords and other mechanisms for restricting unauthorized access and use.

Data sharing The willingness of a research team to make their data and methods freely available to other researchers.

Deception In research, the intentional misleading of research participants about the true purpose and procedures of a study.

Declaration of Helsinki A document written by the World Medical Association in 1964 to provide ethical guidelines for clinicians conducting experimental studies.

Deduction A mode of reasoning in which logical inferences are based on facts or premises, and the conclusions are assumed to be certain.

Degrees of freedom The number of values in the calculation of a statistic that are free to vary.

Deidentification The process of removing potentially identifying information from a data file so that the data can be shared with others without violating the privacy of the individuals whose data are included in the file.

Deliverable A tangible or intangible object produced to fulfill the terms of a contract-funded research project.

Delphi method A structured consensus-building method in which experts complete questionnaires, a facilitator summarizes and shares the responses, and panelists reconsider their perspectives after reflecting on the opinions expressed by others.

Demography The study of the size and composition of populations and of population dynamics, such as birth and death rates.

Denominator The bottom number in a ratio; that is, the "B" in the ratio "A/B."

Dependability An indicator of consistency in a qualitative study that is demonstrated through transparency about data collection, analysis, and interpretation methods.

Dependent variable A variable in a statistical model that represents the output or outcome for which the variation is being studied; also called an *outcome variable*.

Derived variable A new variable created during data analysis from existing variables in the data file.

Descriptive epidemiology Studies that quantify how often various health-related exposures and outcomes occur in a population and characterize the person, place, and time factors associated with adverse health outcomes.

Descriptive statistics Statistics that describe the basic characteristics of quantitative data, such as means and proportions.

Desk rejection The rejection of a manuscript from a journal without external peer review; also called *bench rejection*.

Detection bias A form of bias that occurs when a population group that is routinely screened for adverse health conditions incorrectly appears to have a higher-than-typical rate of disease because more frequent testing enables a higher case detection rate in that population than in the general population; also called *surveillance bias*.

Determinants of health Biological, behavioral, social, environmental, political, and other factors that influence the health status of individuals and populations.

Deterministic model A mathematical model in which the outcomes are the same every time the model is run with the same inputs.

Diagnostic accuracy The percentage of individuals who a diagnostic test correctly classifies as true positives or true negatives.

Dichotomous variable A categorical variable with only two possible answers.

Digital object identifier (DOI) An alphanumeric code assigned to a document by a registration body to allow quick online access to the document or its abstract.

Direct age adjustment A method of age standardization that applies age-specific rates in two or more study populations with different age structures to one standard population so that the rates in the study populations can be more fairly compared.

Direct costs The specific monetary expenses associated with a particular research project.

Directed acyclic graph A model that uses nodes and arrows to illustrate hypothesized causal pathways from distal exposures to proximal exposures to outcomes.

Disability-adjusted life year (DALY) A burden of disease metric that is quantified as the sum of years lived with disability (YLDs) and years of life lost (YLLs) to premature death in a population.

Discordance Disagreement.

Discourse analysis The use of tools of linguistics to evaluate the ordinary use of written and spoken language.

Discrete variable A numeric variable that is not continuous.

Discriminant analysis A statistical method that identifies the set of ratio/interval and/or nominal variables that most accurately predicts group membership in a model with a nominal dependent variable; also called *discriminant function analysis*.

Discriminant function analysis A statistical method that identifies the set of ratio/interval and/or nominal variables that most accurately predicts group membership in a model with a nominal dependent variable; also called *discriminant analysis*.

Discriminant validity Survey instrument validity demonstrated when two items that a construct says should not be related are shown not to be associated.

Discrimination The ability of a statistical model to distinguish between independent groups.

Discussion section The final section of a typical four-part scientific report, which compares the new findings to the prior literature on the topic, acknowledges the limitations of the study, and summarizes the implications and conclusions of the study.

Disease The presence of signs or symptoms of poor health, or the particular adverse health outcome that is the focus of a health science study.

Disorder A functional impairment that may or may not be characterized by measurable structural or physiological changes.

Dispersion A measure that describes the variability and distribution of responses to a numeric variable; also called *spread*.

Distributive justice A principle of research ethics that requires the benefits and burdens of research to be fairly allocated.

DOI An abbreviation for *digital object identifier*.

Double-blind An experimental study design in which neither the participants nor the researchers assessing the participants' health status know which participants are in an active or control group; in publishing, a peer-review process in which the reviewers do not know the identity of the authors and the authors do not learn the identity of the reviewers.

Double-entry A method for ensuring the accuracy of a data file by having two individuals enter the same data into separate computer files, comparing the two files for agreement, and resolving any discrepancies.

Double translation A translation approach in which one person translates a questionnaire from the original language to a new language and a second person then translates the survey instrument in the new language back into the original language to ensure that the correct meanings were conveyed in the translation; also called *back translation*.

Dummy variables Derived variables created by recoding one variable with *n* categorical responses into a set of *n* – *1* dichotomous (0/1) variables.

Duplicate publication A form of research misconduct in which authors publish a second paper that is identical or very similar to a paper they have already had published.

Duplicate submission The act of submitting the same manuscript to two or more journals at the same time, a practice that is not permitted in the health sciences.

Dynamic population A study population with rolling enrollment that allows new participants to be recruited after the study begins collecting data; also called an *open population*.

E

Ecological fallacy The incorrect assumption that individuals follow the trends observed in population-level data.

Ecological study A correlational study that explores an environmental exposure, such as distance from the equator or level of air pollution.

EDPs An abbreviation for exposures, diseases/outcomes, and populations, which can be combined to form study questions using a standard format of "Is [exposure] related to [disease/outcome] in [population]?"

Effectiveness A measure of the success of an intervention under real-world conditions.

Effect modifier A third variable that defines groups of individuals who experience different biological responses to various exposures

Effect size The magnitude of the difference in the value of a statistic in independent populations.

Efficacy A measure of the success of an intervention that is calculated as the proportion of individuals in the control group who experienced an unfavorable outcome but could have expected to have a favorable outcome if they had been assigned to the active group instead of the control group.

Efficiency An evaluation of the cost-effectiveness of an intervention that is based on both its effectiveness and resource considerations.

EHR An abbreviation for *electronic health record*.

Eigenvalues The proportion of variance accounted for by the correlation between pairs of canonical variates during canonical analysis.

Electronic health record (EHR) A digital version of a patient's health information that is designed to be shared among different healthcare providers.

Electronic medical record (EMR) A digital version of a patient's medical history and other details recorded at one healthcare provider's office.

Eligibility criteria The inclusion criteria that must be present for an individual (or, for a systematic review, a research manuscript) to be allowed to participate in a study and the exclusion criteria that require an individual (or manuscript) to be removed from the study population.

Emergent codes In qualitative analysis, concepts that are identified during the early stages of analysis and assigned a label or code that describes them.

Emic perspective In ethnography, a study that aims to develop an insider's view.

Empiricism The assumption that the senses (such as seeing, hearing, and touching) are the best way to measure truth about the world.

EMR An abbreviation for *electronic medical record*.

End matter Information that some journals place between the end of the main text of an article and the start of the reference list, such as acknowledgments of funders and disclosures about possible conflicts of interest.

Environment External natural, physical, social, or political factors that facilitate or inhibit health.

Epidemiology The study of the distribution and determinants of health and disease in human populations.

Epistemology The study of knowledge and how an investigator knows what is real and true.

Equipoise A research principle that requires experimental research to be conducted only when there is genuine uncertainty about which treatment will work better.

Equivalence trial An experimental study that aims to demonstrate that a new intervention is as good as some type of comparison.

Erratum A published correction to a minor error in an article that was introduced during the publishing process.

Error A difference between the value obtained from a study population and the true value in the larger population from which the study participants were drawn that occurs by chance rather than as a result of systematic bias.

Ethnography The systematic study of people and cultures in their natural environments.

Etic perspective In ethnography, a study that aims to develop an outsider's view.

Etiology The cause of a disease or other health disorder.

Evaluation An assessment process that includes a variety of approaches for examining how well a project, program, or policy has achieved its associated goals, processes, and/or outcomes.

Evidence-based medicine (EBM) The use of rigorous research studies to optimize clinical decision making.

Evidence-based practice The integration of research into professional decision-making processes.

Excess risk The absolute difference between the incidence rates in two independent populations (often an exposed group and an unexposed group in a cohort study); also called *attributable risk*.

Exclusion bias A form of bias that occurs when different eligibility criteria are applied to cases and controls, such as when controls with health conditions related to an exposure are excluded but cases with those comorbidities are not excluded.

Exemption from review A determination by an institutional review board that a research protocol does not require review by the full committee because it does not meet the committee's definition of human subjects research.

Expedited review A determination by an institutional review board that a proposal requires review but a review by the full committee is not required because a minor change to a previously approved protocol is being requested or because a new proposal will not expose participants to risks greater than those encountered in ordinary daily life or during routine clinical examinations or procedures.

Experimental study A study that assigns participants to receive a particular exposure; also called an *intervention study*.

Explanatory research Investigations that test hypotheses about causal relationships.

Exploratory research Investigations that aim to discover new ideas and develop hypotheses.

Exposure A personal characteristic, behavior, environmental encounter, or intervention that might change the likelihood of developing a health condition.

External grant A grant funded by an organization outside the researcher's institution.

External validity The likelihood that the results of a study with internal validity can be generalized to other populations, places, and times.

Extraneous variable A third variable that produces an apparent but false association between two other variables that are not causally related.

F

F&A costs Facilities and administrative costs.

Fabrication A form of research misconduct involving the creation of fake data, such as creating fictitious rows of data in a spreadsheet for people who never completed a questionnaire or never participated in an experiment.

Face validity Survey instrument validity demonstrated when content experts and users agree that a survey instrument will be easy for study participants to understand and correctly complete.

Factor analysis A statistical method that uses measured variables to model a latent variable that represents a construct that cannot be directly measured with one question but appears to have a causal relationship with a set of measured variables.

Factorial ANOVA A test that compares the mean values of an interval/ratio variable across groups that are defined by two different variables (such as both sex and smoking status); also called *two-way ANOVA*.

Factorial design An experimental design that tests several different interventions in various combinations within one trial.

False negative rate The proportion of people who actually have a disease (according to a reference standard) who incorrectly test negative with a screening or diagnostic test.

False positive rate The proportion of people who actually do not have a disease (according to a reference standard) who incorrectly test positive with a screening or diagnostic test.

Falsification A form of research misconduct involving the misrepresentation of research results, such as modifying extreme data values to improve the results of statistical tests, manipulating photographs or other images collected during laboratory work, or intentionally misreporting a study's methods to make the study look more rigorous than it was.

Feasibility study An evaluation of the likelihood that a task can be completed with the time, money, technology, and other resources that are available for the activity.

Federalwide assurance (FWA) A status that applies to institutional review boards that are registered with the U.S. federal government.

Field notes Observation records, interview transcripts, and other documents compiled during the qualitative research process.

Figure In a research report, the visual presentation of key findings in the form of a diagram, flowchart, drawing, map, photograph, or other graphic.

Filter question A survey question that determines whether the respondent is eligible to answer a subsequent question or set of questions; also called a *contingency question*.

FINER An acronym reminding a researcher that a good research plan is feasible, interesting, novel, ethical, and relevant.

First author Typically, the person who was the most involved in drafting a manuscript; also called the *lead author*.

Fisher's exact test A statistical test that compares the values of a binomial variable in two independent populations.

Fixed effects model A statistical model for meta-analysis that can be used to create a pooled estimate when there is little variability among the included studies.

Fixed population A prospective or longitudinal study design in which all participants start the study at the same time and no additional participants are added after the study's start date.

Focused coding Qualitative analysis that seeks to identify the most frequent and important categories.

Focus group A qualitative data gathering technique in which approximately 8 to 10 people spend 1 or 2 hours participating in a moderated discussion.

Forest plot A graphical display of the effect sizes of the studies included in a meta-analysis and the pooled statistic calculated from those statistics.

Formal sources Scholarly works that were critically reviewed before being disseminated by a publishing group.

Formative evaluation The needs assessments and feasibility studies conducted as part of developing a new intervention or modifying an existing one.

Free-response questions Survey or interview questions that allow an unlimited number of possible answers; also called *open-ended questions*.

Frequency matching A sampling design that ensures that cases and controls in a case–control study or exposed and unexposed participants in a cohort study have similar group-level demographic characteristics; also called *group matching*.

Frequentist An approach to statistics in which probability is based on the expected frequency of an event occurring over a long time period or if an experiment is repeated many times.

Friedman test A statistical test that compares the values of an ordinal/ratio variable across several time points or in several individually matched populations.

***F* test** A statistical test used to determine whether the mean values of an interval/ratio variable are different or not different across three or more independent populations.

Full review A determination by an institutional review board that the full committee must discuss a study protocol in order to ensure that the requirements for the protection of human subjects are met.

Funnel plot A graphical display of the results of the studies included in a meta-analysis that reveals the likelihood that publication bias has kept relevant studies with null results out of the formal literature.

FWA An abbreviation for *federalwide assurance*.

G

Galley proofs The copyedited and formatted version of a manuscript that is sent to a corresponding author for review prior to publication; also called *page proofs*.

Gantt chart A type of bar chart that visually displays a research timeline and marks critical calendar dates and deadlines.

Gaps in the literature Missing pieces of information in the scientific body of knowledge that a new study could fill.

Gaussian distribution A histogram with a bell-shaped curve that has one peak in the middle; also called *normal distribution*.

Generalizability The external validity of a study that allows its results to be considered applicable to a broader target audience.

General knowledge Information that should be familiar to a typical person working in a research area; also called *common knowledge*.

Geographic information system (GIS) A computer-based platform for mapping the locations of events, identifying spatial clusters, and testing complex spatial associations.

Ghost authorship Failure to include as a coauthor on a manuscript a contributor who has made a substantial intellectual contribution to a research project.

Gift authorship The addition of someone to the list of authors of a manuscript when that individual has not earned authorship according to disciplinary standards, such as those spelled out in the authorship criteria of the International Committee of Medical Journal Editors.

GIS An abbreviation for *geographic information system*.

Gold open access A publishing model in which authors (or their funders) pay to make a journal article freely available to readers on the Internet.

Gold standard A test that shows the actual presence of disease in affected people.

Goodness-of-fit A statistical test for how well real data match the values predicted by a model.

Grant continuation An extension of a grant that provides additional funding to continue the research project and expand it in new directions; also called *grant renewal*.

Grant renewal An extension of a grant that provides additional funding to continue the research project and expand it in new directions; also called *grant continuation*.

Graph An illustration of quantitative results, such as a scatterplot or a line graph that shows the values of a numeric variable over time.

Graphical abstract A single visual representation that displays the most important finding of a study in a format that can be readily disseminated through social media.

Green open access A publishing model in which authors are allowed to post a version of a published article on their personal websites or in institutional repositories after an embargo period of 1 year or longer.

Grey literature Research reports that are available in a format that is not indexed in databases of journal article abstracts.

Grounded theory A qualitative research approach that uses an inductive reasoning process to develop general theories that explain observed human behavior.

Group matching A sampling design that ensures that cases and controls in a case–control study or exposed and unexposed participants in a cohort study have similar group-level demographic characteristics; also called *frequency matching*.

H

Habituation An error that occurs when participants completing a questionnaire or interview become so accustomed to giving a particular response (like "agree . . . agree . . . agree . . .") that they continue to reply with the same response even when that does not reflect their true perspective.

Hand searching A literature review technique that involves scanning every article in the table of contents of selected volumes of relevant journals to see if any of those articles might be eligible for inclusion in a review.

Hawthorne effect A type of bias that occurs when participants in a study change their behavior for the better because they know they are being observed.

Hazard function An equation describing the conditional probability of an individual having an event (such as death) at a particular time given that the person has survived to that time.

Hazard ratio The ratio of two hazard functions, such as a comparison of the durations of time to an event (such as death) in two populations.

Health A state of complete physical, mental, and social well-being, and not merely the absence of disease or infirmity.

Health belief model A theoretical framework that considers health behavior change to be a function of perceived susceptibility to an adverse health outcome, perceived severity of the disease, perceived benefits of behavior change, perceived barriers to change, cues to action, and self-efficacy.

Health informatics The application of advanced techniques from information science and computer science to the compilation and analysis of health data.

Health Insurance Portability and Accountability Act (HIPAA) A set of regulations about patient protection that apply in the United States.

Health-related quality of life A multidimensional construct that captures an individual's perceived physical, mental, emotional, and social well-being and the perceived impact of health status on the quality of daily life.

Health research Investigation of health and disease or any of the biological, socioeconomic, environmental, and other factors that contribute to the presence or absence of physical, mental, and social health and well-being.

Health services research The examination of factors related to the types of health services and providers available to a population, the organization and financing of those health services, and the impact of governments and policies on population health.

Healthy worker bias A form of bias that can occur when participants are recruited from occupational populations and therefore are systematically healthier than the general population.

Hermeneutics The study of the interpretation of texts.

Heterogeneity Dissimilarity.

Heteroscedasticity The heterogeneity of variance among the variables in a linear regression model that is demonstrated when the distribution of residuals from a regression model across the length of the best-fit line is uneven.

Hierarchical model A multilevel, multivariable regression model that adjusts for different levels of exposure, such as adjusting for both census tract and county.

***h*-index** A bibliometric that indicates that an author has at least *h* publications that have each been cited at least *h* times.

Histogram A graphical representation of the distribution of ratio/interval data in which the *x*-axis shows the values of responses and the *y*-axis displays the count of the number of times each response appears in the data set.

Historic cohort study A cohort study that recruits participants based on data about their exposure status at some point in the past and typically also measures outcomes that have already occurred (but happened after the baseline exposures were established); also called a *retrospective cohort study*.

Homogeneity Similarity.

Homogeneous sampling Recruiting participants with similar backgrounds, experiences, or perspectives.

Homoscedasticity The homogeneity of variance among the variables in a linear regression model that is demonstrated by the even distribution of residuals from a regression model across the length of the best-fit line.

Host A human who is susceptible to an infection or another type of disease or injury.

House style A particular publisher's requirements for spelling, citation style, and other formatting details.

Hybrid journal A subscription journal that gives authors the option of paying to make their article freely available to all.

Hypothesis An informed assumption about the likely outcome of a well-designed investigation that can be tested using scientific methods.

I

***i*10 index** A count of the number of publications by an author that have been cited at least 10 times each.

I^2 statistic A statistic used to examine heterogeneity in the studies included in a meta-analysis that adjusts the *Q* statistic based on the number of studies being pooled.

ICD codes International Classification of Diseases codes, more formally called the *International Statistical Classification of Diseases and Related Health Problems*.

Illness A person's perception of his or her own experience of having an adverse health condition.

Impact evaluation The determination of whether an intervention achieved its objectives.

Impact factor An annual determination by the Clarivate Analytics company about the number of times a typical article in a particular journal is cited in its first year or two after publication.

Incidence The number of new cases of disease in a population during a specified period of time.

Incidence proportion The percentage of people at risk in a population who develop new disease during a specified period of time; also called the *cumulative incidence*.

Incidence rate The number of new cases of disease in a population during a specified period of time divided by the total number of people in the population who were at risk during that period.

Incidence rate ratio The most common measure of association for cohort studies, calculated as the incidence rate among the exposed divided by the incidence rate in the unexposed.

Independent populations Groups in which no individual is a member of more than one of the groups being compared; for example, in a case–control study, each participant must be either a case or a control, so the case and control populations are independent.

Independent variable A variable in a statistical model that predicts the value of some outcome variable; also called a *predictor variable*.

Independent-samples *t* test A statistical test that compares the mean values of a ratio/interval variable in two independent populations; also called *two-sample t test*.

In-depth interview A qualitative research technique in which an interviewer spends 1 or 2 hours interviewing a key informant using open-ended questions.

Indexing In qualitative analysis, the use of words or short phrases to briefly summarize the contents, attitudes, processes, or other aspects of each item in a transcript or other qualitative document; also called *coding*.

Indicator A variable used to measure performance, achievement, or change.

Indirect age adjustment A method of age standardization that applies age-specific rates in a standard population to a study population so that a determination can be made about whether the overall rate in the study population is greater or lesser than expected given the population's age distribution.

Indirect costs The general research-related expenses that institutions incur but cannot attribute to specific research projects.

Individual matching A sampling design that links each case in a case–control study or each exposed individual in a cohort study to one or more controls with similar characteristics, such as genetic siblings or community members with the same date of birth; also called *matched-pairs matching*.

Individual-based modeling A type of modeling that uses computers to simulate the actions and interactions of various individuals (agents) in a population; also called *agent-based modeling*.

Induction A mode of reasoning in which inferences are based on observations, and the conclusions are assumed to be likely.

Inferential statistics The use of statistics from a random sample of members of a population to make evidence-based assumptions about the values of parameters in the population as a whole.

Informal sources Webpages, factsheets, blogs, podcasts, and other sources of information that are not peer reviewed.

Information Data that have been processed and presented in a format usable for understanding a situation and making decisions.

Information bias A form of bias in an epidemiological study that arises due to systematic measurement error.

Informed consent An individual's voluntary decision to participate in a research study after reviewing essential information about the project.

Inquiry The process of finding answers to questions that arise from personal experiences.

Institutional Animal Care and Use Committee A committee that oversees research with animals and operates separately from an IRB.

Institutional review board (IRB) A research ethics committee responsible for protecting human subjects who participate in research studies.

Instructions for authors Detailed directives from a journal about how manuscripts should be formatted prior to submission; also called *author guidelines*.

Intention-to-treat analysis An analysis of experimental data that includes all participants, even if they were not fully compliant with their assigned intervention; also called *treatment-assigned analysis*.

Interaction A situation in which the effect of one predictor variable on an outcome variable depends on the presence or absence of a second predictor variable.

Intercorrelation A situation in which two or more related items in a survey instrument measure various aspects of the same concept.

Internal consistency A measure of how well the items in a survey instrument measure various aspects of the same concept; internal consistency can be assessed with Cronbach's alpha, KR-20, and other tests.

Internal grant Research funds provided by the researcher's school or employer.

Internal validity Evidence that a study measured what it intended to measure.

International Committee of Medical Journal Editors (ICMJE) An organization that provides guidelines about manuscript formatting and authorship criteria that are widely used in the health sciences.

Interobserver agreement The degree of concordance among independent raters assessing the same study participants; also called *inter-rater agreement*.

Interpretivism A paradigm in which researchers consider the reality in the social world to be different from reality in the natural world.

Interprofessionalism The ability to work and communicate well with colleagues in different practice areas in order to achieve a shared goal.

Interquartile range The range for the 25th to 75th percentiles of values for a numeric variable, which captures the middle 50% of responses.

Inter-rater agreement The degree of concordance among independent observers assessing the same study participants; also called *interobserver agreement*.

Interval variable A numeric variable for which a value of zero does not indicate the total absence of the characteristic (such as 0°F not meaning the complete absence of heat, since it is possible to measure a temperature lower than 0°F).

Intervention A strategic action intended to improve individual and/or population health status.

Intervention study A study in which participants are assigned to receive a particular exposure; also called *experimental study*.

Interview The process of a researcher verbally asking a participant questions and recording that person's responses.

Interviewer bias A form of information bias that occurs when interviewers systematically question cases and controls or exposed and unexposed members of a study population differently, such as probing individuals they believe to be cases or exposed individuals for more information but not doing the same for participants they believe to be controls.

Introduction section The first section of a scientific report, which presents foundational theories, provides important definitions, and spells out the study goals; also called a *background section*.

IQR An abbreviation for *interquartile range*.

IRB An abbreviation for *institutional review board*.

Iteration A repetitive process.

K

Kaplan-Meier plot A time graph that displays cumulative survival rates in a study population.

Kappa statistic A statistical measure that determines whether two assessors who evaluated the same study participants agreed more often than is expected by chance.

KAP survey A survey instrument that asks participants about their knowledge, attitudes (or beliefs or perceptions), and practices (or behaviors).

Kendall's tau A statistical measure of the degree to which changes in the value of one ordinal/rank variable predict changes in the value of another ordinal/rank variable.

Key informants Individuals selected to participate in a qualitative study because they have expertise relevant to the study question.

Keyword A word, MeSH term, or short phrase used in a database search.

Kinesiology The study of the mechanics, physiology, and psychology of body movement, function, and performance.

KR-20 An abbreviation for the *Kuder-Richardson Formula 20*.

Kruskal-Wallis *H* test A statistical test that compares the median values of an ordinal/rank variable in three or more independent populations.

Kuder-Richardson Formula 20 (KR-20) A measure of internal consistency used with binary variables.

Kurtosis A description of how peaked or flat a normal distribution is.

L

Last author In some fields, the position in the authorship list that designates the senior researcher in whose lab the work was conducted.

Lead author Typically, the person who was the most involved in drafting a manuscript; also called the *first author*.

Lead researcher Typically, the researcher who will do the majority of the work on a project.

Lead-time bias A form of bias that occurs when a screening test that enables early detection of an adverse health condition is incorrectly interpreted as prolonging survival with the condition.

Leptokurtic A bell-shaped distribution curve that is very peaked.

Letter of inquiry A letter a researcher sends to a potential funding organization to ask about whether a particular research idea might be of interest to the funder.

Letter of intent A letter that presents a preliminary research plan to a funding organization and states the intention to submit a full proposal.

Levene's test A statistical test of the homogeneity of the variances across different groups.

Life table An actuarial table that displays conditional and cumulative survival probabilities in a population.

Likelihood ratio tests Probability ratios used to evaluate the accuracy of screening and diagnostic tests.

Likert scale Ordered responses to a questionnaire item that asks participants to rank preferences numerically, such as by using a scale for which 1 indicates strong disagreement and 5 indicates strong agreement.

Linear regression A statistical model that is used when the outcome variable is a ratio or interval variable.

Logic model A visual representation of the hypothesized causal pathways that lead to an outcome of interest.

Logical validity Agreement by subject-matter experts that a set of survey items captures the most relevant information about the study domain; also called *content validity*.

Logistic regression A probability-based regression model used when the outcome variable is binomial; also called *logit regression*.

Logit regression A probability-based regression model used when the outcome variable in a regression model is binomial; also called *logistic regression*.

Log-rank test A statistical test that determines whether survival rates are longer in one population than another.

Longitudinal cohort study A study that follows a group of individuals who are representative members of a selected population forward in time but does not recruit them based on exposure status; also called a *panel study*.

Loss to follow-up Inability to continue tracking a participant in a prospective or longitudinal study because the person drops out, relocates, dies, or stops responding to study communication for another reason.

LR+ An abbreviation for the *positive likelihood ratio test*.

LR– An abbreviation for the *negative likelihood ratio test*.

Lurking variable A third variable that was not measured in a study but is affecting the apparent association between an exposure variable and an outcome variable.

M

M&E Monitoring and evaluation.

Machine learning A method of data analysis derived from artificial intelligence in which a computer "learns" more about patterns in a data set by running and rerunning many rounds of analysis.

Major revision An invitation by a journal to significantly update a manuscript in response to reviewer comments and then to submit the revised manuscript for another round of review by external peer reviewers.

MANCOVA Multivariate analysis of covariance, a statistical test that compares differences in group means across multiple dependent variables while controlling for one or more additional ratio/interval or nominal variables.

Mann-Whitney *U* test A statistical test that compares the median values of an ordinal/rank variable in two independent populations; also called a *Wilcoxon rank sum test*.

MANOVA Multivariate analysis of variance, a statistical test that compares differences in group means across multiple dependent variables.

Mantel-Haenszel A weighting method that adjusts the measure of association for an exposure variable and an outcome variable after using a third variable to stratify the data.

Markov chain Monte Carlo (MCMC) Stochastic processes that use algorithms to take samples from simulated probability distributions.

Masking An experimental design element that keeps participants (and sometimes also some members of the research team) from knowing whether a participant is in the active intervention group or the control group; also called *blinding*.

Matched-pairs matching A sampling design that links each case in a case–control study or each exposed individual in a cohort study to one or more controls with similar characteristics, such as genetic siblings or community members with the same date of birth; also called *individual matching*.

Matched-pairs odds ratio (OR_{mp}) A special kind of odds ratio for a matched-pairs case–control study that compares the number of pairs in which the case had the exposure and the control did not (in the numerator) to the number of pairs in which the control had the exposure and the case did not (in the denominator).

Matched-pairs *t* test A statistical test that compares the values of an interval/ratio variable in members of one population measured twice or among individually matched pairs from two different groups.

Matching The process of recruiting one or more controls who are demographically similar to each case in a case–control study or recruiting one or more unexposed individuals who are demographically similar to each exposed person in a cohort study.

Maximum The greatest (highest) numeric value for a variable in a data set.

Maximum likelihood estimate (MLE) The value of a coefficient in a logistic regression model that gives the model the greatest probability of matching the observed data.

McNemar's test A statistical test used to compare the values of a binomial or nominal variable in one population measured twice or among individually matched pairs from two different groups.

Mean A measure of the average value of a ratio or interval variable that is calculated by adding up all the values for a particular variable and dividing that sum by the total number of individuals with a value for the variable.

Measure of association A number that summarizes the relationship between an exposure and a disease outcome, such as an incidence rate ratio for a cohort study or an odds ratio for a case–control study.

Median A measure of the average value of a numeric variable that is identified by putting all the values for a particular variable in order from least to greatest and then finding the middle number.

Medicine The practice of preventing, diagnosing, and treating health problems in individuals and families.

Memoing The act of documenting personal reflections and impressions about observations, participants, experiences, codes, categories, and themes.

Mentorship A formal or informal relationship in which an experienced mentor offers professional development advice and guidance to a less experienced mentee.

MeSH (Medical Subject Headings) A dictionary that can be used for searches in MEDLINE and other databases.

Meta-analysis The calculation of a pooled statistic that combines the results of similar studies identified during a systematic review.

Meta-synthesis A tertiary analysis that integrates the results from several different qualitative studies.

Methods section The second section of a scientific report, which presents details about the processes used for data collection and analysis.

Metric A composite indicator derived from two or more other measures.

Minimum The least (lowest) numeric value for a variable in a data set.

Minor revision An invitation by a journal to make a limited set of manuscript updates in response to reviewer comments and then to resubmit the manuscript for editorial review and a final decision about acceptance.

Misclassification bias A form of bias that occurs when participants are not correctly categorized, such as when some controls in a case–control study are incorrectly classified as cases.

Mixed methods The use of both quantitative and qualitative techniques in one research study.

MLE An abbreviation for *maximum likelihood estimate*.

Mode The most frequently occurring value for a particular variable in a data set.

Modifiable risk factor A risk factor for a disease that can be avoided or mitigated.

Monitoring Ongoing assessment to ensure that a project or program is staying on track toward achieving predefined targets.

Moran's *I* coefficient A statistical test of spatial autocorrelation.

Morbidity Nonfatal illness.

Mortality Death.

Mortality rate The proportion of members of a population who die of any condition during a specified time period, typically expressed in units such as "per 100,000."

Multicausality A causal pathway in which many different risk factors or combinations of risk factors contribute to a disease occurring.

Multicollinearity A problem that occurs when two or more predictor variables in a multiple regression model are highly correlated, and that redundancy means the coefficients for one or more of those variables are highly inaccurate.

Multilevel model A multivariable regression model that adjusts for different levels of exposure, such as adjusting for both census tract and county.

Multiple linear regression A model that examines the relationships between several ratio/interval and/or nominal predictor variables and one ratio/interval outcome variable.

Multiple logistic regression A statistical method that examines the relationships between several ratio/interval and/or nominal predictor variables and the value of one nominal outcome variable.

Multivariable analysis Statistical tests such as multiple regression models that examine the relationships among three or more variables.

N

Narrative analysis A qualitative analysis method that seeks to understand personal stories.

Narrative inquiry Qualitative research that examines autobiographies, personal letters, family stories, and other records to understand how people frame their identities and social relationships.

Narrative review A tertiary analysis that provides a unique perspective about a topic by using evidence from the literature to support the author's commentary.

Natural experiment A research study in which the independent variable is not manipulated by the researcher but instead changes due to external forces.

Natural language processing A machine learning algorithm that is used in the analysis of qualitative and social media data to examine how people speak and write in real-life situations.

Naturalistic observation A method of qualitative field observation in which a researcher unobtrusively observes study subjects in a natural setting, typically without the knowledge of the subjects.

Negative likelihood ratio (LR–) test A statistic that examines whether a diagnostic test is good at predicting the absence of disease.

Negative predictive value (NPV) The proportion of people who test negative with a screening or diagnostic test who actually do not have the disease (according to a reference standard).

Nested case–control study A case–control study that uses the participants of a large longitudinal cohort study as the source population for both cases and controls.

NNH An abbreviation for *number needed to harm.*

NNT An abbreviation for *number needed to treat.*

No-cost extension An extension of the timeline for spending grant money that moves the closing date to a later time but provides no additional funding.

Nominal variable A categorical variable for which the values have no inherent rank or order.

Noninferiority trial An experimental study that aims to demonstrate that a new intervention is no worse than some type of comparison.

Nonmaleficence The ethical imperative for a research study to do no harm.

Nonmodifiable risk factor A risk factor for a disease that cannot be changed through health interventions, such as a person's age.

Nonparametric test A statistical test that does not make assumptions about the distributions of responses.

Nonrandom-sampling bias A form of bias that occurs when each individual in the source population does not have an equal chance of being selected for the sample population.

Nonrecursive model A causal analysis model in which causal pathways can be bidirectional.

Nonresponse bias A form of bias that occurs when the members of a sample population who agree to participate in a study are systematically different from nonparticipants.

Normal distribution A histogram with a bell-shaped curve that has one peak in the middle; also called a *Gaussian distribution*.

Null hypothesis A statement describing the expected result of a statistical test if there is no difference between the two or more values being compared.

Null result A statistical test that shows no statistically significant differences between populations or over time.

Number needed to harm (NNH) The number of people who would need to receive a particular treatment in order to expect that one of those people would have a particular adverse outcome.

Number needed to treat (NNT) The expected number of people who would have to receive a treatment to prevent an unfavorable outcome in one of those people.

Numerator The top number in a ratio; that is, the "A" in the ratio "A/B."

Nuremberg Code One of the first codes of research ethics, which in 1947 mandated voluntary consent for experimental studies of humans.

Objectivity The unbiased evaluation of facts.

Observational methods Qualitative research techniques that involve systematic observations of human actions and interactions.

Observational study A study in which no participants are intentionally exposed to an intervention or asked to change their behavior.

Observer bias A form of bias that occurs when a researcher intentionally or unintentionally evaluates participants differently based on their group membership, such as systematically evaluating cases and controls in a case–control study differently.

Odds The ratio of the likelihood of an event happening and the likelihood of that event not occurring; for example, when a person has a 25% likelihood of developing a particular disease and a 75% likelihood of not developing that disease, the odds are 25%/75% = 0.33.

Odds ratio (OR) A ratio of odds in which the denominator represents the reference group; for a case–control study, the ratio of the odds of exposure among cases (in the numerator) to the odds of exposure among controls (in the denominator).

OLS An abbreviation for *ordinary least squares*.

One-sample *t* test A statistical test that compares the mean value of a ratio/interval variable to a selected value.

One-sided *p* value The probability value that is used for a statistical test when a direction is specified in the alternative hypothesis.

One-way ANOVA A statistical test that compares the mean values of one interval/ratio variable across three or more independent populations.

Ontology The study of the nature of reality and truth.

Open-access fee A charge that authors pay to a journal in order to make their articles freely available to online readers.

Open-access journal A journal that mandates that authors pay a publication fee before their manuscripts are published.

Open-ended questions Survey or interview questions that allow an unlimited number of possible responses; also called *free-response questions*.

Open population A study population with rolling enrollment that allows new participants to be recruited after the study begins collecting data; also called a *dynamic population*.

OR An abbreviation for *odds ratio*.

Oral consent Informed consent for participation in a study that is spoken and witnessed rather than requiring a participant's signature; also called *verbal consent*.

Oral history The audiovisual recording of historical information about individuals, families, groups, or events.

Oral presentation Conference presentation in which one individual speaks to a group, typically for about 15 minutes.

ORCID Open researcher and contributor identifier, a 16-digit number that serves as a unique persistent identifier for a researcher.

Ordinal variable A variable with responses that span from best to worst, most to least, or always to never or that are expressed using other types of ranked scales.

Ordinary differential equation (ODE) An equation that includes one or more functions of one independent variable along with the derivatives of those functions.

Ordinary least squares (OLS) A linear regression modeling approach that finds the line that minimizes the average vertical distance from each point in a data set to the fitted line.

Originality The aspects of a new research project that are novel and will allow it to make a unique contribution to the health science literature.

Outcome An observed event such as the presence of disease in participants in an observational study or the measured endpoint in an experimental study.

Outcome evaluation Processes that examine whether an ongoing intervention is making good progress toward achieving stated objectives.

Outcome variable A variable in a statistical model that represents the output or outcome whose variation is being studied; also called a *dependent variable*.

Outlier A value in a numeric data set that is distant from other observations and outside the expected range of values.

Overhead The institutional costs of maintaining research infrastructure, operating research facilities, and administering compliance activities.

Overmatching The recruiting challenges and possible statistical bias that can result from matching too many characteristics of the cases and controls in a case–control study or the exposed and unexposed participants in a cohort study.

Overt observation Qualitative research in which the participants are aware that they are being observed and the researcher is transparent about the goals and methods of the study.

P

Page charge A per-page fee that some subscription journals mandate prior to an accepted article being published; also called *page fee*.

Page fee A per-page charge that some subscription journals mandate prior to an accepted article being published; also called *page charge*.

Page proofs The copyedited and formatted version of a manuscript that is sent to a corresponding author for review prior to publication; also called *galley proofs*.

Paired data Variables linked together for analysis because they were gathered from individuals who were matched on specific characteristics (such as genetic siblings) or they were gathered from one individual at two or more points in time (such as at baseline and after an intervention).

Panel study A research study that measures participants or samples of participants at multiple points in time.

Parameter A measurable numeric characteristic of a population.

Parametric test A test that assumes the variables being examined have particular distributions, often requiring the variables to have normal or approximately normal distributions.

Parsimony The principle that when two models are equally good, the one that is simpler or more economical should be used.

Participant–observation A method of qualitative field observation in which a trained investigator seeks to understand a community by engaging with its members and immersing in its practices.

Participation rate The percentage of members of a sample population who are included in the study population.

Passive surveillance The compilation of reports of notifiable disease diagnoses submitted by medical laboratories.

Path analysis A recursive causal analysis strategy that uses regression models to examine unidirectional causal patterns among variables.

Pattern coding Qualitative analysis that seeks to group codes into a limited number of categories.

Pearson correlation coefficient A statistical measure of the degree to which changes in the value of one ratio/interval variable predict changes in the value of another ratio/interval variable.

Percentage A proportion or other type of ratio presented in units of "per 100."

Percentile The percentage of all observations in a data set that are below a particular individual's value for a variable.

Period prevalence The proportion of a population with a particular characteristic at one point in time.

Person–time A way of accounting for individual members of a study population participating in the study for different lengths of time that uses units like person-years, person-months, or person-days to quantify how long participants in a study were observed.

Phenomenology A qualitative research approach that seeks to understand how individuals understand, interpret, and find meaning in their own unique life experiences and feelings.

Phenomenon The central concept being studied during a qualitative research project.

Phi coefficient A statistical measure of the degree to which changes in the value of one binomial variable predict changes in the value of another binomial variable.

Photovoice A qualitative research technique in which participants take photographs that they feel represent their communities and then they share what aspects of their lived experiences they intended to capture in those images.

PICOT A framework of patient/population, intervention, comparison, outcome, and time frame that is helpful for developing clinical research questions and designing intervention studies.

Pie chart A circle in which each wedge or slice displays the percentage of participants who provided a particular answer to one question; the sum of the percentages for the slices must add up to 100%.

Pilot test A small-scale preliminary study conducted to evaluate the feasibility of a full-scale research project.

Placebo An inactive comparison that is similar to the therapy being tested in an experimental study, such as a sugar pill used as a control for a pill with an active medication, a saline injection used as a control for an injection of an active substance, and a sham procedure that is designed to look and feel like a real clinical procedure used as a control for that active procedure.

Plagiarism The use of someone else's ideas, words, images, or creative work without proper attribution.

Platykurtic A bell-shaped distribution curve that is relatively flat.

Pluralism In qualitative research, drawing on more than one theoretical framework to guide the design, analysis, and interpretation of a research project.

Point estimate The value of a statistic in a study population, which is typically presented along with a corresponding 95% confidence interval that provides additional information about the likely value of the statistic in the source population.

Point prevalence The proportion of a population with a particular characteristic at some point during a defined time period.

Policy A set of principles and procedures defined by governments or other groups to guide decision making and resource allocation.

Population A group (or subgroup) of individuals, communities, or organizations; research studies usually carefully identify a target population, source population, sample population, and study population.

Population at risk People who do not have the disease being tracked in a cohort study.

Population attributable risk The rate of new disease in a population that can be attributed to some people in the population having an exposure.

Population attributable risk percentage (PAR%) The proportion of incident cases among the total population that can be attributed to some people having the exposure.

Population-based study A study that uses a random sampling method to generate a sample population that is representative of a well-defined larger population.

Population health The health outcomes and determinants of health in groups of humans.

Population health research Health research that examines health outcomes at the community, regional, national, and worldwide levels.

Positive likelihood ratio (LR+) test A statistic that examines whether a diagnostic test is good at predicting the presence of disease.

Positive predictive value (PPV) The proportion of people who test positive with a screening or diagnostic test who actually have the disease (according to a reference standard).

Positivism A paradigm in which researchers apply a realist perspective that assumes that reality is knowable and that inquiry should be logical and value-free.

Poster session A designated time during an academic or professional conference when selected researchers display printed placards and are expected to be available to talk about their posters with other attendees.

Post hoc test A test that examines paired comparisons after an omnibus (overall) test comparing three or more populations shows differences among the populations.

Post-positivism A qualitative research paradigm in which researchers aim to experimentally test theories about how the world works, but they acknowledge that the unpredictability of human behavior limits the validity of some empirical methods.

Power In statistics, the ability of a test to detect significant differences in a population when differences really do exist; the power of tests is increased when the number of participants included in the analysis is large.

PPTs An abbreviation for person, place, and time, which are components of comprehensive case definitions and descriptive epidemiology studies.

Pragmatism A qualitative research paradigm in which researchers assume that reality is situational, and it is acceptable to use any and all research tools and frameworks to try to understand a particular problem so it can be solved.

Précis A concise one- or two-sentence summary of a research study's key finding.

Precision In a survey instrument, diagnostic test, or other assessment tool, a quality that is demonstrated when consistent answers are given to similar questions and when an assessment yields the same outcome when repeated several times; also called *reliability*.

Predatory open-access journal An exploitative journal that does not provide the quality editorial and peer-review services associated with legitimate journals.

Predictive validity Survey instrument validity demonstrated when a new test is correlated with subsequent measures of performance in related domains.

Predictor variable A variable in a statistical model that predicts the value of some outcome variable; also called an *independent variable*.

Preprint An unformatted version of a manuscript that is posted online by the authors or in the advance access section of a journal's website.

Preproposal A brief research plan required by a funding organization that wants to confirm alignment between the funder's vision and the proposed research plan before inviting a full proposal to be written and submitted.

Pretest A small-scale preliminary study conducted to evaluate the utility of a new survey instrument; also called a *pilot test*.

Prevalence The percentage of members of a population who have a given trait at the time of a study.

Prevalence ratio (PR) A statistic that compares the prevalence of a characteristic in two independent populations by taking a ratio of their prevalence rates.

Prevalence study A study that measures the proportion of members of a population who have a particular exposure or disease at a particular point in time; also called a *cross-sectional study*.

Prevention science The study of which preventive health interventions are effective in various populations, how successful the interventions are, and how well they can be scaled up for widespread implementation.

Primary investigator (PI) The researcher who accepts principal responsibility for a research project, guaranteeing that the protocol is being followed, the budget is properly managed, and any adverse outcomes are immediately reported to the institution's research ethics committee; the PI for a project conducted by a new researcher is often a professor or senior employee.

Primary prevention Health behaviors and other protective actions that help keep an adverse health event from occurring.

Primary study The collection of new data from individuals.

Principal component analysis (PCA) A statistical method that creates one or more index variables (called components) from a larger set of measured variables.

Privacy The assurance that research participants get to choose what information they reveal about themselves.

Probability The likelihood that an event will happen.

Probability-based sampling Methods for ensuring that all members of a source population have an equal likelihood of being invited to participate in a research study.

Probing An interviewing technique that prompts an interviewee to provide a more complete or specific response.

Process evaluation The systematic analysis of an ongoing intervention in order to ensure that procedures are being implemented as planned.

Processing charge A per-article fee that some subscription journals mandate prior to an accepted article being published; also called a *processing fee*.

Processing fee A per-article charge that some subscription journals mandate prior to an accepted article being published; also called a *processing charge*.

Professional development An ongoing and intentional process of establishing short- and long-term professional goals, identifying and completing activities that enable systematic progress toward achieving those goals, and routinely evaluating performance, competencies, and growth.

Program An ongoing group of projects.

Program evaluation The systematic collection and analysis of information to answer questions about the effectiveness and efficiency of a program.

Project A specific, time-limited set of activities.

Project narrative The section of a research proposal that describes the methods that will be used to answer the research question.

Proofreading The process of confirming the quality of the almost-final version of a soon-to-be-printed manuscript.

Propensity score matching A statistical technique for predicting the probability of group membership while adjusting for covariates.

Proportion A ratio in which the numerator is a subset of the denominator.

Proportionate mortality rate (PMR) The proportion of deaths in a population during a particular time period that were attributable to a particular cause.

Proposal A written request for approval of or funding for a research project.

Prospective cohort study A cohort study that recruits participants because they have or do not have an exposure of interest and then follows them forward in time to look for incident cases of disease.

Prospective study A study that follows participants forward in time.

Protected health information The information about an individual's health history or health status that by law must be kept confidential.

Protective factor An exposure that reduces an individual's likelihood of subsequently experiencing a particular disease or outcome.

Protocol A detailed written description of all the processes and procedures that will be used during participant recruitment (if relevant), data collection, and analysis.

Provisional acceptance Notification from a journal that a manuscript will be accepted once minor adjustments are submitted.

Publication bias A form of bias in a systematic review or meta-analysis that occurs when articles with statistically significant results are more likely to be published than those with null results.

Public health The actions taken to promote health and prevent illnesses, injuries, and early deaths at the population level.

PubMed A service of the U.S. National Library of Medicine that provides access to more than 28 million abstracts of journal articles.

Purposiveness Research designed to answer one well-defined research question.

Purposive sampling A nonprobability-based sampling method that recruits participants for a qualitative study based on the special insights they can provide.

p value The probability value, which is the likelihood that a test statistic as extreme as or more extreme than the one observed would occur by chance if the null hypothesis were true; a very small *p* value means that the observed test result is highly unlikely to have occurred by chance.

Q

QALY An abbreviation for a *q*uality-*a*djusted *l*ife *y*ear.

Qualitative research A research approach that uses in-depth interviews, focus group discussions, participant–observation, and other unstructured or semi-structured methods to explore attitudes and perceptions, identify themes and patterns, and formulate new theories.

Quality-adjusted life year (QALY) A metric used in health economics to estimate the additional duration of life and quality of life conferred to populations by successful public health interventions.

Quality of life A construct that captures an individual's perceived position in life in the context of that person's expectations, goals, values, and concerns.

Quantitative research A research approach that uses structured, hypothesis-driven approaches to gather data that can be statistically analyzed.

Quartiles The division of a data set into four ordered parts of equal size.

Quasi-experimental design An experimental study that assigns participants to an intervention or control group using a nonrandom method.

Questionnaire A series of questions used as a tool for systematically gathering information from study participants; also called a *survey instrument*.

R

R&R An abbreviation for *revise and resubmit*.

Random-digit dialing Calls made to a computer-generated list of unscreened telephone numbers.

Random effects model A statistical model for meta-analysis that can be used to create a pooled estimate when there is considerable variability among the included studies.

Randomization Assignment of participants to an exposure group in an experimental study using a chance-based method that minimizes bias.

Randomized controlled trial (RCT) An experimental study in which some participants are randomly assigned to an active intervention group, the remaining participants are assigned to a control group, and all participants from both groups are followed forward in time to see who has a favorable outcome and who does not.

Range The difference between the minimum and the maximum values of a variable in a data set.

Ranked variable A variable with responses that span from best to worst, most to least, or always to never, or that are expressed using other types of ordered scales; also called an *ordinal variable*.

Rapid diagnostic test (RDT) A test that can detect the presence of a pathogen (or markers for a pathogen) in a small drop of blood (or another body fluid) within 15 to 30 minutes.

Rate A ratio in which the numerator and denominator have different units; the denominator typically expresses a measure of time.

Rate difference The absolute difference between the incidence rates in two independent populations (often an exposed group and an unexposed group in a cohort study).

Rate ratio (RR) A ratio of two rates, with the reference (comparison) group in the denominator; may also be called the *relative rate*, the *risk ratio*, or the *relative risk*.

Ratio A comparison of two numbers.

Ratio variable A numeric variable that can be plotted on a scale on which a value of zero indicates the total absence of the characteristic (such as 0 inches meaning no height).

RCR An abbreviation for *responsible conduct of research*.

RCT An abbreviation for *randomized controlled trial*.

Realism The assumption that one reality exists and it can be understood.

Realist synthesis A qualitative analysis technique that uses a systematic process to find and interpret evidence for the complex reasons why some programs succeed and others fail.

Recall bias A form of bias that occurs when cases and controls in a case–control study systematically have different memories of the past.

Receiver operating characteristics (ROC) curve A graphical plot of the true positive rate against the false positive rate for the different possible cutoff points of a screening or diagnostic test.

Recoding The process of generating values for a new variable based on one or more existing columns of data in a file.

Reconciliation The process of resolving any discrepancies between a researcher's financial records and the reports produced by the institution hosting the researcher's grant or contract accounts.

Recursive model A causal analysis model in which all causal pathways are unidirectional.

Reference population A group that is used as a comparison for another population; in rate ratios, the rate for the comparison population is placed in the denominator of the ratio.

Reference standard The test used for comparison when examining the validity of a new screening or diagnostic test.

Registry A centralized database containing information about people who have had a particular exposure or been diagnosed with a particular disease.

Regression model A statistical model that seeks to understand the relationship between one or more independent (predictor) variables and one dependent (outcome) variable.

Reject and resubmit A rejection letter from a journal editor that invites the authors to revise a manuscript and resubmit it to the same journal for consideration as a "new" article; most rejection letters specify that a different version of the same manuscript will not be considered by the journal, but this unusual outcome is akin to a "revise and resubmit" offer.

Relative rate A ratio of two rates, with the reference (comparison) group in the denominator.

Relativism The assumption that multiple realities exist and they cannot be fully understood.

Reliability In a survey instrument, diagnostic test, or other assessment tool, a quality that is demonstrated when consistent answers are given to similar questions and when an assessment yields the same outcome when repeated several times; also called *precision*.

Repeated cross-sectional study A series of cross-sectional studies that resample and resurvey representatives from the same source population at two or more different time points.

Repeated-measures ANOVA A statistical test that compares the values of an interval/ratio variable across several time points or in several individually matched populations.

Replicability The principle that a study protocol implemented in a new study population should generate results similar to those of the original study, as long as the same protocol is used.

Replication studies Research studies that repeat a study protocol in a new population as part of attempting to confirm that the original findings were not due to chance.

Reporting bias A form of bias that occurs when members of one study group systematically underreport or overrerport an exposure or outcome.

Representativeness The degree to which the participants in a study are similar to the source population from which they were drawn.

Reproducibility The ability of an independent researcher to implement another researcher's data analysis protocol and generate the same results as the original researcher if given access to the original data set.

Request for applications (RFA) A notice distributed by a funding organization seeking applications from researchers who want to conduct research on topics of interest to the funder; also called a *request for proposals*.

Request for proposals (RFP) A notice distributed by a funding organization to inform researchers of their desire to receive grant proposals for research on topics of interest to the funder; also called a *request for applications*.

Research The process of systematically and carefully investigating a topic in order to discover new insights about the world.

Research ethics committee An institutional review board responsible for protecting human subjects who participate in research studies.

Residual The difference between the observed value in a data set and the value predicted by a regression model.

Respect for persons A research principle that emphasizes autonomy, informed consent, voluntariness, and protection of potentially vulnerable individuals.

Responsible conduct of research (RCR) A concept encompassing research ethics, professionalism, and best practices for collaborative research.

Results section The third section of a typical four-part scientific report, which contains key findings with text as well as tables and/or figures.

Retraction Removal of a published article from the accepted scientific literature due to major errors or author misconduct.

Retrospective cohort study A cohort study that recruits participants based on data about their exposure status at some point in the past and typically also measures outcomes that have already occurred (but happened after the baseline exposures were established); also called a *historic cohort study*.

Revise and resubmit A term used by journal editors when authors are invited to edit a manuscript in response to reviewer comments and then submit the revised manuscript back to the journal for another round of consideration; often abbreviated *R&R*.

Rigor The careful design, implementation, interpretation, and reporting of an exacting, unbiased, and ethical research protocol that answers a clearly defined scientific question.

Risk The probability of an individual in a population becoming a case during a defined period of time.

Risk factor An exposure that increases an individual's likelihood of subsequently experiencing a particular disease or outcome.

ROC curve An abbreviation for *receiver operating characteristics curve*.

RR An abbreviation for *rate ratio*.

Salami publication The practice of inappropriately writing two or more similar manuscripts about the same research finding rather than telling the complete story in one manuscript.

Sample population Individuals from a source population who are invited to participate in a research study.

Sample size In statistics, the number of observations in a data set (that is, the number of individuals in the study population).

Sample size calculator A tool used to identify an appropriate number of participants to recruit for a quantitative study based on a series of guesses about the expected characteristics of the sample population.

Sampling bias A form of bias that occurs when the individuals sampled for a study are not representative of the source population as a whole; also called *ascertainment bias*.

Sampling frame A well-defined subset of individuals from the target population from which potential study participants will be sampled; also called a *source population*.

Scholarship of Teaching and Learning (SoTL) The process of using systematic investigations to improve the quality of education.

Scope For a journal, the subject areas the publication covers.

Screening A type of secondary prevention in which all members of a well-defined group of people are encouraged to be tested for a disease based on evidence that members of the population are at risk for the disease and early intervention improves health outcomes.

Secondary analysis A study in which a researcher analyzes data collected by another entity.

Secondary prevention The detection of health problems in asymptomatic individuals at an early stage when the conditions have not yet caused significant damage to the body and can be treated more easily.

Secondary study A study that analyzes an existing data set or existing health records.

Selection bias A form of bias that occurs when the members of the study population are not representative of the source population from which they were drawn.

Self-administered survey Questionnaire forms that participants complete for themselves, either using a paper-and-pencil version or an online version of the survey instrument.

Semiotics The study of signs and symbols.

Semi-structured interview A qualitative interview with a key informant that covers a range of preselected topics using open-ended questions, probing for clarifications about verbal responses, and observations of body language and other nonverbal communication.

Senior author An experienced researcher, often the head of a research group or the primary research supervisor for a student, who may choose to be listed last in the order of authors of manuscripts produced under that individual's leadership or supervision.

Senior researcher An experienced researcher who guides the work of a newer investigator.

Sensitivity The proportion of people who actually have a disease (according to a reference standard) who test positive with a screening or diagnostic test.

Sensitivity analysis The process of examining the robustness of statistical methods and the results of models.

Sentinel surveillance The continuous collection and analysis of high-quality data from a limited number of clinics or hospitals so that public health officials will be able to detect changes in health status in the larger population from which the sentinel sites were sampled.

Sickness The way in which a person with an adverse health condition is regarded by his or her community.

Sign An objective indication of disease that can be clinically observed, such as a rash, cough, fever, or elevated blood pressure.

Significance level The p value (usually $p = .05$) at which the null hypothesis is rejected and a statistical result is considered statistically significant.

Significant figures The number of digits in a number that are known to be accurate.

Simple linear regression A model that examines whether there is a linear relationship between one ratio or interval predictor variable and one ratio or interval outcome variable.

Simple randomization The use of a coin toss, a random number generator, or some other simple mechanism to randomly assign each individual in an experimental study to one of the exposure groups.

Simultaneous multiple regression A model that includes all predictor variables in the model rather than fitting the model using a stepwise approach.

Single-blind An experimental study design in which the participants do not know whether they are in an active group or a control group; in publishing, a peer-review process in which the reviewers are provided with the authors' names, but authors are not provided with reviewers' names.

SIR model A mathematical model of infection transmission that describes how the susceptible (S) individuals in a population may become infected (I) and then eventually recover (R) with immunity.

Skewness A description of how asymmetrical a nearly normal distribution is.

Skip logic Codes used in computer-based surveys to automatically hide irrelevant questions from participants based on their answers to filter questions.

SMART An acronym describing a good goal statement as being specific, measurable, attainable, relevant, and timely.

Snowball sampling A literature searching technique that involves looking up every article cited by eligible articles in order to identify additional sources that might be relevant even though they are not indexed in the selected databases.

Social ecological model A theoretical framework that considers individual health and health behaviors to be a function of the social environment, which includes intrapersonal (individual), interpersonal, institutional (organizational), community, and public policy dimensions.

Social media analytics The process of compiling and analyzing data from social networking services like Instagram and Twitter.

Solicited proposal A request for funding submitted by a researcher after a funder has contacted the researcher to invite that person to submit a proposal.

Source population A well-defined subset of individuals from the target population from which potential study participants will be sampled; also called a *sampling frame*.

Spearman rank-order correlation A statistical measure of the degree to which changes in the value of one ordinal/rank variable predict changes in the value of another ordinal/rank variable; usually called "Spearman's rho (ρ)" or listed as r_s.

Specific aim A carefully described action that will help a researcher make progress toward achieving the big-picture goal; also called a *specific objective*.

Specific knowledge Information that is specific to a particular study, such as a particular statistic or a particular laboratory finding.

Specific objective A carefully described action that will help a researcher make progress toward achieving the big-picture goal; also called a *specific aim*.

Specificity The proportion of people who do not have a disease (according to a reference standard) who test negative with a screening or diagnostic test.

Spread A measure that describes the variability and distribution of responses to a numeric variable; also called *dispersion*.

Spreadsheet A file that stores data in the cells of a row-by-column table.

Spurious A term used to describe results that are false or invalid.

Stakeholder A person who has an interest in the success or failure of a group and can influence or be affected by that group's decisions or actions.

Standard deviation A measure of the narrowness or wideness of a normal distribution that is calculated as the square root of the variance; in a normal distribution, 68% of the responses will fall within one standard deviation above or below the mean and 95% of the responses will fall within two standard deviations above or below the mean.

Standard error A measure of the narrowness or wideness of a normal distribution that is calculated by dividing the variance by the total number of observations and then taking the square root of that number.

Standardized mortality ratio (SMR) In indirect age adjustment, a statistic that compares the number of deaths observed in the study population to the number of deaths expected in the study population based on the age-specific mortality rates in the standard population.

Standard of care An existing therapy (such as the best therapy currently available or the therapy that is used most often in the location where the study is being conducted) that is used as a comparison for a new therapy being experimentally tested.

Statistic A measured characteristic of a sample population.

Statistical significance A classification based on a test result having a *p* value less than a preselected significance level (typically $\alpha = 0.05$).

Stepwise multiple regression A model that systematically adds or removes predictor variables to a regression model to find the most parsimonious model that provides a good fit.

Stochastic model A mathematical model with inputs that vary according to a probability distribution, so the outcomes differ slightly every time the model is run.

Stratified randomization The division of a population into subgroups prior to randomly but systematically assigning each individual within each subgroup to one of the exposure groups in an experimental study.

Structural equation modeling A nonrecursive causal analysis strategy that can be used to examine complexities in the directionalities of the path diagram.

Structured abstract A research summary that uses subheadings like objective, methods, results, and conclusion.

Study goal The single overarching objective of a research project or the main question that a research project seeks to answer.

Study population The eligible members of the sample population who consent to participate in the study and complete required study activities.

Subjectivity Interpretation of claims and experiences that is based on an evaluator's beliefs, perceptions, and feelings.

Submission fee A charge that some journals mandate for all manuscripts prior to review.

Subscription journal A journal that covers its costs from library and/or individual subscriptions and advertising and does not charge any author fees.

Superiority trial An experimental study that aims to demonstrate that a new intervention is better than some type of comparison, not merely as good as the comparison.

Surveillance The process of continually monitoring health events in a population so that emerging public health threats can be detected and appropriate control measures can be implemented quickly.

Surveillance bias A form of bias that occurs when a population group that is routinely screened for adverse health conditions incorrectly appears to have a higher-than-typical rate of disease because more frequent testing enables a higher case detection rate in that population than in the general population; also called *detection bias*.

Survey The gathering of data from individuals using a list of questions.

Survey instrument A series of questions used as a tool for systematically gathering information from study participants; also called a *questionnaire*.

Survival analysis Statistical evaluation of the distribution of the durations of time that individuals in a study population experience from an initial time point (such as the time of enrollment in a study or the time of diagnosis of a particular condition) until some well-defined event, which can be death, discharge from a hospital, or some other outcome.

SWOT An evaluation method that identifies the strengths, weaknesses, opportunities, and threats of a program.

Symbolic realism A paradigm in which researchers treat individuals' realities as being real to those individuals.

Symptom A subjective indication of illness that is experienced by an individual but cannot be directly observed by others.

Syndrome A collection of signs and symptoms that occur together.

Syndromic surveillance The process of tracking potential outbreaks or other disease events based on reports of symptoms rather than relying solely on counts of laboratory-confirmed diagnoses.

Synthesis research The integration of existing knowledge from previous research projects, typically in the form of a narrative review, a systematic review, or a meta-analysis.

Systematic review The use of a predetermined and comprehensive searching and screening method to identify relevant articles during a tertiary analysis.

Systems thinking The process of identifying the underlying causes of complex problems so that sustainable solutions can be developed and implemented.

T

Table In a research report, the concise presentation of key findings in a grid.

Talk-aloud protocol A research technique in which participants in a qualitative study are asked to describe their thoughts and actions while they complete a task; also called *think-aloud protocol*.

Target journal The journal a researcher intends to submit a manuscript to first.

Target population The broad population to which the results of a study should be applicable.

Temporality The timing of events.

Tertiary prevention Interventions that reduce impairment, minimize pain and suffering, and prevent death in people with symptomatic health problems.

Tertiary study A research analysis that reviews and synthesizes the existing literature on a topic, such as a narrative review, systematic review, or meta-analysis.

Testability The ability of a research question to be answered using experiments or other types of measurements.

Test–retest reliability In a survey instrument or other assessment tool, a condition that is demonstrated when people who complete a baseline assessment and then retake the test later have about the same scores each time they are tested.

Test statistic A value calculated from study data for a hypothesis test, such as the *t* stat used for *t* tests or the *F* stat used for one-way ANOVA.

Text recycling Also called "self-plagiarism," duplicate publication that occurs when one's own words from one publication are copied into a new manuscript.

Theme In qualitative analysis, a concept that encompasses one or several categories.

Theoretical framework A set of established models in the published literature that can inform the components and flows of the conceptual framework for a new research study.

Theoretical sampling In grounded theory, the use of an emerging theory to guide the selection of new data sources.

Theory In qualitative analysis, a construct that provides a systematic explanation about a phenomenon.

Think-aloud protocol A research technique in which participants in a qualitative study are asked to describe their thoughts and actions while they complete a task; also called *talk-aloud protocol*.

Third variable A variable that is associated with an exposure variable and an outcome variable but is not part of the causal pathway from an exposure to an outcome.

Threshold A value that divides a numeric variable into separate categories; also called a *cutpoint*.

Time series study A research study that measures participants or samples of participants at multiple points in time.

Tolerance The inverse of the variance inflation factor, a test used to examine whether the independent variables in a regression model have reasonably independent errors.

Transferability An indicator of quality assurance that is present when the interpretation of qualitative data is likely to be applicable in other circumstances.

Transformative paradigm A qualitative research framework in which researchers assume that reality can be changed when research addresses a social justice issue.

Translational research Bench-to-bedside studies that bridge basic research and clinical research by applying basic science discoveries to improvement of clinical outcomes.

Transparency The quality of being open and clear about the methods used for a research study.

Treatment-assigned analysis An analysis of experimental data that includes all participants, even if they were not fully compliant with their assigned intervention or comparison protocol; also called *intention-to-treat analysis*.

Treatment-received analysis An analysis of experimental data that includes only the participants who were fully compliant with their assigned intervention or comparison protocol.

Triangulation The process of using multiple different types of data, methods, and theories to better understand a phenomenon.

True negative rate The proportion of people who actually do not have a disease (according to a reference standard) who test negative with a screening or diagnostic test; also called *specificity*.

True positive rate The proportion of people who actually have a disease (according to a reference standard) who test positive with a screening or diagnostic test; also called *sensitivity*.

Tukey's test A post hoc test that examines all of the possible pairwise comparisons across three or more populations included in an ANOVA.

Two-by-two (2×2) table A row-by-column table that displays the counts of how often various combinations of events happen; in epidemiological analysis, the columns typically display disease status (yes/no) and the rows typically display exposure status (yes/no).

Two-sample *t* test A statistical test that compares the mean values of a ratio/interval variable in two independent populations; also called *independent-samples t test*.

Two-sided *p* value The probability value that is used for a statistical test when a direction is not specified in the alternative hypothesis.

Two-way ANOVA A test that compares the mean values of an interval/ratio variable across groups that are defined by two different variables (such as both sex and smoking status); also called *factorial ANOVA*.

Type 1 error An error that occurs when a study population yields a significant statistical test result even though a significant difference or association does not actually exist in the source population.

Type 2 error An error that occurs when a statistical test of data from a study population finds no significant result even though a significant difference or association actually exists in the source population.

U

Understood consent Evidence that a potential study participant comprehends the study benefits, risks, and procedures and his or her rights as a study participant prior to agreeing to participate.

Uniform distribution A histogram for a numeric variable that appears rectangular because approximately equal numbers of responses were provided for each allowable value of the variable.

Unimodal A numeric variable with a one-peaked distribution.

Univariate analysis Statistical analysis that describes one variable in a data set.

Unstructured abstract A narrative research summary that does not use section titles like objective, methods, results, and conclusion to divide the content of the paragraph.

V

Validity In a survey instrument, diagnostic test, or other assessment tool, a condition that is established when the responses or measurements in a study are shown to be correct; also called *accuracy*.

Variability The extent to which the values for a particular variable deviate from the average value of that variable in the data set.

Variable A characteristic that can be assigned to more than one value.

Variance A measure of the narrowness or wideness of a normal distribution that is calculated by adding together the squares of the differences between each observation and the sample mean and then dividing by the total number of observations.

Variance inflation factor (VIF) A test for whether the independent variables in a regression model have reasonably independent errors and are not too intercorrelated.

Verbal consent Informed consent for participation in a study that is spoken and witnessed rather than requiring a participant's signature; also called *oral consent*.

Vignette A brief written or pictorial scenario designed to elicit a response from participants in a qualitative study.

Vital signs Physiological measurements that provide clinical information about an individual's essential body functions.

Vital statistics Population-level measurements related to births, deaths, and other demographic characteristics.

Voluntariness A decision made of an individual's own free will without undue outside influence.

Vulnerable populations Special populations whose members might have limited ability to make an autonomous decision about volunteering to participate in a research study.

W

Waiver of consent Permission from an institutional review board not to provide an informed consent form to the individuals who will have their data included in an analysis as well as permission not to give those individuals the opportunity to opt into or out of a study.

Waiver of consent documentation Permission from an institutional review board not to collect signed consent forms from participants because they could be harmed by being able to be linked to participation in a study on a sensitive topic.

Washout period A time between arms of an experimental study when patients receive no treatment.

Weighting Statistical methods that adjust for sampling methods, demographic differences between a study population and a source population, varying sample sizes in a meta-analysis, or other circumstances.

White space Blank areas between printed content on a page.

Wilcoxon rank sum test A statistical test that compares the median values of an ordinal/rank variable in two independent populations; also called a *Mann-Whitney U test*.

Wilcoxon signed-rank test A statistical test that compares the values of an ordinal/rank variable in one population measured twice or among individually matched pairs from two different groups.

Writer's block Sustained struggles with writing that an author might experience due to fear of failure or other barriers to productivity.

Yates correction A method for improving the validity of a chi-square test statistic when the sample size for the test is small.

Years lived with disability (YLDs) A burden of disease metric used to quantify the population-level reductions in health status attributable to nonfatal conditions.

Years of life lost (YLLs) A burden of disease metric used to quantify the population-level reductions in health status due to premature mortality.

YLD An abbreviation for a *y*ear *l*ived with *d*isability.

YLL An abbreviation for a *y*ear of *l*ife *l*ost to premature mortality.

Z

z score A number that indicates how many standard deviations away from the sample mean an individual participant's response is.

Index

D

E

F

G

H

J

K

L

M

N

O

P

Q

R

S

T

V

Z